Atomic Absorption Spectrometry in Occupational and Environmental Health Practice

Volume II
Determination of Individual Elements

Author

D.L. Tsalev, Ph.D.

Associate Professor of Analytical Chemistry
Head, Atomic Spectrometry Laboratory
Faculty of Chemistry
University of Sofia
Sofia, Bulgaria

CRC Press, Inc.
Boca Raton, Florida

Library of Congress Cataloging in Publication Data

Tsalev, D. L. (Dimiter L.)
Atomic absorption spectrometry in occupational and environmental health practice.

Bibliography: v. 1, p.
Includes index.
Contents: v. 1. Analytical aspects and health significance — v. 2. Determination of individual elements.
1. Trace elements—Analysis. 2. Trace elements in the body. 3. Atomic absorption spectroscopy. 4. Industrial hygiene—Technique. 5. Toxicology—Technique. I. Zaprianov, Z. K. (Zaprian K.) II. Title. [DNLM: 1. Environmental health. 2. Occupational medicine. 3. Spectrophotometry, Atomic absorption. 4. Trace elements. 5. Monitoring, Physiologic—Methods. WA 400 T877a]
QP534.T73 1983 615.9'02 83-2741
ISBN 0-8493-5603-2 (v.1)
ISBN 0-8493-5604-0 (v. 2)

Direct all inquiries to CRC Press, Inc., 2000 Corporate Blvd., N.W., Boca Raton, Florida, 33431.

Second Printing, 1985

International Standard Book Number 0-8493-5603-2 (Volume I)
International Standard Book Number 0-8493-5604-0 (Volume II)

Library of Congress Card Number 83-2741
Printed in the United States

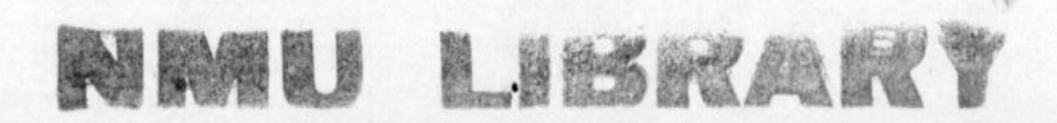

PREFACE

This volume is intended to cover the application of Atomic Absorption Spectrometry (AAS) to the determination of 34 elements in biological materials. All the available information has been treated with a view to its usefulness in occupational and environmental health practice and science, but the text may hopefully be of use in other branches of science as well: biology, medicine, forensic toxicology, pharmacology, etc.

AAS applications have been subjected to a thorough literature search, covering 1955 to 81 (and partly up to mid-1982), but — as the number of relevant references counted to over 3000 — the final list had to be confined to about 1600 references.

AAS is currently the technique-of-choice for more than 20 of the discussed elements; but applications of other analytical techniques are also mentioned and cited, so as to help select a reliable independent method for the verification of AAS analytical results.

The text is illustrated by numerous tables whose purpose is to classify the available AAS procedures, to give selected reference figures of "normal" levels, to present important preconcentration and/or speciation techniques, and briefly to outline some selected AAS procedures.

A general characteristic of AAS as an analytical technique, as well as a compressed reference information on the occupational and environmental health significance of the elements has been given in Volume I.

Acknowledgments are due to Dr. I. P. Havezov, Dr. G. Georgiev, Prof. P. R. Bontchev, Dr. D. Nonova, Dr. Z. K. Zaprianov, and Dr. I. I. Petrov for their valuable comments on the manuscript. Mr. A. Rizov edited the English text, Mr. I. Tchomakov has given useful linguistic advice, and Mrs. M. Dimitrieva typed the text. Many individual authors and companies sent reprints of their papers, reports, as well as other relevant information. John Hunter, senior Editor and staff of CRC Press provided excellent editorial and composition services. The preparation of the volume would have been impossible without the constant encouragement and help of my wife Eugenia and my son Andrej.

Dimiter L. Tsalev

THE AUTHOR

D. L. Tsalev, Ph.D., is a staff member of the Faculty of Chemistry of the University of Sofia, Bulgaria. His professional duties include the laboratory training of students in qualitative and quantitative analysis and instrumental analytical methods, laboratory training of specialists in postgraduate courses in atomic spectrometry, and lecturing in postgraduate courses on atomic aborption spectrometry and flame emission spectrometry.

Dr. Tsalev received his higher education at the Higher Institute of Chemical Technology, Sofia, Bulgaria and the Moscow Higher Institute of Steel and Alloys, USSR. He received his Ph.D. from Moscow State University, in 1972.

Dr. Tsalev has published 33 papers in Bulgarian, Russian and international journals. He holds two patents, has co-authored two books, and is currently preparing a third.

Dr. Tsalev's current research interests include atomic aborption spectrometry, liquid/liquid extraction, analysis of ecological materials (blood, urine, tissue, soil, plants, and water), and trace analysis.

TABLE OF CONTENTS

Chapter 1

ALUMINIUM

I. DETERMINATION OF ALUMINIUM IN BIOLOGICAL SAMPLES

Aluminium is ubiquitous in nature, but its normal concentrations in biological materials are relatively low — usually sub-micrograms per gram[1-28] * (Table 1). In the assays of trace Al in biological samples there are two major problems: (1) pronounced external contamination during specimen collection storage, and handling and (2) a limited choice of sensitive analytical methods. Most existing instrumental techniques require expensive equipment: electrothermal atomic absorption spectrometry (ET-AAS) — which is the method of choice at present — inductively coupled plasma emission spectrometry (ICP-ES), the arc-source atomic emission spectrometry (AES), neutron activation analysis (NAA), and mass spectrometry (MS).

The importance of ICP-ES is steadily growing, as this technique is sufficiently sensitive (detection limits down to 0.2 to 5 ng/mℓ[1,29-31]), and all that is needed is a dilution of the biological liquids or digests. Applications of ICP-ES to analyses of blood,[1,30,32,33] diet,[33] feces,[33] serum/plasma,[30,34] urine,[1,33] etc.[35] have already been described.

The arc-source AES is a classical technique for trace Al in biological samples, and usually involves dry ashing of specimens: blood,[36,37] bone,[38] feces,[39] milk,[14,40] saliva,[41] serum/plasma,[36,41,42] teeth,[26] tissues,[36,43,44] etc.[36,37,39,41,42]

The NAA[45] is not a sufficiently sensitive and convenient method for Al owing to the short lifetime of the ^{28}Al isotope, but its inherent absence of reagent blanks renders it attractive. Applications to blood,[45] bone,[46] diet,[45,47] hair,[9,11,45,48] nails,[15] serum/plasma,[45,49] teeth,[23,24,45,50,51] tissues,[12,45] and other biologicals have been described.

Only a few analytical laboratories have been able to perform MS analyses of blood,[52,53] cerebrospinal fluid (CSF),[52] hair,[54] nails,[16] plasma,[55] teeth,[25] etc.

II. ATOMIC ABSORPTION SPECTROMETRY OF ALUMINIUM

The flame AAS of Al is relatively selective but not sensitive. In the reducing nitrous oxide-acetylene flame (with a red feather of approximately 2 cm), at the 309.3-nm line, the characteristic concentration is from 0.3 to 1 μg/mℓ and the detection limits are down to 0.01 to 0.03 μg/mℓ.[56-59] An ionization buffer (e.g., 0.1 to 0.2% of KCl) is added to both the sample and the standard solutions. The stoichiometry of the flame (i.e., the acetylene flow rate) and the observation height should be carefully optimized in order to obtain a good signal-to-noise ratio and to reduce interferences. Suitable media for the flame AAS of Al are the dilute acids ($HClO_4$, HCl) and the organic solvents/extracts.

ET-AAS is the most sensitive and fairly selective technique for nanogram amounts of Al. The characteristic concentration is from 0.004 to 3 ng/mℓ (0.4 to 70 pg),[57,59-61] and detection limits down to 0.001 ng/mℓ are claimed.[59] In practice, these limits are restricted to a few tenths of nanograms per milliliter (a few picograms), due to random contamination and reagent blanks.

The electrothermal atomization (ETA) of Al, as well as interferences and their elimination, have been studied by Persson et al.[61a] and Matsusaki et al.[62] Charring up to the maximum tolerable temperature of approximately 1250°C, as well as adding of matrix/analyte modifiers, are generally recommended in order to expel most of the volatile chlorides and other matrix consistuents before the atomization step and to convert the analyte into an Al_2O_3 which is

* All references appear at the end of the volume.

Table 1
ALUMINIUM CONCENTRATIONS IN HUMAN BODY FLUIDS AND TISSUES

Sample	Unit	Selected reference values	Ref.
Blood	μg/ℓ	12.5 ± 4	1
Bone	μg/g	2.39 ± 1.18;[a,b,c] 3.88 ± 1.73[a,b,d]	2
Brain	μg/g	0.5 ± 0.1	3
		1.9 ± 0.7 (1.0—3.9)[a]	4
CSF	μg/ℓ	10 ± 2.5	5
		19.8 (16.5—25.5)	6
Erythrocytes	μg/ℓ	(42—80)	7
Feces	mg/day	43	8
Hair	μg/g	$\tilde{x}$, (3.4—5.6);[e] (0.8—30)	9
		$\tilde{x}$, 5 (1—17)	10
		$\tilde{x}$, 10.3 (1.8—74)	11
Kidney	μg/g	0.4 ± 0.1	3
Liver	μg/g	2.6 ± 1.3	3
Lung	μg/g	18.2 ± 9.7	3
		$\tilde{x}$, 21 (10—100)	12
Milk	mg/ℓ	0.33	8
		0.330 ± 0.042	13
		(0.023—0.872)	14
Muscle	μg/g	0.5 ± 0.2	3
		1.22 ± 0.72[b]	2
Nails (finger)	μg/g	$\tilde{x}$, ♂ 13.2; $\tilde{x}$, ♀, 15.7; (2.1—98.6)[a]	15
		(16—260)[a]	16
Plasma or serum	μg/ℓ	2.1 ± 2.2	17
		<4 (<2.5—7)	18
		6 ± 3	19
		$\tilde{x}$, 14 (3—39)	20
Sweat	mg/day	6.1 (3—10)	21
Tooth (dentine)	μg/g	65.9	22
		68.6 ± 22.5	23
		150 ± 50;[f] 210 ± 110[g]	24
Tooth (enamel)	μg/g	$\tilde{x}$, 5.6 (1.5—70)	25
		66.6 (8—325)	26
		86.13 ± 4.54	23
		89.8	22
		240 ± 80;[f] 330 ± 280[g]	24
Urine	μg/ℓ	3.7 ± 1.4	27
		4.7 ± 2.5	1
		(3.5—31)	28

Note: Wet weight unless otherwise indicated. Mean ± SD (range in parentheses). $\tilde{x}$, median.

[a] Dry weight
[b] Defatted
[c] Trabecular
[d] Cortical
[e] Range of medians.
[f] Permanent teeth.
[g] Primary teeth.

then thermally stable until the proper atomization temperature is attained. Between 1250 and 1750°C, the analyte may be partially lost, probably as volatile gaseous species (AlOH, $AlCl_3$, AlCl, etc.).[61a] The effectiveness of several matrix modifiers on elimination of chloride interference has been found to decrease in the order: $CH_3COONH_4 > HNO_3 > (NH_4)_4EDTA > H_2SO_4$.[62]

The following matrix/analyte modifiers, either alone or in combination with wetting agents (Triton® X-100,[17,18,20,63,64] Teepol®,[6] etc.), have proved useful in ET-AAS of Al: a dilute HNO_3 (0.01 *M*[18,61] to 5% v/v HNO_3[65,66]), a dilute H_2SO_4,[67,68] aqueous NH_4NO_3,[20] aqueous NH_3,[6] Na_2EDTA,[2,69] Na_3PO_4,[70] and Na_2SO_4.[71] The addition of 0.1% Ca^{2+} (as nitrate) resulted in a sevenfold sensitivity enhancement.[72]

In situ pyrocoating,[72] as well as coatings with molybdenum[17] or tungsten carbide,[68] have proved useful in extending the lifetime of the graphite tube. The use of pyrolytically coated tubes and "maximum power", temperature-controlled heating resulted in twice greater sensitivity.[60]

The recommended purge gas in the ET-AAS of Al is argon (or, incidentally, Ar + CH_4[72]); the use of N_2 should be avoided. Due to the formation of aluminium nitride and cyanide, in presence of N_2 the sensitivity is lower (2 ×)[61] and the lifetime of the graphite tubes is shorter.

HCl and $HClO_4$ media are not advisable in the ETA of Al. Even small concentrations of $HClO_4$ (>0.01 *M*) strongly depress aluminium absorbance,[65,66] while the presence of 0.5 to 1 *M* $HClO_4$ causes a 100% depression.[66]

Higher concentrations of other acids (HNO_3, H_2SO_4) or their combinations (HNO_3 + HCl, HNO_3 + Fe, HNO_3 + HCl + Fe)[61a] may also interfere, the extent depending on the age and the condition of the graphite-tube surface as well as on the furnace programming. Standard addition checks are always advisable in the ET-AAS of Al.

III. AAS METHODS FOR ANALYSIS OF BIOLOGICAL SAMPLES

The flame AAS can only be applied to samples which are relatively high in Al, i.e., to food/feces digests. The published flame AAS procedures for CSF and tissues[65,73,74] cannot be recommended. Al has been determined in food after wet digestion with HNO_3 and H_2SO_4 and subsequent extraction with 10% acetylacetone in butylacetate at pH 12.[75] A number of extraction procedures for Al have been compiled and discussed by Cresser,[76] and among these, the extractions with 8-hydroxyquinoline[39] and with cupferron-methylisobutyl ketone (MIBK) have been applied to biological materials.

The sensitivity of ET-AAS is adequate, and its various applications to biological liquids or digests have been described: bile,[70] blood,[77] bone,[2,46,64,68] CSF,[5-7] hair,[78-80b] serum or plasma,[5-7,17-20,28,63-65,70,71,81,82] serum protein fractions,[68] teeth,[50,83,84] tissues,[2,4,7,64,65,67,68,74,85-88] and urine.[5,7,28,46,64,70,71,82] Selected procedures are briefly outlined in Table 2.

The main problem in trace aluminium assays is how to avoid/control the extraneous aluminium additions throughout specimen collection, storage, preparation, and measurement. Accordingly, the ideal procedure should involve no or a minimum of handling steps, as well as ultrapure reagents and thoroughly precleaned laboratory ware. Contamination from syringes (in particular those incorporating glass beads or heparin,[82] Vacutainer® tubes,[89] polycarbonate tubes (unless specially precleaned[17]), new Vitreosil® crucibles,[90] disposable pipet tips (especially when acidic solutions are pipetted),[71,82] plastic foam discs used to collect biopsies of surface tooth enamel,[83] reagents,[33,71] air particulates, glass, porcelain, cigarette ash, talc, stoppers, filter paper, etc. have been encountered, to mention only a few published examples. This problem may explain the by-orders-of-magnitude discrepancies in the "normal" values published for Al in biological materials. Versieck and Cornelis,[49] in reviewing the data on serum aluminium, concluded that "the reported Al content is going up with the increasing complexity of the procedure before the actual detection."[49]

Recently Alderman and Gitelman[17] as well as Oster,[18] under stringent precautions against contamination, obtained blank values of approximately 2 ng/mℓ in their analyses of blood serum — i.e., the same order of magnitude as the normal levels of serum aluminium appear to be.

Thornton et al.[89] studied and encountered both contamination and losses of the serum Al during storage in four containers: Becton Dickinson Vacutainer,® trace metal tubes, BD red top tubes, polystyrene tubes. A slow decrease after 12 hr of storage was found for the first type of tubes, followed by a sharp increase after 144 hr. These authors concluded that the serum Al analyses should be made within 12 hr of collection or, if necessary, the samples can be either refrigerated for up to 48 hr in precleaned polystyrene or polypropylene tubes or stored frozen for a longer period of time.[89] Other reliable container materials for Al are PTFE, polycarbonate, and quartz after thorough precleaning.

Organic matter can be decomposed by either wet[7,65,73,75,88,90,91] or dry ashing.[4,65,69,88] Dry ashing at temperatures between 450 and 650°C should be safe, provided that temperature rises slowly and airborne contamination is restricted. At higher ashing temperatures and/or in presence of higher amounts of silica (e.g., in food, feces, urine), it may be difficult to quantitatively put the resulting oxides into solution. Platinum crucibles have been successfully employed in the dry ashing of tissues[2,4,88] and blood, feces, feed, and urine,[33] at 500 to 650°C.

Single-acid treatment of the tissues and bone with HNO_3[64] and of the tooth enamel with HNO_3[50,83] or HCl[84] is preferable to the common acidic mixtures, owing to its simplicity and low blanks. Pressurized PTFE tubes may conveniently be used to decompose blood[77] and other biologicals with HNO_3.

The Al can be extracted from small defatted biopsy samples of bone or muscle with saturated aqueous Na_2EDTA, prior to ET-AAS determination.[69]

Two common preparation techniques — precipitation of serum proteins[82] and solubilization of muscle biopsies with tetraalkylammonium hydroxide (TAAH) (Lumatom®)[92] — have been studied and found inappropriate for ET-AAS of Al in the indicated samples.

Wherever possible, the preparation of biological liquids for ETA should be confined to a simple dilution with water or with an aqueous matrix/analyte modifier (Table 2). In some cases it has even been possible to directly analyze intact samples, thus greatly reducing both the preparation time and contamination risk; specimens of blood,[77] CSF,[5,7] hair,[79,80] serum/plasma,[5,71,81] water mussels,[86] and urine[5,64,71,82] have been analyzed by means of these simplified procedures. Although some drawbacks are intrinsic with most direct procedures (complicated calibration, longer temperature programs, residue accumulation, sensitivity drift, etc.), the application of the direct ET-AAS still seems to be the best practical way of avoiding the marked positive bias so typical in ultratrace Al assays.

IV. CONCLUSION

ET-AAS is the best analytical technique and ICP-ES is its most promising alternative at submicrograms-per-gram levels. Precautions should be taken to avoid external contamination during specimen collection, storage, and preparation. Direct procedures with versatile equipment (high heating rates, and perhaps, the L'vov platform and pretreated graphite surfaces) are therefore preferable. Matrix/analyte modification is often required; composite diluents, comprising combinations of a surfactant and aqueous NH_3, HNO_3, H_2SO_4, and/or EDTA are useful.

Table 2
SELECTED ET-AAS PROCEDURES FOR ALUMINIUM DETERMINATION IN BIOLOGICAL SAMPLES

Sample	Brief procedure outline	Ref.
Blood	2 μℓ of heparinized, completely hemolyzed (+Triton® X-100) blood injected into a Varian Techtron® CRA-63 graphite cup; ramp dried to 90°C for 40 sec, ramp ashed to 400°C for 30 sec, ashed at 1200°C for 30 sec, and atomized at 2500°C; calibration vs. preanalyzed blood	77
	1 + 10 dilution with water; 90 (ramp), 500 (ramp), 1200, and 2700°C (gas stop); standard addition calibration	77
Bone	HNO_3 decomposition; dilution; 5-μℓ portions injected into CRA-90 graphite tube; 100°C for 30 sec, 900°C for 30 sec, and 2500°C; standard addition calibration	64
Bone/ muscle biopsies	10—25 mg of dried, defatted bone powder extracted for 2—4 hr with 5 mℓ of saturated aqueous solution of Na_2EDTA; ET-AAS: 100°C for 15 sec, 1550°C (ramp for 20 sec), 2650°C (gas stop); standards in aqueous Na_2EDTA	2, 69
CSF, serum	Dilution with aqueous detergent (Teepol® 710) and aqueous NH_3; ET-AAS: 120, 1700, and 2800°C; background correction	6
Plasma	1 + 1 dilution with an aqueous 2.5% NH_4NO_3 in 0.01% Triton® X-100; ET-AAS: 200 (ramp), 1500 (ramp), and 2700°C (gas stop); background correction; standard addition calibration	20
Serum	1 + 3 dilution with water, 100, 600 (ramp!), 1400, and 2650°C (gas stop); background correction; standard addition calibration	82
	1:25 dilution with an aqueous 2% Triton® X-100 in 1% v/v HNO_3; aerosol deposition via FASTAC® 254 into a 555 temperature-controlled furnace (Instrumentation Laboratory) at 175°C; temperature program: 300° (ramp), 1300 (ramp), and 2500°C (ramp); serum-based standards	63
	1:5 dilution with an aqueous 0.01 *M* HNO_3 in 0.1% Triton® X-100; 100 (ramp), 140 (ramp), 1350 (ramp), and 2700°C (gas stop, "maximum power"); background correction; standards in the same diluent; see also Reference 64	18
	1:1 to 1:4 dilution, the diluent containing 20 m*M* Na_2EDTA, 1% v/v H_2SO_4, 1% v/v Triton® X-100, and 2% v/v of aqueous NH_3; ET-AAS with a Mo-treated graphite tube: 400, 1580, and 2600°C (gas stop); standards in diluent	17
Serum protein fractions	Gel filtration; ET-AAS with a pyrolyzed L'vov platform and H_2SO_4-matrix modification	68
Tissues	Wet[8,64,65] or dry[4,88] ashing; digests (in diluent HNO_3) analyzed by ET-AAS with background correction and standard addition calibration	
Tooth enamel (biopsies)	Pressurized digestion with 6 *M* HNO_3; dilution (10 ×); Varian Techtron® CRA furnace: 100, 350, and 2500°C	83
Urine	1 + 2 dilution; 100°C for 60 sec, 300°C for 60 sec, ramp ashing to 1400°C (30°C/sec), atomization at 2650°C (gas stop); background correction; standard addition calibration; see also References 5, 64, and 71	82

Chapter 2

ANTIMONY

I. DETERMINATION OF ANTIMONY IN BIOLOGICAL SAMPLES

The normal levels of Sb in biological materials are around or below 1 ng/mℓ in body fluids and some tens of nanograms per gram in tissues[93-115] (Table 1). Accordingly, only a few highly sensitive techniques, such as hydride generation (HG)- and ET-AAS, NAA, and differential pulse anodic stripping voltametry (DPASV) are suitable for biological analyses.

NAA, a recognized method for ultratrace Sb, has been widely applied to various biological materials: blood,[45,103] bone,[97,103] diet,[45,47,116] hair,[11,98,99,101-104,117,118] milk,[110] nails,[103] platelets,[119] serum/plasma,[111,120] serum protein fractions,[120] skin,[112] teeth, tissues, urine, etc.[45,121,122]

DPASV, with its detection limits down to 0.1 μg/ℓ,[123] has been found suitable for urine screening assay.[124,125] The classical polarographic procedures are applicable to monitor Sb in urine after acute intoxication.[114]

The sensitivity of X-ray fluorescence (XRF) is limited,[127-129] even after an efficient preconcentration by, e.g., Sb coprecipitation as a sulfide[127] or dithiocarbamate.[129]

The AES techniques should also be preceded by a preconcentration step (extraction,[95,130] stibine generation[131]); the potential of both microwave-induced plasma (MIP)-AES[95,131] and ICP-ES should be further explored.

Spark-source mass spectrometry (SSMS) analyses of nails,[16] plasma,[55] teeth,[25] tissues,[132] etc.[132] have been reported.

There are spectrophotometric and fluorimetric techniques for Sb, most of which are based on Rhodamine B reaction[133-136] and extractions of ion-associate complexes; these procedures are tedious and have neither sensitivity nor selectivity.

II. ATOMIC ABSORPTION SPECTROMETRY OF ANTIMONY

The AAS determination of Sb in a stoichiometric air-acetylene flame is very selective but not sensitive. The characteristic concentration is between 0.02 and 0.5 μg/mℓ, and the detection limits range between 0.02 and 0.06 μg/mℓ.[56-59]

The spectral bandpass should be as narrow as 0.2 nm, so as to isolate the resonance line of Sb at 217.6 nm from the two neighboring lines, Sb 217.0 and Sb 217.9 nm, and thus to improve the linearity of the working curve as well as to eliminate spectral interference by the absorption lines of Cu 217.9 nm and Pb 217.0 nm. The second line of antimony, 231.2 nm, though twice lower in sensitivity vs. the Sb 217.6-nm line, has been incidentally preferred in ET-AAS for better baseline stability.[137]

An electrodeless discharge lamp (EDL) light source is recommended for its higher intensity (8 ×) and lower detection limits attainable (2 ×), as compared with the hollow cathode lamp (HCL).[138]

In ETA, the characteristic concentration ranges from 8 to 90 pg, and detection limits are down to 0.08 to 4 ng/mℓ (5 to 15 pg).[57,59-61] Ashing temperatures between 300 and 1000°C, and atomization temperatures from 2000 to 2900°C have been reported, depending on the chemical form of the analyte and the particular matrix, heating rates, matrix/analyte modifier(s) added, graphite-tube coating, etc. Fernandez and Iannarone[60] found that the "maximum power" temperature-controlled heating of the Perkin-Elmer® HGA-2200 graphite furnace resulted in much lower atomization temperatures: 2000 vs. 2800°C (for pyrolytically coated tubes) and 2300 vs. 2900°C (for uncoated tubes), as compared with the "normal" (much slower) heating mode. During the ashing stage, premature volatization of the analyte

Table 1
ANTIMONY CONCENTRATIONS IN HUMAN BODY FLUIDS AND TISSUES

Sample	Unit	Selected reference values	Ref.
Blood	μg/ℓ	(<0.3—1.8)	93
		0.8	94
		1 (0—5.3)	95
		(<0.1—5.9)	96
Bone	μg/g	x̄, 0.007 (0.007—0.1)[a]	97
Brain	ng/g	7 ± 0.4	3
Feces	μg/day	3 (0.6—4.7)	96
Hair	μg/g	x̄, <0.01; 0.26 (<0.01—2.9) (Italy)	98
		x̄, 0.031 (0.016—1.30) (Japan)	11
		x̄, 0.027—0.057;[b] (0.005—187) (Austria)	99
		x̄, 0.064 and 0.212; (0.02—1.0) (U.S.A.)	9
		(<0.01—0.111)[c]	100
		0.12 ± 0.15 (0.003—1.53) (India)	101
		(<0.02—3) (U.S.S.R)	102
		x̄, 0.32 (0.073—0.602) (U.K.)	103
		(1.2—50)[d] (Canada)	104
Kidney	ng/g	x̄, 4.0; 13 ± 7 (1.0—42)	105, 106
		6 ± 1	3
		10 ± 30 (up to 130)	12
		x̄, 18.5 (2.5—445)	103
Kidney cortex	ng/g	5 ± 0.4	3
		39	107
Liver	ng/g	x̄, 6.0; 13 ± 6 (1.0—36)	105, 106
		10 ± 2	3
		10 ± 30 (up to 130)	12
		11 (2—20)	108
		17	107
Lung	ng/g	x̄, 29; 33 ± 8 (11—67)	105, 106
		(35—84)	109
		60 ± 5	3
		x̄, 62 (7—497)	103
		x̄, 250; 310 ± 230	12
Milk	ng/g	<0.2[d]	110
Muscle	ng/g	9 ± 3	3
Nails (finger)	ng/g	x̄, 75; 163 ± 178 (16—523)	103
Plasma or serum	μg/ℓ	0.75 ± 0.51	111
Skin	ng/g	x̄, 39; 78 ± 115 (5—665)	103
		x̄, 200; 200 ± 100[d]	112
Tooth (dentine)	μg/g	0.69 ± 0.41	23
Tooth (enamel)	μg/g	x̄, 0.038; 0.076 ± 0.112 (0.0049—0.646)	103
		0.13 ± 0.01 (0.02—0.34)	25
		0.96 ± 0.69	23
Urine	μg/ℓ	<0.05	113
		≤0.4	114
		(<1—5, generally <1)	115
	μg/day	(Up to 1.1)	98
		1.5 ± 0.57 (0.52—2.6)	111
		<0.1[c] (<0.1—925)	96

Table 1 (continued)
ANTIMONY CONCENTRATIONS IN HUMAN BODY FLUIDS AND TISSUES

Note: Wet weight unless otherwise indicated. Mean ± SD (range in parentheses). x̃, median.

[a] Femur.
[b] Range of medians.
[c] Variation over the same head.
[d] Dry weight.
[e] Most mean values <0.1 (Italy).

(mainly as volatile chlorides) is to be expected. The salts of Ni and Mo, as well as tartaric acid, act as matrix/analyte modifiers. Strong matrix effects (usually depression) and background absorption are typical. Gas-stop facility as well as atomization from a L'vov platform should be very useful.

The HG-AAS is the most popular technique for nanogram amounts of Sb. By adding a strong reductant, typically aqueous 1 to 5% $NaBH_4$, the analyte is reduced to stibine, SbH_3, which is stripped out of solution and introduced into a flame or furnace-heated quartz tube. The temperature of pyrolysis should be at least 850°C[139] or higher (up to 1000°C). The detection limits are between 0.1 and 2 ng/mℓ (1 to 6 ng).[126,139-142] These figures are comparables with the detection limits of several other instrumental methods for the quantitation of Sb after stibine generation: atomic fluorescence spectrometry (AFS) (0.1 ng);[143,144] ICP-ES (0.5 to 1 ng/mℓ);[142,145] MIP-AES (0.5 ng/mℓ and 10 ng);[131] and gas chromatography (GC) (10 to 20 ng/mℓ and 10 ng).[146,147] The lowest detection limits seem to have been obtained so far by Andreae et al.:[148] the antimony species have been selectivity reduced to SbH_3 (from Sb(III) + Sb(V), at pH ~1 and in presence of KI) or to SbH_3 (from Sb(III)) at near-neutral pH) and alkylstibines (from "monomethylantimony" and "dimethylantimony", at pH < 4). The hydrides have been trapped in a U-shaped tube at liquid nitrogen temperature, and subsequently released and purged into a graphite tube heated to 2200°C. The detection limits have been as low as 30 to 60 pg absolute, i.e., 0.3 to 0.6 ng/ℓ in 100-mℓ samples of environmental waters.[148]

Stibine is generated in a broad range of acidity: from 5 *M* HCl to pH = 8.[131,139,140,143,147-153] The optimum acidity appears to be around 1 *M* HCl,[139,153] but higher concentrations of HCl should also be tolerable.[140,150]

In strongly acidic media, i.e., ≥1 *M* HCl, both oxidation states of antimony, Sb(III) and Sb(V), are reduced to stibine but the peaks of Sb(V) are lower (1.3- to 8-fold, depending on acidity, collection time, etc.[139,142,143]) and broader. Peak-area measurements may improve the response of Sb(V), or else any Sb(V) will have to be prereduced to Sb(III). This prereduction can be easily performed at pH 0 to 1 adding KI (or NaI),[139,141,143,148,151,153,154] or ascorbic acid + KI,[136] or Na_2SO_3 and then KI.[154]

Sb(III) is selectively reduced to SbH_3 in presence of Sb(V) at pH ≥ 2 (from citric acid media)[153] or at pH ≥ 4 (from tartaric acid media)[153] and up to pH ~7[148] or pH = 8 (borate buffer).[152] The pH dependence of stibine generation has been studied in detail,[148,152,153] and has been applied to speciation studies of Sb in environmental water samples.[148,152,153] Two more procedures for the selective evolution of SbH_3 from Sb(III) in presence of excess Sb(V) are worth mentioning here: Nakashima[151] determined Sb(III) in presence of a fourfold excess of Sb(V) in 0.35 *M* HCl containing 4 mg/mℓ Zr(IV) (as zirconium oxychloride); Tsujii[143] selectively reduced Sb(III) to SbH_3 in presence of 1 *M* HCl and 0.1 *M* HF, the latter being added to mask the Sb(V) (Caution: paraffin wax-coated cell used!).

The stibine-generation AAS is prone to interferences from acids,[142,154] strong oxidants (HNO_3, NO_3^-, NO_2^-, and certain ions.[153,155] H_2SO_4, $HClO_4$, and H_3PO_4 only slightly depress the analyte signal,[142] while HNO_3, NO_3^-, and NO_2^- are strong depressants and must be removed from the solution by proper completion of the digestion procedure.[154] As a rule, the concentration of any acid or buffer should be matched in both sample digests and standard solutions.

Smith,[155] in an extensive interference study, found very strong interference (>50% suppression) by Ag, Co, Ge, Ni, Pt, Pd, and Ru, and strong depression (10 to 50%) by Ag, As, Cr, Cu, Re, Se, and Sn, each ion at 1000-fold excess in 1 *M* HCl. Yamamoto et al.[153] found significant interference (>10% depression) by diverse ions at the following excess over Sb(III): Ag^+, Bi^{3+}, Cu^{2+}, Se(IV), Te(IV) (10 ×), As(III) (200 ×), S^{2-} (3000 ×), Sn^{2+} (5000 ×), and As(V) (9000 ×); all present in citrate buffer at pH = 2. Fortunately, most of these potential interferents are not normally present in biological digests.

III. AAS METHODS FOR ANALYSIS OF BIOLOGICAL SAMPLES

The flame AAS is not applicable to biological specimens owing to its limited sensitivity. The flameless AAS techniques generally involve a preconcentration/separation of Sb from the matrix, either by solvent extraction[7,93,134,156,157] or by stibine evolution.[126,136,141,154,158-160] Analyses of blood,[7,93,136,156,157,160] dried milk,[154] food,[141] tissues,[7,126,158] urine,[7,159,160] and yeast[134] have been described. Selected procedures for the determination of Sb in biological samples are briefly outlined in Table 2.

Antimony may be expected to be rather nonhomogeneously distributed in solid biological specimens, as observed, e.g., in samples of hair[161] and liver.[108] The exogenously deposited Sb may be very difficult to wash from hair specimens unless an EDTA washing is applied.[162]

External contamination of biological samples during collection and handling is likely to occur, as nanogram and subnanaogram levels of the analyte are involved. The dust fallout in a normal and in a dust-free laboratory, as measured by Lievens et al.,[108] proved to be equivalent to 0.013 and 0.001 ng Sb/cm^2/day, respectively. Karin et al.[163] found as much as 7.5 ng Sb/g in polyethylene, and 2.6 ng/g of this was leached-out by 8 *M*HNO_3 in 3 days.

Irreproducible wet-ashing blanks arising in NAA from Sb impurity in quartz have recently been discussed by Versieck and Cornelis (see Reference 49 and references therein).

Analyte losses are even more likely than contamination to occur at all stages of the assay: storage, drying, ashing, separation from the matrix, etc. Both volatization and retention[164] losses of Sb have been observed, which is not surprising in view of the pronounced volatility of some Sb species (halides, methylated forms, etc.) as well as of its strong hydrolysis in solution. Antimony has been found in the precipitate formed during storage of spiked human urine.[165] Iyengar et al.[166] studied losses of metabolized ^{124}Sb in rat blood, serum, erythrocytes, brain, heart, kidney, liver, lung, muscle, nail, ovary, spleen, testis, thyroid, and uterus during freeze drying up to 72 hr at 0.05 Torr, and during 3-day oven drying at 80°C, followed by 1 day at 105°C and 1 day at 120°C. Freeze-drying and oven drying at 80°C proved to be safe for all specimens, while small but significant and tissue-specific losses of Sb were found after an oven drying of blood samples at 105 and 120°C (recovery (R) of 97.1 and 95.4%, respectively), and rat brain, kidney, lung, and spleen at 120°C (R = 94 to 96%).[166]

Behne and Matamba,[167] by instrumental NAA (INAA), were able to quantitatively recover the endogenous Sb in human serum (0.28 ng Sb/g) after freeze-drying or oven drying at 90°C for 3 days, but lower results (0.18 ng Sb/g) were obtained after low temperature ashing (LTA) of the lyophilized serum samples. By using a modified LTA procedure (with a liquid-nitrogen cooled condenser, as described by Kaiser et al.[168]), Van Montfort et al.[95] ashed dried-blood samples; the ashes were subsequently dissolved in HCl under reflux, extracted

Table 2
SELECTED AAS PROCEDURES FOR ANTIMONY DETERMINATION IN BIOLOGICAL SAMPLES

Sample	Brief procedure outline	Ref.
Blood	A 3-mℓ sample mixed (1 + 1) with an ashing aid [aqueous 50% w/v $Mg(NO_3)_2$], dried at 60°C for 12 hr and ashed at 550°C for 6 hr; ashes dissolved in 5 mℓ of 6*M* HCl. KI added to reduce Sb(V) to Sb(III); Sb(III) extracted twice with 5 mℓ of Amberlite® LA solution in xylene; back extraction of Sb(III) into 2 mℓ of 0.01 *M* tartaric acid; ET-AAS; background correction	93,157
Blood, urine	Blood (1 mℓ) and urine (50 mℓ) wet ashed with HNO_3 + H_2SO_4;[a] digests diluted to give ca. 9 N H_2SO_4; HG-AAS	136
Fish, food, tissue	Wet digestion with HNO_3-H_2SO_4-$HClO_4$ (4:1:1);[a] HG-AAS	141
Food, milk	HNO_3-H_2SO_4 digestion (general procedure) or HNO_3-H_2SO_4-$HClO_4$ digestion (milk);[a] Na_2SO_3 and KI added [→ Sb(III)]; HG-AAS in 3.6 *M*HCl	154
Food, tissue	2-g sample wet ashed with 30 mℓ of a (9:1:2) HNO_3-H_2SO_4-$HClO_4$ mixture;[a] finally completed with (9:2) HNO_3-$HClO_4$;[a] HG-AAS in presence of 4 *M*HCl, $Na_2P_2O_7$ (to mask iron), and KI [→Sb(III)]	126
Tissue	Samples (0.4 to 5 g) wet digested with HNO_3 + H_2SO_4; digestion completed[a] with HNO_3 and H_2O_2; HG-AAS in presence of EDTA, KI, ascorbic acid, and 1 *M*HCl	158
	1—5 g samples wet ashed with 20 mℓ of HNO_3 and 2 mℓ of H_2SO_4; SDDC-MIBK extraction at pH 5.5 to 6.5; ET-AAS: 100°C for 20 sec, 300°C for 20 sec, and 2400°C for 6 sec; background correction; standards extracted similarly	7

Note: [a] Gradual, temperature-controlled heating and safety precaution mandatory with most ashing procedures.

with DDDC in CCl_4, back extracted into 0.1 *M*HCl, and finely analyzed by MIP-AES (the recovery of Sb in the whole procedure was, however, as low as 40.5%).[95]

Dry-ashing techniques are even more troublesome for Sb: in presence of NaCl and organic matter, both retention on silica crucibles and volatization of the analyte have been documented.[91,169] Dry ashing in presence of $Mg(NO_3)_2$ ashing aid has proved more successful.[91,93,157]

In wet-digestion techniques, it is essential to (1) maintain strong oxidizing conditions by using excess HNO_3, (2) start digestion slowly (e.g., at room temperature for a few hours), (3) apply gradual, temperature-controlled heating, and (4) avoid charring conditions. The most successful digestion mixtures obviously contain H_2SO_4;[7,91,114,115,154,164] the HNO_3-H_2SO_4-$HClO_4$ digestion appears to be the best choice,[114,164] while the digestion with HNO_3 and H_2SO_4 may often require proper completion (with caution!) by the dropwise addition of H_2O_2[129,170] or H_2O_2 + HNO_3[158] or $HClO_4$ + HNO_3.[114,170]

Wet-digestion procedures involving H_2SO_4 and H_2O_2[115] or HNO_3 and $HClO_4$[156] are prone to losses of Sb.[129,164] The decomposition of organic matter prior to Sb determination has been detailed in References 115, 154, and 164.

The stibine-generation AAS is sensitive and relatively selective, provided that:

1. Oxidants such as NO_3^- or NO_2^- are removed from the digest by proper completion of the digestion procedure;[154,164] otherwise the calibration must be performed by standard additions and peak-area measurements.
2. Acid-matched standard solutions are used.
3. Any Sb(V) is prereduced to Sb(III) as already discussed.
4. Masking agents: EDTA,[152,158] or citrate,[153] or pyrophosphate ($Na_4P_2O_7$)[114] are added to both sample digests and standard solutions, as required in some procedures in order to mask some interfering ions (e.g., excess iron,[114] copper,[153] selenium,[153] etc.).

Major limitations are intrinsic with the direct ET-AAS of Sb in biological liquids or digests: limited sensitivity, severe background, volatization losses, and pronounced interferences.

Only one direct ET-AAS procedure appears to have been described; Ward et al.[137,171] studied the concentration levels and distribution of Sb(III) and Sb(V) drugs (Triostam® and Pentostam®, respectively), either in free or in liposome-entrapped forms, both in vitro (after incubation with human blood) and in vivo (in mice experiments). Samples of tissue homogenates (1:10, in 10 m*M* Tris buffer at pH 7.2) or (1:100) diluted plasma or (1:1000) diluted erythrocytes have been directly injected, ramp ashed up to 900°C, and atomized under gas-stop conditions at 2200°C. The Sb(V) bound to gluconic acid residues in the Pentostam drug has proved to be thermally stable so that matrix-matched calibration is required.

Smaller amounts of Sb are most conveniently separated/preconcentrated from biological digests by means of liquid-liquid extraction.[76,172] In these procedures it is essential to have the analyte in a definite oxidation state, either Sb(III) or Sb(V) (see References 129, 172, 173, and 164). Prereduction to Sb(III) can be performed by KI, Na_2SO_3, or hydrazin sulfate, whereas oxidation to Sb(V) can be done by $KMnO_4$, $Ce(SO_4)_2$, or other strong oxidants. A few extraction systems seem to have been successfully applied to biological samples: Sb(III)—dithiocarbamates at pH 4 to 6.5;[7,95,115,130,173] Sb(III) chloride—6 M HCl — 5% Amberlite® LA-xylene;[93,157] Sb(III)chloride—6 *M*HCl-MIBK;[156] and Sb(V)chloride—Rhodamine B dye complex.[134,135] The dithiocarbamate extracts are readily analyzed by ET-AAS,[7,115,172,173] whereas the extracted chloride complexes are volatile and should therefore be backextracted (e.g., with a small volume of 0.01 *M* tartaric acid[93]). Direct extractions of Sb with Na diethyldithiocarbamate (SDDC)-methylisobutyl ketone (MIBK) from heparinized hemolyzed blood at pH 6.5 and from pH-adjusted urine (pH 5.5 to 6.5),[7] as well as ammonium pyrollydinedithiocarbamate (APDC)-MIBK extraction from undigested urine at pH 4 to 5[115] have been described; however, the complete recovery of all analyte forms is questionable.[115]

The APDC-MIBK extracts of Sb(III) have proved to be stable during storage.[172,173] A suitable temperature program should include ashing at 300 to 500°C and atomization at 2400 to 2600°C, with simultaneous background correction.

Other effective preconcentration procedures for trace Sb(III) are precipitation with dibutyldithiocarbamate (from biological digests, at pH 3.5 to 4.0)[129] and batch sorption on a hydrophilic glycomethacrylate gel with bound thiol groups (Spheron Thiol®, Lachema, Brno), from 1 to 3 *N* solutions of HCl or H_2SO_4.[174]

IV. CONCLUSION

NAA appears to be the most reliable analytical method at nanograms-per-gram Sb levels; the use of ET-AAS, HG-AAS, and ASV, as more accessible techniques, may be expected to increase. Sample preparation generally involves: (1) wet ashing with HNO_3 + $HClO_4$ + H_2SO_4 (better) or with HNO_3 and H_2SO_4, under carefully controlled conditions; (2) reducing the analyte to Sb(III); and (3) separation/preconcentration from digests by liquid-liquid extraction or stibine generation. Care should be taken to avoid retention and volatization losses. Matrix effects are encountered in both ET- and HG-AAS.

Chapter 3

ARSENIC

I. DETERMINATION OF ARSENIC IN BIOLOGICAL SAMPLES

In most human biological materials, the concentration levels of As are well below 1 μg/g, and down to a few nanograms per milliliter in biological liquids[175-183] (Table 1). The flameless AAS techniques (HG and, to a lesser extent, ETA) and NAA are among the best techniques available.

NAA has been widely applied to As assays in blood,[45,103,170,178,179] bone,[97,103] diet,[45,116] erythrocytes,[176] hair,[11,45,48,99,103,104,118,179,184] milk,[40,110,182,185] nails,[103] serum/plasma,[45,111,176,186] skin,[112] teeth,[103] tissues,[12,45,103,108,179] urine,[45,111,165,179] and other biological samples.[45,121] This technique has a number of assets: high sensitivity, low blanks, small sample-size requirements, as well as suitability to instrumental performance (INAA),[11,12,99,104,118,121] with no wet chemistry involved.

Most spectrophotometric procedures[40,133,188-190] involve volatization of the analyte as AsH_3 which is then trapped and reacted with AgDDC to form a color product; they are, however, neither sensitive nor selective.

Applications of the SSMS technique to hair,[54] nails,[16] plasma,[55] teeth,[25] tissues,[132] etc.[132] have been described.

Sensitive AES procedures with plasma-source excitation have been developed;[131,191-194] the analyte has been separated from the matrix as a volatile bromide[192] or hydride,[131] and then introduced into an ICP[192] or MIP excitation source.[131] The classical arc-source AES method also needs preconcentration (by coprecipitation or dithiocarbamate extraction[130]).

XRF lacks sensitivity,[128] unless preceded by an efficient enrichment (e.g., by coprecipitation[129]).

The PIXE technique[128,195] has been used to measure the As distribution in very small sample areas, e.g., across the diameter of a single human hair.[195]

Incidentally, the high-speed ASV[195a] and DPASV[196] have been applied to biological specimens.

Various sensitive analytical techniques have been involved in speciation studies of As in biological/environmental samples: AAS,[177,197-215] MIP-AES,[193,194] ICP-ES,[191,216] AgDDC-spectrophotometry,[217-219] GC-MS,[220-221] and thin layer chromatography (TLC)-XRF.[222] See also Lauwerys et al.[205]

II. ATOMIC ABSORPTION SPECTROMETRY OF ARSENIC

The flame AAS of As suffers from poor sensitivity: the characteristic concentration and detection limits (DLs) are within the ranges of 0.15 to 0.8 and 0.04 to 0.3 μg/mℓ, respectively, depending on the flame type and light source used.[56-59]

Significant light losses are typical, owing to the short wavelength of the As resonance line, 193.7 nm. Hence, an EDL light source is recommended, for its better DLs (3×), higher intensity (9×), and longer lifetime.

In ETA, the characteristic concentration is as low as 0.06 to 0.4 ng/mℓ (6 to 38 pg), and the DLs are between 0.02 and 0.2 ng/mℓ (about 20 pg).[57,59-61]

Major problems are encountered in the practical application of the ET-AAS to complex solutions containing low amounts of As:

1. Volatization losses of the analyte during the ashing

Table 1
ARSENIC CONCENTRATIONS IN HUMAN BODY FLUIDS AND TISSUES

Sample	Unit	Selected reference values	Ref.
Blood	μg/ℓ	♀, 1.5;[a] 2.3[b]	175
		2.5;[c] 21.6[d]	176
		3.8 (0.5—9.4)	177
		4 ± 12[e]	178
		x̄, 7.6; 29 ± 54 (0.2—184)	103, 179
Bone	ng/g	x̄, 57; 80 ± 68 (10—240)[f]	103
		(<5—7)[g]	97
Brain	ng/g	x̄, 2.8; 3.4 ± 2.1 (0.21—7.7)	103
Erythrocytes	μg/ℓ	2.7 ± 1.3;[c] 32.7[d]	176
Hair	μg/g	x̄, 0.047—0.108; (0.01—19.43) (Austria)	99
		x̄, 0.058 (0.0079—0.67) (Japan)	11
		0.083 ± 0.14 (0.001—0.93) (India)	101
		0.27 ± 0.30 (<0.15—1.6) (Iraq)	118
		0.40 ± 0.26 (U.S.A.)	180
		(0.32—2.8)[f](Canada)	104
		x̄, 0.46 (0.02—8.17)[f] (U.K.)	103, 179
Kidney	ng/g	5 ± 1	105
		x̄, 7.4; 11.3 ± 16.9 (0.45—82)	103,179
Kidney (cortex)	ng/g	x̄, 4—5 (3—14)	106,181
Liver	ng/g	x̄, 3.0 (1—13)	106, 181
		6 ± 2	105
		9 ± 14	109
		x̄, 7.8; 16 ± 16 (1.4—143)	103, 179
Lung	ng/g	x̄, 7.0 (1.0—18)	106,181
		x̄, 18; 25 ± 22 (1.3 —113)	103,179
		20 ± 10	3
Milk	μg/ℓ	1.2[e,f]	110
		3.2 (1.6—6)	182
Muscle	ng/g	2 ± 1	3
Nails	μg/g	x̄, 0.3; 0.36 ± 0.31 (0.02—2.9)[f]	103, 179
Nails (finger)	μg/g	(0.16—2.7);[f,i] (0.37—2.2)[f,j]	16
Plasma or serum	μg/ℓ	2.4 ± 1.9;[c] 15.4 ± 14[d]	176
		8.9 ± 6.2	111
Skin	ng/g	x̄, 90;[f] 124 ± 119 (9—590)[f]	103
		x̄, 100;[f] 180 ± 160[f]	112
Tooth (enamel)	μg/g	<0.02	25
		x̄, 0.05;[f] (0.003—0.635)[f]	103
Urine	μg/ℓ	<10	183
		x̄, 42;[f] 114 ± 164 (6—770)[f]	179

Note: Wet weight unless otherwise indicated. Mean ± SD (range in parentheses). x̄ median.

[a] Nonsmokers.
[b] Smokers.
[c] Denmark.
[d] Taiwan (environmental exposure).
[e] ng/g.
[f] Dry weight.
[g] Femur.
[h] Range of medians.
[i] Adults.
[j] Children.

2. Strong dependence of arsenic behavior and sensitivity on its chemical form
3. Severe background absorption
4. Interferences by numerous concomitants

In view of the volatility of some arsenic species, ashing temperatures should be restricted to as low as 100 to 200°C (in presence of organoarsenic compounds and halides[203]) or 500 to 700°C (for inorganic arsenic).[223-225] Thus, the analyte has to be thermally stabilized by adding a suitable matrix/analyte modifier.[223] Typically, this is a nickel salt — 0.1 to 0.2% of Ni^{2+} as sulfate or nitrate[206,219,223-226] — the addition of which results in raising the maximum ashing temperatures up to 1000 to 1400°C. Silver can also serve as a thermal stabilizer (up to 1150°C).[225] Both Ni and Ag improve the analytical sensitivity as well (by a factor of up to 2 for inorganic As[225] and even more for methylated arsenic species[203]). Accuracy and precision are also improved. The matrix-modifier solution can be directly added to the analyzed solution into the graphite tube by using versatile autosampling devices:[61,227] this permits one not only to save time and reagents, but also to use Ag^+ solution as matrix/analyte modifier (which otherwise gives a precipitate of AgCl when mixed with biological liquids/digests).

The next problem in the ET-AAS of As is that the analyte may be present in various chemical forms (e.g., arsenite, arsenate, methylated arsenicals, etc.) in intact samples or in incompletely digested samples; these forms exhibit a quite different behavior on charring and atomization.[203,224] Odanaka et al.[202,203] successfully dealt with this problem by adding solutions of 0.01 *M* Ni^{2+} (587 μg Ni/mℓ) and 0.01 *M*NaOH to their extracts and TLC concentrates from biological and environmental materials which contained arsenite, arsenate, methylarsonate (MA), and dimethylarsinate (DMA). The alternative way is to completely digest the samples and thus to convert the analyte into a definite chemical form and oxidation state, i.e., into an arsenate.

In spite of the addition of a matrix/analyte modifier and the careful optimization of the heating program, the background absorption by the As 193.7-nm line is severe and may well exceed the correction ability of the deuterium background corrector. This problem is even more pronounced with older AA spectrophotometer models as well as with slowly heated graphite furnaces. A radical solution is to use the Zeeman ET-AAS, which can correct for a much higher, structured, and transient background.

Many concomitants such as acids, ions, and salts may depress or enhance the As absorption signal.[224,226,228] Severe depression by small amounts of K^+, K^+ plus SO_4^{2-}, and Na^+ plus SO_4^{2-}, all at concentrations of only a few micrograms per milliliter, has been observed.[224] Though to a lesser extent, some other ions may also interfere: Na^+, Fe^{3+}, Ca^{2+}, Mg^{2+}, Cl^-, SO_4^{2-}, PO_4^{3-}, MoO_4^{2-}, etc.[224,226,228] The matrix effects appear to be less marked in presence of a matrix modifier and under a versatile furnace programming. It is advisable to apply faster heating rates during the atomization step — the "maximum power" heating allowed so as to decrease the optimum atomization temperature to 2200 vs. 2900°C as well as to extend the useful lifetime of the tube).[60] The atomization from the L'vov platform,[229] the modification of the tube surface,[230] and the "gas-stop" mode of atomization are to be recommended as well.

For the above reasons, the ET-AAS of As in biological samples is — at least now — restricted to analyses of relatively simple aqueous or organic solutions obtained after a preconcentration/separation of the analyte from the bulk salt matrix. Standard additions or matrix-matched calibration modes are often mandatory.

The hydride-generation AAS (HG-AAS) appears to be the best current technique for the determination of nanogram amounts of As. The lowest values of the characteristic concentration and the DLs are in the range of 0.02 to 1 (0.09 to 2 ng) and 0.025 to 0.8 ng/mℓ (0.25 to 10 ng), respectively.[57,139-142,150,198,231,232] These figures are quite favorable compared with the DLs of some other instrumental techniques involving arsine evolution and subsequent excitation of the atomized arsenic: AFS, 0.1 ng;[144] AFS (with electrochemical reduction to AsH_3), 0.015 ng;[233] MIP-AES, 0.35 ng/mℓ;[131] and ICP-ES, 0.4 to 0.8 ng/mℓ.[142,145]

The HG-AAS is well documented; there are recent reviews,[232,234] optimization,[235] and interference studies[120,140,155,198,199,226,236-238] whose results and conclusions are not always consistent. In general, this technique is fairly selective, relatively rapid and simple, and can be semiautomated or automated.[141,198,239] There are, however, many instrumental parameters and experimental factors which affect the analytical sensitivity, the form of the analyte peaks, the nature and extent of the interferences, and the reproducibility of the assays: (1) the design of apparatus; (2) the acidity or pH of the sample solution; (3) the presence of (added) buffer components or masking agents; (4) the oxidation state and the chemical form (e.g., the presence of methylated species) of the analyte; (5) the volume of the sample solution as well as the total volume of the system; (6) the amount and concentration of the reductant, the $NaBH_4$, its age, purity, etc.; (7) the time of hydride collection; (8) the purge-gas flow; (9) the interferent: arsenic concentration ratio as well as their absolute amounts; (10) the oxidation state of the interferent; (11) the presence of foaming and antifoaming agents; (12) the pyrolysis temperature of AsH_3/arsines, and so on. Just a few of these factors will be discussed.

The arsine (AsH_3) evolution takes place within a broad range of acidity: from 10 *M*HCl to pH ~6.[140,209,211,237,240] Concentrations of HCl between 1 and *4* M appear to be most convenient for the generation of AsH_3 from both oxidation states, As(III) and As(V). At pH >0 (<1 *M*HCl), the peak height of the As(V) (arsenate) gradually decreases, and at pH 4.5 to 5.0 the As(III) (arsenite) is selectively determined in the presence of a 30 to 100-fold excess of arsenate.[153,193,198,209,211,240] Methylated arsenic species (MA, DMA, etc.) are also reduced by $NaBH_4$ to their corresponding alkylarsines (e.g., to CH_3AsH_2, $(CH_2)_2AsH$, etc.) which, however, exhibit a behavior (rate of evolution, stripping-out of solution, etc.) quite at variance with that of inorganic arsenic/arsine.[198,211]

Depending on the experimental conditions, the peaks of the arsenate may be significantly lower (e.g., by 10 to 30%) than the corresponding peaks of arsenite. This difference should not be neglected; it is accounted for either by using standard solutions containing As in the same oxidation state in which it is believed to be present in sample solutions, or by prereducing any As(V) to As(III) before quantitation. Unfortunately, this prereduction is not as straightforward as it may appear. There is evidence now that the most common reductant, KI (or NaI), may require up to 1 hr to reduce arsenate to arsenite.[129,139,172]

The interfering effects (depression) of the following concomitants are considered as most serious: strong oxidants ($Cr_2O_7^{2-}$, MnO_4^-, VO_3^-, $S_2O_8^{2-}$, MoO_4^{2-}, NO_3^-),[226] ions of noble metals which are easily reduced to their elemental state (Ag, Au, Pd, Pt),[155] and other hydride-forming elements (Se, Te, Sb, Sn),[153,155] as well as certain ions which react with the analyte (Fe, Co, Ni, Cu, S^{2-}, etc.).[153,155,214,238]

Finally, the $NaBH_4$ reagent is worth a couple of words, too. Aqueous $NaBH_4$ solutions are not stable on storage and therefore KOH or NaOH (0.5 to 10% w/v) has to be added to stabilize. Depending on the particular lot of the reductant used, as well as on the further storage conditions, the $NaBH_4$ solutions may contain a carbonaceous precipitate,[241,242] variable amounts of interfering impurities,[242] and different blank levels of As.[243,244] This may result in interferences varying in degree,[242] lower reducing power, impairment of precision,[241] and variable blank levels (e.g., from 0.1 to 90 ng As[243,244]). The shift of the pH when an alkaline $NaBH_4$ is added may not be negligible, either.[211,245]

Freshly prepared $NaBH_4$ solutions should be vacuum-filtrated through a 0.45-μm membrane filter.[241] Most authors prefer to store the reductant solution under refrigeration or to prepare it fresh daily.

Although the typical performance of the HG-AAS should be relatively simple, i.e., against acid-matched standard solutions and without a simultaneous background correction, every newly adopted procedure and every application to a new matrix should be checked for background and matrix effects.

III. AAS METHODS FOR ANALYSIS OF BIOLOGICAL SAMPLES

Both flameless AAS techniques, HG- and the ET-AAS, have been widely applied to analyses of biological materials. The former is of greater practical value, at least at present, and has been applied to samples of blood,[236,246,247] feces, [248] food,[141,249-252] fish,[238,239] hair,[238,247,253] milk,[247] tissues,[238,239,247,254-256] and urine.[27,159,183,209,236,246,247,253,254,257,259-261] ET-AAS has been used in analyses of blood,[7,175,177,230,262,263] feces,[230] fish,[264] tissues,[7,247,262] and urine.[7,201,205,230]

Both techniques are also suitable to speciation studies of As in biological specimens and are therefore attracting growing attention.[177,197-215]

There are three main groups of problems which have to be solved adequately in these assays:

1. To perform an adequate mineralization step, i.e.: (a) to avoid losses of the analyte while (b) ensuring complete cleavage of the As -C bond (in MA and DMA which are especially difficult to oxidize, yet are rather volatile) and (c) keeping acceptably low blank levels
2. To quantitatively convert the analyte into a definite oxidation state in which form it is separated/preconcentrated from the matrix [often, that is to reduce all the arsenic to As(III)]
3. To properly account for possible interferences

Due to interferences and other restrictions of the flameless techniques already discussed above, the direct methods are practically inapplicable to determinations of the total As. Accordingly, the typical As assays involve successive steps of complete ashing, separation of the analyte by hydride generation or solvent extraction, and a final measurement by means of AAS. Only a few exceptions from this general scheme seem to have been reported.[246,258,260,265]

Ashing techniques which most successfully conform to the above requirements of complete ashing are summarized in Table 2. Unfortunately, many of the published procedures have not been thoroughly tested by means of endogenously incorporated radioisotopes or at least by exhaustive studies of recoveries of inorganic and methylated arsenic compounds.

Losses and/or incomplete digestion are typical of this element in many common ashing procedures. Moreover, these two major drawbacks are obviously matrix and technique dependent. Inadequate are the plain dry ashing,[91,266] and even some oven-drying procedures.[91] Most bomb decompositions likewise yield lower results,[265,267] probably due to incomplete digestion and/or matrix effects, although they may be used only as a pretreatment technique and then completed in an open vessel.[250] There are conflicting opinions on the applicability of LTA.[91,169,265,275,276] Further work is called for on validating wet-digestion techniques with HNO_3 + H_2SO_4 + $KMnO_4$[267] and H_2SO_4 + $K_2S_2O_8$.[91,209] The binary mixtures of HNO_3 + $HClO_4$[141,209,248] and HNO_3 + H_2SO_4[209,256,267] may not be adequate either (both losses and incomplete ashing have been experienced), unless certain especial precautions are taken: close temperature control,[129,170,256,269] use of long-neck flasks[129,170,249] or tall test tubes,[183] addition of catalysts (molybdate[248,270] or vanadate[264] or chromate[91]), assistance of microwave radiation,[277] proper completion of the digestion with H_2O_2,[129,170,247] etc. Digestion should be properly started, preferably by leaving the sample with HNO_3 at room temperature for

Table 2
SELECTED ASHING PROCEDURES FOR BIOLOGICAL MATERIALS PRIOR TO AAS DETERMINATION OF THE TOTAL ARSENIC

Oxidation mixture	Matrix
HNO_3, $HClO_4$, and H_2SO_4	Blood,[230] feces,[230] fish,[141,238,239] food,[141,230,252] tissues,[141,230,238] urine[230,268,269]
HNO_3, $HClO_4$, H_2SO_4, and Na_2MoO_4	(Pig) feces,[248] feed,[248] urine[270]
Dry ashing in presence of $Mg(NO_3)_2$	Blood,[163,177] food,[251,271] tissues[214,263]
Dry ashing in presence of $Mg(NO_3)_2$ + MgO	Blood,[262] feces,[261] food,[261] tissues,[262] urine[27,190,205,261]
HNO_3, H_2SO_4, and V_2O_5	Fish[264]
HNO_3 and $HClO_3$	Fish, meat, milk, serum, etc.[272]
HNO_3 (autoclave), then completed by dry ashing in presence of $Mg(NO_3)_2$	Food, (evaporated) milk, tissues, etc.[250]
Combustion in O_2 (closed vessel)	Food, milk, tissues, etc.[265,273,274]

a few hours or overnight and excess of HNO_3 until all organic matter is oxidized should be ensured.

In order to reduce the reagent blanks as well as the consumption of high-purity reagents (which should contain only a few nanograms As per milliliter or less), it is recommended to use long-neck digestion vessels, predigestion with HNO_3 alone (e.g., in closed PTFE-lined tubes), and/or catalysts.

There are several methods for preconcentration/separation of trace As from biological digests: hydride evolution, liquid-liquid extraction,[7,76,172,177,201,205,230,263,278] distillation as a volatile bromide[192] or chloride,[195a,278] and coprecipitation.[129] The first two methods appear to be most convenient and are widely applied.

As(III) can be extracted as dithiocarbamate (e.g., with APDC-MIBK, at pH between 1.0 and 6.8[172]),[7,230,278] as a chloride[177,201,263,278] or other halide complex (e.g., from HCl-HBr-$HClO_4$-KI-medium[205]), as well as with the NH_4 sec-butyldithiophosphate.[280]

The dithiocarbamate extracts are directly injected into a (pyro-coated[172]) graphite-tube furnace, rampashed at 100 to 325°C and atomized at 2500 to 2700°C.[7,172,230] Background correction is required. If the organic phase is not completely separated and contains occluded tiny droplets of the aqueous phase, there is a strong background absorption of the occluded Na^+,[7] as well as an enhancement of the As absorbance by the iron.[230] Both interferences can be eliminated by rinsing the extracts with high-purity water; this step should also help to increase the time stability of the chelate from 3 to 9 days.[172] An excess of Cu(II) (e.g., 50 ×, as present in liver digests), has been found to depress the As(III) extraction with SDDC-$CHCl_3$; thus Cu(II) has to be pre-extracted from the digests.[230]

In halide-extraction procedures, the analyte must be back extracted from the organic phase with dilute aqueous solutions of $Mg(NO_3)_2$ (0.025% w/v),[177,263] or $Co(NO_3)_2$,[205] or $K_2Cr_2O_7$ (0.005 *M*),[201] or simply with water.[201]

Inasmuch as most digestion procedures leave the analyte in both oxidation states, As(III) + As(V), all liquid-liquid procedures must be preceded by a step of prereduction to As(III) (As(V) *is not* extracted in most cited procedures!). As already mentioned, prereduction with KI or NaI may be slow (1 hr) and/or incomplete.[129,172] Depending on the experimental conditions (e.g., the concentration of the reductant, duration the presence of oxidants in excess, etc.), the reductions with ascorbic acid[129] or with Na_2SO_3[129] may likewise be incomplete. Successful reduction procedures involving hydrazine sulfate,[129] $Na_2S_2O_3$,[172] KI + $Na_2S_2O_3$,[172,279] $NaHSO_3$ + HCl + $Na_2S_2O_3$,[280] and KI + $Na_2S_2O_5$[230] have been described recently.

The arsine generation is undoubtedly the best separation technique for nanogram amounts of As. Its main assets are efficiency, speed, "on-line" combination with the next AAS quantitation, and relative freedom from interferences. Another very useful feature of this technique is its ability to differentiate between the two oxidation states of the analyte as well as between its inorganic and methylated forms. However, while useful in speciation studies of As in untreated biological and environmental specimens, this feature tends to create serious problems when attempting to determine the total As in homogenates, untreated biological liquids, and even in incompletely digested samples.

In most commercially available apparatus, the evolved arsine is collected for a preset time and then purged into a flame or furnace-heated (850 to 950°C) quartz tube. There are also other techniques in which the arsine (or the arsines, in speciation studies) is (are) additionally concentrated. These techniques may offer certain advantages but at the expense of modifications and complications of the apparatus and/or the duration of the assay. The most effective approach consists of freezing-out the AsH_3/arsines into a cold trap at liquid nitrogen temperature, with a subsequent release at higher temperature.[193,198,214,235,247] The arsine is thus separated from the bulk H_2 evolved during the reduction, the sensitivity, the DL and the accuracy are significantly improved, and selective volatization of different arsenic species becomes possible.[193,198,281]

Another technique involves trapping the evolved AsH_3/arsines into a suitable absorbing solution which is subsequently analyzed by means of ET-AAS. This absorbing solution can be AgDDC in an organic solvent or aqueous I_2-KI-$NaHCO_3$.[245]

The evolved arsine can also be introduced into a heated graphite-tube furnace.[214,282]

The HG-AAS is relatively selective. As a rule, the quantitation is performed against acid-matched standard solutions, but some potential interferents are to be considered when present in excess in the sample digests: HNO_3,[226,236,253] viscous acids,[226,236] Fe,[198,214] Cu,[155,214] as well as some components of the ashing aids or the catalysts (if added during the ashing step) — molybdate, vanadate, dichromate, permanganate, $Mg(NO_3)_2$, etc. Excessive foaming due to incompletely oxidized matrix constituents may be also be encountered.[260] These adverse effects can reportedly be eliminated or accounted for by using peak-area measurements,[198,236,244] optimizing the HCl concentration,[211] adding of masking (EDTA,[198,214] oxalate,[235,258] etc.) and antifoam[246,260] agents, or by combinations of these approaches. Obviously more research work is called for in this field, as recently demonstrated in the AOAC collaborative study on the determination of As and Se in food.[249] Standard addition checks should be frequently made.

And, finally, the most difficult, challenging, and increasingly important[197,205] analytical task is the determination of the specific chemical forms and oxidation states of arsenic in biological and in related environmental materials (Tables 3 and 4). These speciation techniques are often based on selective hydride-evolution procedures[153,198,209,212,213,257,281] or liquid-liquid extractions,[197,201,205,207,208,278,279] though some other separation techniques such as selective distillation of inorganic As (from 6.6 *M*HCl),[278] TLC,[202,203] GC,[283] ion-exchange chromatography,[206,215,284,285] ion chromatography (IC),[204] high performance liquid chromatography (HPLC),[210] etc. may be used as well. In all these procedures, the separated As species are atomized in a graphite furnace[201-206,210,283-285] or in a flame- or furnace-heated quartz tube.[198,209,212,213]

Some difficulties in speciation studies of As are to be mentioned. The preparation step is a serious concern, as it should leave intact the concentration, the oxidation state, and the chemical form of the species of interest. The relatively mild homogenation/digestion/extraction procedures with HCl [198,207,208,278] or with aqueous NaOH[202,203,212,213,215] are widely used at present. Diamondstone and Burke[217] found no suitable acid digestion system that would effectively digest organic and biological materials without altering the original valence state of the As; during the H_2SO_4-HNO_3 treatment some As(III) was oxidized to As(V), whereas the H_2SO_4 digestion led to a partial reduction of As(V) to As(III).[217] Lauwerys et al.[205] found that As(III) added to urine was partially oxidized to As(V) on storage. Other

Table 3
AAS PROCEDURES FOR ARSENIC SPECIATION IN BIOLOGICAL MATERIALS

Sample	Species determined	Brief procedure outline	Ref.
Algae, clam tissue	DMA, MA, inorganic As	Extraction with 0.1 *M* NaOH; evaporation in vacuo; residue dissolved in 8.5 *M* HCl; 1 *M* KI saturated with ascorbic acid added; toluene extraction; back extraction in water; HCl and $K_2Cr_2O_7$ added; chromatography on Dowex® AG 50-X8 resin; eluents: 0.5 *M* HCl for inorganic As, water for MA, and aqueous 1.5 *M* NH_3 for DMA; HG-AAS	215
Algae, molluscs, water	DMA, MA, inorganic As	Overnight extraction with HCl at 65—70°C; filtration; dilution; pH-controlled HG-AAS in presence of EDTA; cold trapping and then selective volatization of AsH_3, CH_3AsH_2, and $(CH_3)_2AsH$	198
Algae, shark muscle	Inorganic As, organic As	Solubilization with 6 *M* HCl; *inorganic* As extracted into toluene and back extracted in water; *total* As determined after a (HNO_3-H_2SO_4-$HClO_4$) digestion	208
Bile, blood, feces, tissue, urine (all from exposed lab animals)	DMA, MA, inorganic As	Extraction with aqueous 0.2 *M* NaOH; benzene extraction; back extraction with water (according to Reference 220); TLC; extraction of the scrapped fractions with water; ET-AAS (in presence of Ni(II) and NaOH)	202, 203
Fish	Total inorganic As	*Option A:* distillation of the inorganic As from 6.6 *M* HCl; HG-AAS. *Option B:* dissolution in aqueous NaOH, extraction with APDC-MIBK, back extraction with HNO_3; digestion with H_2SO_4; HG-AAS	278
Lobster, mussels, urine	DMA, MA, TMA, inorganic As	CH_3OH extraction, digestion in aqueous NaOH, neturalization; HG-AAS with cold trapping followed by fractional volatization of corresponding arsines	212, 213
Phospholipids	DMA, MA, arsenite, arsenate	HPLC and Zeeman ET-AAS	210
Tissues	DMA, MA, inorganic As	HCl extraction of tissues; species selectively extracted from HCl-solutions; GLC or ET-AAS	207
Urine	Hydrolyzable As, total As	ET-AAS	200
	Inorganic As(III)	Toluene extraction; back extraction with aqueous $Co(NO_3)_2$; ET-AAS	205
	Inorganic As, total As	12 *M* HCl added; left for 1 hr at 60—70°C; KI added; As(III) extracted in $CHCl_3$; back extraction with water (inorganic As) or with 5 m*M* $K_2Cr_2O_7$ (total As); ET-AAS	201
	DMA, MA, inorganic As, total As	HG-AAS at pH 0; cold trapping and fractional volatization	197, 209

Table 4
SELECTED AAS PROCEDURES FOR ARSENIC DETERMINATION IN BIOLOGICAL SAMPLES

Sample	Brief procedure outline	Ref.
Blood, feces, food, serum, urine	Predigestion with HNO_3, completion with $HClO_4$ and H_2SO_4;[a] reduction to As(III) with KI and $Na_2S_2O_5$; SDDC-$CHCl_3$ extraction; ET-AAS: 100°C for 25 sec, 525°C for 45 sec, and 2700°C for 7 sec; background correction; standards extracted similarly	230
Blood, tissues,	Extraction of $AsCl_3$ from HCl digests with $CHCl_3$; back extraction in aqueous 0.025% $Mg(NO_3)_2$; ET-AAS	177, 263
Feces, feed	Predigestion with HNO_3,[a] and then digestion with $HClO_4$-H_2SO_4-Na_2MoO_4;[a] HG-AAS; acid-matched standards (in 5% v/v H_2SO_4)	248
Feces, urine, food	Dry ashing in presence of $Mg(NO_3)_2$ and MgO; ashes dissolved in 6 *N* HCl; KI added; HG-AAS	261
Fish, tissues	Predigestion[a] with HNO_3 (overnight, at room temperature); complete digestion with $HClO_4$; semiautomated HG-AAS	239
Food	Dry ashing in presence of $Mg(NO_3)_2$ at 500°C for 2—5 hr; ashes dissolved in HCl; HG-AAS	251
	HNO_3-$HClO_4$-H_2SO_4 digestion;[a] HG-AAS	141, 252
Food, milk, tissues	Bomb digestion[a] with HNO_3; final dry ashing at 450°C in presence of $Mg(NO_3)_2$; ashes dissolved in 8 *M* HCl; KI added; HG-AAS; standards contained $MgCl_2$, KI, and HCl	250
Tissue (kidney, liver)	Ashing in O_2 in an autoclave; HG-AAS; acid-matched standards	255
Tissue (NBS SRMs "Bovine Liver", Oyster Tissue", etc.)	Dry ashing in presence of $Mg(NO_3)_2$ and α-cellulose (up to 600°C, gradually!); ashes dissolved in 5 *M* HCl; HG-AAS in 1 *M* HCl-EDTA medium	214
Urine	Dry ashing in presence of $Mg(NO_3)_2$ and MgO; HG-AAS	27
	Dry ashing in presence of $Mg(NO_3)_2$ and MgO; ashes treated with HCl-HBr-$HClO_4$ (10 + 3 + 3), then extracted with toluene in presence of KI; back extraction of As(III) in aqueous $Co(NO_3)_2$; ET-AAS	197, 205
	Wet ashing with HNO_3, $HClO_4$, and H_2SO_4;[a] KI added; HG-AAS; acid-matched standards	269

Note: [a] Gradual, temperature-controlled heating and safety precautions mandatory with most wet-ashing procedures. See text.

researchers have also observed transformations of arsenic into urine during storage[165,209] or treatment,[205] so that this matter should be further studied.

The most popular speciation procedures are based on the original work of Braman and Foreback;[193] the arsenic species are reduced to their corresponding arsines; arsenite and arsenate — to AsH_3; MA — to CH_3AsH_2; and DMA — to $(CH_3)_2AsH$, which are dried, frozen-out in a cold trap (at −196°C), and then fractionally volatized into a heated tube or hydrogen flame. A double HG procedure may be needed:[218] at first, AsH_3 is generated at pH 4 to 5 (from arsenite) together with $(CH_3)_2AsH$ (from DMA), and then, after acidification with HCl (0.1 to 1 *M*), the arsenate and the MA are reduced to AsH_3 and CH_3AsH_2, respectively.[198,209,218,281] However, it would seem that these separations may not be as selective as expected.[198,211,240]

IV. CONCLUSION

Arsine-generation AAS appears to be the most suitable routine technique for nanogram amounts of As; NAA is its most reliable alternative. Volatization losses of the analyte are typical of certain digestion procedures/conditions. The most reliable ashing techniques are tabulated (Table 2). Preconcentration/separation from the matrix (digests) is involved; this is accomplished by liquid-liquid extraction of As(III) (as dithiocarbamate or halide) or by arsine evolution. Direct flameless techniques (both HG- and ET-AAS) are impaired by severe background absorption (in ET-AAS only) and matrix effects, as well as by pronounced species-dependent behavior of the analyte, but may be used in speciation studies. Important instrumental and procedural parameters should be considered and optimized. Thermal stabilization of the analyte (with Ni^{2+} or, incidentally, with Cu^{2+} or Ag^{+}) and efficient background correction are a sine qua non; Zeeman ET-AAS is highly appropriate.

Chapter 4

BARIUM

I. DETERMINATION OF BARIUM IN BIOLOGICAL SAMPLES

The concentration of Ba in most biological liquids and soft tissues is some tens of nanograms per gram or less (Table 1). The most sensitive and suitable techniques seem to be ICP-ES (with DLs ranging from 0.01 to 0.5 ng/mℓ[29,35,286,289] and ET-AAS (DLs between 0.04 and 0.15 ng/mℓ[59,60]).

Arc-source AES procedures have been applied to samples of blood,[36] hair,[288] milk,[14,40] teeth,[26] tissues,[36,44] and other biologicals.[41,42,44]

The NAA technique has been used in assays of blood,[45,287] bone,[45] diet,[45,116] hair,[45,290] nails,[290] serum or plasma,[287] teeth,[2,3] tissues,[45,291] urine,[45] etc.[45,121]

SSMS has been applied to hair,[54] nails,[16] plasma,[55] teeth,[25] and tissues,[132] with DL as low as 2 ng/g.[123]

Though not sensitive enough, XRF may be applied to some biological specimens (e.g., dry-ashed hair or tissues[128,292,293]).

II. ATOMIC ABSORPTION SPECTROMETRY OF BARIUM

In a fuel-rich nitrous oxide-acetylene flame, the characteristic concentration and detection limits are in the range of 0.15 to 0.4 μg/mℓ and 0.008 to 0.02 μg/mℓ, respectively, depending on the apparatus used.[56-59] It is essential to optimize some instrumental parameters: flame stoichiometry, burner height, lamp current, and slit width, in order to obtain a better signal-to-noise ratio.

An ionization buffer — typically 0.1 to 0.5% K^+ — is added to both sample and standard solutions.

The significant emission noise arising from the intensive emission of a nitrous oxide flame or an incandescent graphite tube or excess CaOH is to be expected. This adverse effect can be attenuated by increasing the intensity of the light source (HCL), by using a narrower bandpass or by proper programming of the graphite-tube furnace.[294]

In ETA, the characteristic concentration is in the range of 0.04 to 3 ng/mℓ (4 to 190 pg), and the DLs are from 0.04 to 0.15 ng/mℓ (4 to 50 pg),[59-61] depending strongly on the graphite-tube coating and heating rate.[60] Pyrolytically coated tubes provide a sensitivity enhancement by one order of magnitude,[60] and the addition of 1% CH_4 to the Ar purge-gas improves the analyte signal by a factor of 1.5 to 2.5.[72,295] Argon is to be preferred to N_2 as an inert gas for better sensitivity provided (3 ×).[61]

In order to lower the barium carbide formation and thus improve sensitivity, it is advantageous to apply very high heating rates (''maximum power'' heating[60,296]) during the atomization step. Metal or metal-carbide coatings may also prove useful.[297] The addition of 1% Ca^{2+} (as nitrate) has resulted in a marked signal improvement (10 ×),[72] but the excess of Ca^{2+} may also give rise to background problems (from CaOH).

III. AAS METHODS FOR ANALYSIS OF BIOLOGICAL SAMPLES

The literature on AAS of Ba in biological materials is scant. Flame AAS procedures[298-300] include a preconcentration step by coprecipitation of Ba^{2+} as a sulfate[298] or phosphate[299,300] and by ion exchange.[299,300] The final solution, containing, e.g., 2 *N* NH_3, 1% EDTA, and an ionization buffer, is sprayed into a nitrous oxide-acetylene flame. These procedures are

Table 1
BARIUM CONCENTRATIONS IN HUMAN BODY FLUIDS AND TISSUES

Sample	Unit	Selected reference values	Ref.
Blood	μg/ℓ	<1	286
		100 ± 60	3
Bone	μg/g	18.0 ± 2.8[a-c] and 19.3 ± 2.0[a,b,d]	3
Brain	ng/g	6 ± 0.3	3
Erythrocytes	μg/ℓ	7 ± 2	287
Feces	mg/day	0.69	8
Hair	μg/g	♂, 0.82;[e] ♀, 2.41[e] (0.121—29.0)	288
		2.1 (0.46—5.2)[f]	54
Kidney	μg/g	0.01 ± 0.001	3
Liver	μg/g	0.01 ± 0.003	3
Lung	μg/g	0.03 ± 0.008	3
Milk	μg/ℓ	20	13
		6—42[f,g] (2—170)	14
Muscle	μg/g	0.02 ± 0.006	3
Nails (finger)	μg/g	(1.5—43);[f,h] (2.0—10)[f,i]	16
Plasma or	μg/ℓ	60 (<30—290)	42
serum		66 ± 5	287
Tooth (dentine)	μg/g	129.0 ± 54.7	23
Tooth	μg/g	4.2 ± 0.60 (0.8—13)	25
(enamel)		15 (4.2—44)	26
		125.1 ± 23.7	23
Urine	μg/ℓ	4.8 ± 1.4	286

Note: Wet weight unless otherwise indicated. Mean ± SD (range in parentheses).

[a] Rib.
[b] Ash.
[c] Soft water area (U.K.).
[d] Hard water area (U.K.).
[e] Geometric mean.
[f] Dry weight.
[g] Range of mean values.
[h] Adults.
[i] Children.

tedious, time consuming, and not sensitive enough to analyses of biological fluids. A standard addition check is advisable.

The direct ET-AAS has been applied to the microanalyses of freeze-dried pancreatic islets:[301] samples of 1 to 15 μg have been dried for 50 sec at 90°C, ashed for 60 sec at 1030°C, and atomized for 3.5 sec at 2250°C; the background absorption has proved to be negligible.

Most ashing procedures except those involving H_2SO_4-containing mixtures should be useful for sample preparation.[91]

Barium can be extracted as thenoyltrifluoroacetylacetonate (TTA) or as a hexafluoroacetylacetonate complex, but these procedures do not seem to have been applied to biological samples.

Inasmuch as the ET-AAS determination of Ba is prone to serious interferences (e.g., from $MgSO_4$, $MgCl_2$, Na, K, Ca, Fe),[297,301] the calibration mode chosen should be verified by standard additions. The background absorption at the Ba line of 553.6 nm, though not very pronounced, should be checked by means of tungsten-halide continuum source.[296] The lines of Mo 553.5 nm or Y 557.6 nm may be used as well.[302]

The graphite-tube performance may be subject to long-term sensitivity drift, memory effects, etc.; temperature-controlled heating, autosampling, and reslope facilities should therefore prove to be useful.

The key to a successful direct assay of Ba in a heavy matrix — sea water — was found in a recent paper by Beaty and Cooksey:[296] use of a pyro-coated tube, versatile programming including high ashing temperatures (900°C), and very rapid heating in the atomization step, as well as simultaneous background correction (by tungsten-iodide lamp) and standard addition calibration. These findings should be a most useful guide to analyses of biological samples as well.

IV. CONCLUSION

ET-AAS and ICP-ES, both of which are sensitive down to a few tenths of nanograms per milliliter, may be considered as the most promising techniques for trace Ba. In ET-AAS, there are problems due to carbide formation, emission noise, persistent background, and matrix effects; hence, versatile apparatus and standard addition calibration are required.

Chapter 5

BERYLLIUM

I. DETERMINATION OF BERYLLIUM IN BIOLOGICAL SAMPLES

The concentration of Be in most biological materials is generally less than a few nanograms per gram and is often <1 ng/g (Table 1). Just a few analytical techniques are sensitive enough in this low concentration range: ET-AAS, chelate-GC,[304] fluorimetry,[133,306,307] and SSMS.[25,55] Incidentally, the arc-source AES has been applied to biological specimens as well.[26,36,44,288] The ICP-ES, with its impressive DLs of 0.003 to 0.5 ng/mℓ, is a promising analytical technique.[29,35]

II. ATOMIC ABSORPTION SPECTROMETRY OF BERYLLIUM

The flame AAS of Be is sensitive and selective. At the resonance line of 234.9 nm, the characteristic concentration and the DL range between 10 and 30 ng/mℓ and 1 and 2 ng/mℓ, respectively.[56-59] Instrumental parameters such as the stoichiometry of the nitrous oxide-acetylene flame and the burner height significantly affect the signal-to-noise ratio and should be optimized. The absorbance is enhanced by organic solvents, H_3PO_4, HF, CH_3COOH, and NH^+_4, whereas an excess of Al, Mg, and Si causes signal depression.[57,308]

In ETA, the characteristic concentration and the DL are down to 0.001 to 0.2 ng/mℓ (0.1 to 3 pg) and 0.0003 to 0.03 ng/mℓ (0.5 to 3 ng), respectively.[59,61,294]

The addition of CH_4 (10% by volume) to the Ar purge gas as well as the addition of a calcium salt (0.1% of Ca^{2+} as nitrate) to the analyzed solutions both improve the sensitivity by a factor of 2.[72] A pretreatment of the graphite tube with a Zr salt has been found to enhance the analyte signal by factors of 3 and 9, for benzene extracts of Be(II)acetylacetonate and aqueous standard solutions of Be, respectively.[309]

Matrix effects are common in ET-AAS of Be; pronounced interferences by HCl, HF, $HClO_4$, Al, Ce, Cr, La, Mo, Mn, Si, W, etc. have been reported.[310] Small amounts of acids (≤0.1 *M*) are tolerable, the H_3PO_4 being the best in view of a thermal stabilization of the analyte.[311] The neutralization of the acidic solutions with ammonia results in both an additional thermal stabilization (e.g., to 1600 vs. 900°C) and a minor (by 10 to 50%) sensitivity improvement.[294]

Depending on the particular matrix involved, the ashing temperatures are between 600 and 1900°C, and thus many matrix constituents can be eliminated from the graphite tube before the start of the atomization step.[294,305,309,311,312] Still a simultaneous background correction and a standard addition check are often called for.

III. AAS METHODS FOR ANALYSIS OF BIOLOGICAL SAMPLES

AAS is undoubtedly a very successful technique for the determination of Be in biological materials. Applications to analyses of blood,[7,309] blood/serum fractions,[303,313] bone,[314,315] feces,[303,310] hair,[310,314] nails,[310] tissues,[7,309,312,315] and urine[7,303,305,308-310,316] have been described (Table 2).

The flame AAS can only be applied to specimens containing high Be concentrations, e.g., to detect more than 2 μg/ℓ Be in urine[308] or to analyze tissue digests from exposed lab animals.[315] A background correction and a standard addition check are advisable.

The direct ET-AAS of Be in biological liquids or digests is possible,[305,312,316] although some difficulties have been encountered. Hurlbut[310] found both a positive bias of analytical

Table 1
BERYLLIUM CONCENTRATIONS IN HUMAN BODY FLUIDS AND TISSUES

Sample	Unit	Selected reference values	Ref.
Blood	μg/ℓ	1.0 ± 0.4	303
		3.6 (2—5)	304
Feces	μg/day	10?[a]	8
Hair	ng/g	(5—8)	304
Nails	ng/g	<10	133
Plasma or serum	μg/ℓ	<4	55, 113
Tooth (enamel)	μg/g	<0.01[b]	25
		x̃ <0.2; 0.3 (<0.2—2.5)[b]	26
Urine	μg/ℓ	0.9 ± 0.4 (0.40—0.9)	305
		0.9 ± 0.5[c]; 2[c,d]	303
		1.1	113

Note: Wet weight unless otherwise indicated. Mean ± SD (range in parentheses). x̃, median.

[a] "Reference Man".
[b] Dry weight.
[c] ng/g.
[d] Heavy smoker of nonfilter cigarettes.

Table 2
SELECTED ET-AAS PROCEDURES FOR BERYLLIUM DETERMINATION IN BIOLOGICAL SAMPLES

Sample	Brief procedure outline	Ref.
Blood, tissues, urine	Bomb decomposition of 1-mℓ samples with 0.5 mℓ of HNO_3 for 1.5 hr at 170°C; acetylacetone-benzene extraction at pH 5—7 in presence of acetate and EDTA; ET-AAS: 39, 138, 1490, and 2660°C, and Zr-treated graphite tubes	309
Blood/serum fractions	Separation by centrifugation, ultrafiltration, and electrophoresis; ET-AAS as above, and some isotachophoretic fractions may be directly injected into the graphite furnace, as well, and quantitated by standard additions	313
Blood	Protein precipitation of heparinized blood (5 mℓ) by 10 mℓ of 5% TCA, precipitate washed by TCA, the two supernatants combined; cupferron-MIBK extraction at pH 6—6.5; ET-AAS: 100, 300, and 2400°, and maximum temperature clean-out necessary; background correction; standards extracted similarly	
Bone, hair, tissue	Dry ashing at 500°C, and ashes dissolved in HCl; acetylacetone-MIBK extraction at pH 6 in presence of EDTA (0.5 *M*); back extraction with 1 *M* HCl; ET-AAS; standard addition calibration	
Feces	Wet digestion with HNO_3 and H_2O_2[a] in presence of 60 mg of Fe(III) (as chloride), followed by gradual dry ashing from 300 to 700°C; ashes dissolved in 5% v/v HNO_3 containing 100 μg/mℓ of La^{3+}; 100°C for 50 sec, 1000°C for 60 sec, and 2600°C for 10 sec (gas stop); background correction	310
Hair, nails	HNO_3-$HClO_4$ digestion;[a] final solutions contained 100 μg La^{3+}/mℓ in 5% v/v HNO_3; ET-AAS as above	310
Tissues	HNO_3-H_2SO_4 digestion; cupferron-MIBK extraction at pH 6—6.5; ET-AAS as above (Reference 7)	7
Urine	Dilution with 0.5% v/v HNO_3 and 4% v/v H_2SO_4; ET-AAS: ramp drying and ashing to 1200°C, atomization at 2700°C; background correction; standard addition calibration	305
	Coprecipitation of Be from a 100-mℓ sample by aqueous NH_3; precipitate dissolved in 5 mℓ of 5% v/v HNO_3 containing 100 μg La^{3+}/mℓ; ET-AAS; 100 samples per day	310

[a] Safety precautions mandatory.

results and bent standard additions curves in direct analyses of urine. The excess of Ca in a urine matrix (>1 mg/mℓ) resulted in a 6 to 11% enhancement of the Be signal.[305] Ownes and Gladney[312] experienced pronounced memory effects in their direct assays of Be in solid tissue samples (2 to 8 mg samples of the NBS SRMs "Bovine Liver" and "Orchard Leaves").

In spite of the indicated problems, the direct injection of biological fluids/digests is quite acceptable for screening purposes, provided that the following conditions are observed: a simultaneous background correction, a standard addition (or matrix-matched) calibration, a high-temperature decontamination step after each atomization cycle, and a suitable analyte/matrix modification.

Additions of lanthanum (100 μg/mℓ) La^{3+} as nitrate, in 5% v/v HNO_3)[310] or acids (0.5% v/v HNO_3 plus 4% v/v H_2SO_4),[305] have been recommended.

Analytical results of high accuracy and precision are obtained only after major preparation steps involving ashing and preconcentration.[309,310,313,314]

The decomposition of organic matter poses few problems, if any;[91,266] formation of a very insoluble beryllium oxide after prolonged high-temperature ashing (at 750°C for 17 hr) has been observed.[91] Various ashing techniques have been applied with an apparent success: dry ashing at 500°C, followed by HCl dissolution of the ashes;[314] LTA with subsequent dissolution with aqua regia;[304,315] pressure decomposition with HNO_3 (caution!);[304,309,313] wet digestion with HNO_3 and $HClO_4$ (caution!),[310] or with HNO_3 and H_2SO_4,[7] or with HNO_3, $HClO_4$, and H_2SO_4 (caution!).[312] Hurlbut[310] recommended a wet-digestion procedure for feces (see Table 2) as faster than the other common preparation procedures. Reagent blanks are very low (e.g., less than 0.01 ng Be/mℓ in Merck Suprapur® acids[311]).

Beryllium can be separated from the matrix by chelate extraction[7,309,313,314] or coprecipitation.[310] Several extraction systems are available,[76] the extractions of Be(II)acetylacetonate at pH 5 to 7 in presence of EDTA[309,313,314] or Be(II)cupferronate at pH 6 to 6.5[7] being most suitable to biological materials. Berman[7] was able to perform cupferron-MIBK extractions directly from pH-adjusted urine and from TCA supernatants of deproteinized blood.

A detailed study of the distribution of Be between blood and serum protein fractions, both in vivo and in vitro, was conducted by Stiefel et al.;[303,313] the separation scheme involved centrifugation, ultrafiltration, and electrophoresis.

IV. CONCLUSION

ET-AAS is a very suitable method for the determination of nanograms-per-gram concentrations of Be in biological samples offering certain advantages over other current techniques: chelate GC, fluorimetry, and ICP-ES. Direct ET-AAS calls for versatile apparatus (Zr-treated/pyro-coated graphite tubes, high heating rates, efficient background correction) and methodology (additional thermal stabilization of Be, standard additions). Therefore, the presently adopted procedures preferably involve ashing and extraction (as Be(II) acetylacetonate or cupferronate).

Chapter 6

BISMUTH

I. DETERMINATION OF BISMUTH IN BIOLOGICAL SAMPLES

The Bi content is very low in most biological samples, ranging from a few nanograms per milliliter in body fluids to dozens of nanograms per gram in tissues (Table 1). At present, there are only two analytical techniques, DPASV[124,125,317,321] and flameless AAS,[7,318,322-324] which are sensitive and practically useful at these nanogram levels of Bi; the DLs are down to 0.005 to 0.05 and 0.01 to 0.02 ng/mℓ, respectively.

A few SSMS assays of Bi and other trace elements in nails,[16] plasma,[55] teeth,[25] and other biologicals[325] have been described.

After ashing and preconcentration, AES[44,320,326] and spectrophotometry[327] have been used. The DLs claimed for some of the indicated procedures are, however, well above the normal levels of Bi in biological samples which have been analyzed.[25,55,321,323,326]

II. ATOMIC ABSORPTION SPECTROMETRY OF BISMUTH

The flame AAS of Bi is very selective but not sensitive. The characteristic concentration and DLs are within the ranges of 0.16 to 0.6 and 0.015 to 0.05 μg/mℓ, respectively.[56-59] An EDL is the preferred light source providing about 8 times higher intensity and better DL (2.5 ×) as compared with a HCL.[138]

In ETA, the characteristic concentration is 0.05 to 1.4 ng/mℓ (5 to 40 pg), and the DLs range from 0.01 to 0.1 ng/mℓ (~10 pg).[57,59-61] The behavior of Bi in real analytical matrices should be a serious concern. Certain bismuth species, in particular the metal itself and its chloride, may be partially lost during ashing above 300 to 400°C. Owing to these relatively low ashing temperatures, the background absorption at 223.1 nm Bi line is high. Several matrix/analyte modifiers have proven to thermally stabilize Bi during ashing at higher temperatures: 0.05% w/v of Ni(II) (up to 1000°C),[328] aqueous ammonia (up to 800°C),[329] 1 to 2% v/v H_2SO_4 (to 470 to 550°C),[322,330] and 1 m*M* EDTA (tetraacid) up to 650°C in O_2.[1030]

Inui et al.[331] found that coextracted copper (in 250-fold excess over Bi) thermally stabilized bismuth in DDDC-xylene extracts during ashing at 650 to 800°C; minor enhancement of Bi absorbance was caused by the same excess of Fe, Mn, and Ni, and was controlled by peak-area calibration.[331]

Although Bi is readily atomized, starting at temperatures as low as 1200°C,[60] its atomization pattern and the extent of matrix effects may dramatically depend on parameters such as heating rate and tube coating (or condition). As reported by Fernandez and Iannarone,[60] the optimum atomization temperature of Bi, at normal heating rates, was 2600°C in an uncoated graphite tube and 2200°C in a pyrolytically coated tube. With a much faster heating mode (''maximum power'' heating), the corresponding figures were as low as 1600 and 1200°C, respectively.[60]

Using the L'vov platform[229] and/or metal-carbide coatings should prove advantageous, as well.

The hydride-generation AAS of Bi is sensitive and relatively selective; DLs of 0.02 to 0.2 ng/mℓ (0.1 and 2 ng) have been reported.[142,149,232,242] Bismuthine, BiH_3, is evolved by reducing the acidified sample solution with aqueous (freshly prepared![323]) $NaBH_4$. The concentration of HCl appears not to be critical (e.g., from 0.1 to 5 *M* HCl);[140,142,149,150] small amounts of H_2SO_4 or H_3PO_4 (<2 *M*) are also tolerable,[142] while HNO_3 and $HClO_4$ depress hydride evolution.

Table 1
BISMUTH CONCENTRATIONS IN HUMAN BODY FLUIDS AND TISSUES

Sample	Unit	Selected references values	Ref.
Blood	μg/ℓ	<3[a]	3
		<5	317
		<10	318
Bone	μg/g	<0.2[b,c]	3
Brain	μg/g	0.01	3
Feces	μg/day	18?[d]	8
Kidney	μg/g	0.4 ± 0.1	3
Kidney cortex	μg/g	0.4 ± 0.04	3
Liver	μg/g	~0.004	3
		0.2 ± 0.26 (<0.1—0.9)[e]	319
Lung	μg/g	0.01 ± 0.001	3
Plasma or serum	μg/ℓ	<0.6[f]	55
Tooth (enamel)	μg/g	x̃, <0.02; 0.006 (<0.02—0.07)	25
Urine	μg/day	1.6 (<1—2.4)	320
	μg/ℓ	<5	317
		<10	318

Note: Wet weight unless otherwise indicated. Mean ± SD (range in parentheses). x̃, median.

[a] Preliminary data.
[b] Rib.
[c] Ash.
[d] "Reference Man".
[e] Dry weight.
[f] N = 1.

There are two extensive interference studies. Smith,[155] in 1 *M* HCl, found a significant depression of bismuthine evolution by a 2000-fold excess of many ions (each a 1000 μg/mℓ level): Ag, Au, Co, Cu, Ni, Pd, Pt, Rh, Ru, Se, and Te (more than 50% depression) and As, Cd, Cr, Fe, Ge, Ir, Mo, Sb, and Sn (10 to 50% depression). Thompson et al.,[149] in 5 *M* HCl, found depressive interference by 100 μg/mℓ of Cu and by 1000 μg/mℓ of Cd, Co, Cr(VI), Hg, Ni, Pb, V, and Zn. It should be added here that the extent of these interferences will also depend on acidity, hydride collection time, the concentration and age of the $NaBH_4$ solution, and other experimental factors. Fortunately, most of these interferents are not present in normal biological liquids or digests at such high levels.[324]

III. AAS METHODS FOR ANALYSIS OF BIOLOGICAL SAMPLES

AAS has been applied to the determination of Bi in blood,[7,318,323,324,332] CSF,[318] milk (powder),[333] plasma,[330] shellfish,[334] serum,[1030] tissues,[7,319,322,324,335-337] urine,[7,319,322,324,335-338,1030] and other biological[327,331] or organic[328] samples.

Several points appear to be of major concern in these analyses:

1. Choice of an effective preconcentration method
2. Control of reagent blanks (from acids, $NaBH_4$, etc.)
3. Elimination of possible analyte losses during preparation and storage of dilute sample or standard solutions (e.g., due to hydrolysis, coprecipitation, adsorption, volatization, etc.)

4. Proper background correction
5. Elimination of or accounting for matrix effects

The direct flame AAS[319,337] can be applied only to samples obtained from exposed/treated lab animals. In most other flame AAS procedures, an effective preconcentration — typically by chelate extraction — is needed, in order to determine Bi in biological liquids[332,335,338,339] or tissue digests.[334-336] With minor modifications, these procedures can be adapted to ET-AAS.

The flameless AAS techniques — both ETA[7,331,333,336] and bismuthine-generation AAS[323,324] — require ashing and/or a preconcentration step. Only two exceptions seem to have been described[322,330] which tend to support rather than refute this view, due to their unacceptable DLs of 3 to 100 ng/g. Bourdon et al.[330] injected directly 5-$\mu\ell$ portions of (1 + 1) diluted plasma or (1 + 9) diluted urine into a Varian Techtron® Carbon Rod Atomizer. The diluent was 2% v/v H_2SO_4, and the temperature program included ashing at approximately 550°C and atomization at approximately 1300°C. In another direct ET-AAS procedure, Djudzman et al.[322] decomposed tissue samples by LTA, and dissolved and diluted the ashes with 2% v/v of H_2SO_4. (See also Reference 1030.)

All common decomposition techniques should be applicable in the oxidation of organic matter prior to Bi determinations, except for the dry ashing at temperatures above 450 to 500°C. Apparently successful procedures involving dry ashing at 450°C, LTA,[322] HNO_3 (urine),[338] HNO_3 and HCl (urine),[339] HNO_3 and H_2SO_4,[7] HNO_3 and $HClO_4$ (caution!),[323,331] HNO_3, $HClO_4$, and H_2SO_4 (caution!),[332] as well as acid extraction of liver tissue with HNO_3 and HCl at 100°C[319] have been described. The choice of a digestion method should be in accord with the next analytical steps, e.g., dilute acids (1 or 2% v/v of HNO_3 or H_2SO_4) are suitable media in ETA, whereas 1 to 4 *M* HCl or H_2SO_4 media should better suit HG-AAS. Nitric acid and/or nitrogen oxides must be absent from the final solution if hydride generation or chelate extraction[340] are to be attempted; to a lesser extent this should refer to $HClO_4$, as well.

Among the preconcentration methods for trace amounts of Bi: bismuthine evolution,[323,324] chelate extraction,[76,331-336,338,339] coprecipitation (with $La(OH)_3$,[149] $Fe(OH)_3$, $Al(OH)_3$, $Zr(OH)_4$ or with sulfides), and ion exchange,[174,317,326,327,341] the first two are most widely explored.

Bismuth is readily extracted from digested biologicals as dithiocarbamate[7,318,331-335,338] or xanthate.[336] Other extraction systems (see, e.g., Cresser[76]) may prove useful, as well.

In APDC-MIBK extractions, the pH value between 2.8 and 6.5 appears not to be critical.[324,334,335,338,339] Direct extraction of Bi from pH-adjusted urine, following the original work of Willis,[339] is possible,[7,318,339] but persistent emulsions are usually formed and centrifugation is mandatory. Allain[318] directly extracted Bi(III) from hemolyzed blood samples with APDC-hexane, and Berman[7] extracted Bi(III) at pH 5.5 to 6.5 with SDDC-MIBK from TCA-treated blood samples. The determination of Bi in organic extracts, with simultaneous background correction, is straightforward in both flame and ET-AAS. The peak-area mode of calibration has been advocated[331] in order to minimize the effect of coextracted ions.

Finally, the blank contribution in trace Bi analyses may restrict the practical detection limits.[243,322,323] Knudson and Christian[243] found about 2 ng of Bi per pellet of $NaBH_4$ in all lots examined. Reportedly, very high-purity $NaBH_4$ pellets have been obtained by Aldrich Chemical Co.[323] Acid digestion procedures may introduce an intolerable blank contribution.[322]

In practical work, the limited lifetime of both the $NaBH_4$ solutions (the useful life of 1% w/v $NaBH_4$ being only 1 hr[323]) and dilute bismuth solutions (hydrolysis, adsorption losses) should not be overlooked.

A few selected procedures are briefly outlined in Table 2.

Table 2
SELECTED AAS PROCEDURES FOR Bi IN BIOLOGICAL MATERIALS

Sample	Brief procedure outline	Ref.
Blood, urine	5-mℓ sample digested with 5 mℓ of HNO_3 and 5 mℓ of $HClO_4$, evaporated to almost dryness, dissolved in 2 mℓ of (1 + 1) HCl, and diluted to 5 mℓ; HG-AAS; standards in 20% v/v HCl.	323
Blood, CSF, urine	APDC-hexane extraction at pH 6; ET-AAS: 100°C for 10 sec, 230°C for 20 sec, and 1200°C for 8 sec (gas stop); background correction; standards on blood/urine matrix	318
Blood, tissues	TCA precipitation of blood proteins or wet ashing of tissue with HNO_3 and H_2SO_4 or pH adjustment of urine samples; SDDC-MIBK extraction at pH 5.5—6.5; centrifugation; ET-AAS: 100°C for 20 sec, 300°C for 20 sec, and 2400°C for 6 sec; background correction; standards extracted similarly	7
Milk (powder)	SDDC-MIBK extraction from milk digests; ET-AAS	333
Tissues	LTA for 15 hr; ashes moistened with 0.2 *N* HCl, re-ashed for 3 hr, and then dissolved in 2% v/v H_2SO_4; ET-AAS: 100°C for 45 sec, 470°C for 30 sec, and 2650°C for 10 sec; background correction; standards in 2% v/v H_2SO_4	322
Serum, urine	1 + 1 dilution with aqueous 1 m*M* EDTA (tetraacid); ET-AAS: 120°C (ramp!), 650°C (120 sec ramp in O_2 as alternate gas), 650°C (20 sec in Ar), and 2500°C (gas stop); background correction; standard addition calibration	1030

IV. CONCLUSION

The current method of choice at nanograms-per-gram Bi levels is ET-AAS; ASV and HG-AAS should be further studied as promising alternatives. Preconcentration by (dithiocarbamate) extraction or else by BiH_3 evolution is required. Analyte losses due to hydrolysis/retention/volatization are probable.

Chapter 7

BORON

I. DETERMINATION OF BORON IN BIOLOGICAL MATERIALS

The concentration of B in most biological materials is about or less than 1 μg/g (Table 1).

Optical emission methods[29,36,342,348-351] are very sensitive and useful in trace B analyses. The DLs of ICP-ES can be as low as 0.1 to 7 ng/mℓ;[29,35,289] this is the most promising technique of the future.

Hollow-cathode emission, after an ion-exchange preconcentration, has been successfully applied to analyses of hair, tissues, and other biological specimens.[346]

Molecular emission of BO_2 at 518 nm in a flame[351] or in a flame-heated cavity—molecular emission cavity analysis (MECA)[349,350] is another sensitive and simple instrumental technique for nanogram amounts of boron (e.g., 5 to 10[349] or 10 to 400 ng[350]).

The arc-source AES, though not as sensitive as the above-mentioned methods, has also found application to analyses of biological specimens.[36,44,288,342]

Incidentally, mass spectrometry has been used in multielement (including boron) assays of blood[52,53] and CSF.[52]

Several sensitive fluorimetric[344] and spectrophotometric[188,345,352] procedures have been described, as well. See also Kliegel.[353]

II. ATOMIC ABSORPTION SPECTROMETRY OF BORON

Among the elements discussed in this volume, boron is least sensitive in both flame and ET-AAS. Its determination is restricted, due to the formation of very stable and refractory boron species, in particular, oxides, carbides, and nitrides.

The flame AAS of B is relatively selective but the sensitivity is extremely poor. The characteristic concentration ranges from 7 to 15 μg/mℓ and the DLs are between 0.5 and 2 μg/mℓ.[56-59] Organic solvents and HF markedly enhance the sensitivity. Of utmost importance are (1) the quality of HCL and the lamp current applied; (2) the stoichiometry of the nitrous oxide-acetylene flame (e.g., red zone of about 2 cm); and (3) the burner height. Bent calibration curves are often observed, both in flame and ET-AAS, due to the doublet 249.7/249.8 nm measured as well as to carbide formation.

A progress in graphite-furnace atomization of B was achieved only recently by using pyrolytically coated tubes,[354,355] and/or applying *in situ* pyro-coating,[354] heating at a very high rate during the atomization stage,[354,355] and adding suitable matrix/analyte modifiers.[354,355] The characteristic concentration dropped to 6 to 50 ng/mℓ (0.04 to 1 ng), and the DLs went down to 5 to 10 ng/mℓ (about 0.5 ng).[58,354,355]

The temperature program consist of ashing at 500 to 1500°C and atomization to maximum available temperatures and with as high a heating rate as possible. Background correction may be necessary. Due to the high temperatures applied, the lifespan of the graphite tube is restricted.[354] A gradual sensitivity drift should be expected. Using high-purity argon (not nitrogen!) and an autosampling device is advisable. Coating the graphite tube with other refractories may likewise prove advantageous.

In recent procedures for analysis of water[354,355] and plant tissue digest,[354] the addition of 1000 μg/mℓ of $Ba(OH)_2$,[354] or 100 μg/mℓ of Ca^{2+} + 200 μg/mℓ of Mg^{2+},[355] in dilute HNO_3 media (0.01 to 0.1 *M* HNO_3), improved sensitivity and other analytical parameters.

Table 1
BORON CONCENTRATIONS IN HUMAN BODY FLUIDS AND TISSUES

Sample	Unit	Selected reference values	Ref.
Blood	mg/ℓ	x̃, 0.0985; 0.114 (0.039—0.365)	342
		0.25 (0—1.25)[a]	343
		0.74	344
Bone	μg/g	6.2 ± 2.1[b–d] and 10.2 ± 5[b,c,e]	3
Brain	μg/g	0.06[f]	3
		0.85	345
SF	mg/ℓ	0.01	52
Feces	mg/day	0.27	8
Hair	μg/g	0.21	346
		♂, 0.90;[g] ♀, 1.04;[g] (0.037—25.0)	288
Kidney	μg/g	0.6[f]	3
Liver	μg/g	0.2[f]	3
Lung	μg/g	0.6[f]	3
Milk	μg/ℓ	88 (Pool)	347
Muscle	μg/g	0.1[f]	3
Tooth (enamel)	μg/g	5.0 ± 1.5 (0.5—39)	25
		19 (<2—141)	26
Urine	mg/ℓ	x̃, 0.715; 0.919 (0.040—6.6)	342
	mg/day	1.0	113

Note: Wet weight unless otherwise indicated. Mean ± SD (range in parentheses). x̃, median.

[a] Children.
[b] Rib.
[c] Ash.
[d] Hard water area (U.K.).
[e] Soft water area (U.K.).
[f] Preliminary data.
[g] Geometric mean.

The efficiency of several tested modifiers decreased in the order: $Ba^{2+} > Ca^{2+} > Mg^{2+} = Sr^{2+}$.[355]

In a graphite furnace of open-type design (Varian Techtron® CRA-90), the addition of H_2 to the Ar sheath gas has proven advantageous; yet the useful lifetime of the graphite tube was only about 30 firings.[354]

The years to come will most probably yield new developments in ETA of boron.

III. AAS METHODS FOR ANALYSIS OF BIOLOGICAL SAMPLES

The literature on AAS of B in biological materials is, for the reasons discussed, scant.[354,356-358]

Bader and Brandenberger[356] directly determined lethal boron concentrations (0.2 to 0.4 mg/g) in post-mortem samples of CSF, serum, and digested tissues and blood by flame AAS; the DL was 15 μg/mℓ.

Holak[357,358] described an official procedure for the determination of boron added to foods as H_3BO_3. Digestion of samples with H_2SO_4 and HNO_3 under reflux, extraction with 2-ethyl-1,3-hexane-diol-MIBK, and flame AAS were involved.

Recently, Szydlowski[354] determined boron in plant tissues by ET-AAS. Samples were dry ashed at 600°C for 10 hr, dissolved in HNO_3, diluted to 1% final concentration of HNO_3, and finally, an addition of 1000 μg/mℓ of $Ba(OH)_2$ (as analyte modifier) was made. The Perkin-Elmer® HGA-2100 furnace was programmed as follows: drying at 125°C (20 sec ramp + 20-sec hold), ashing at 1500°C (15 + 15 sec), and atomization at 2950°C for 8 sec. Alternatively, a Varian Techtron® CRA-90 was used (100°C for 35 sec, 1100°C for 25 sec, and 2600° for 3 sec). Simultaneous background correction and standard curve calibration were applied.[354]

The preparation step in B assays should be considered with caution: losses of B during certain common wet- or dry-ashing procedures (e.g., as H_3BO_3, BH_3, esters, etc.) and major contamination from borosilicate glass (Pyrex®, etc.) are to be expected. Lab ware made of polyethylene, polypropylene, PTFE, quartz, and platinum should be useful. Accidental contamination by cosmetics and $NaBH_4$ (a common reagent in AAS laboratories) should not be allowed, either.

Apparently suitable decomposition procedures proved to be LTA,[346,352] wet ashing under reflux,[356,358] and dry ashing with an excess of ashing acid — $Ca(OH)_2$[344] or Li_2CO_3.[342,356] Pressurized decomposition vessels may be used, as well.

Trace boron can be preconcentrated from the matrix by extraction,[76,351,352,357,358] ion exchange,[346,352] and by distillation as a volatile BF_3 or $B(CH_3O)_3$.

Of liquid-liquid extraction procedures available (see, e.g. Cresser[76]), extraction with 2-ethylhexane-1,3-diol[352,357,358] or with 2,2,4-trimethylpentane-1,3-diol[351] proved to be applicable to biological digests. An effective ion-exchange separation of trace B on boron-specific XE-243 resin (Rohm and Haas Co.) has been applied to biological materials.[346,352]

IV. CONCLUSION

ICP-ES appears to be the best analytical technique. ET-AAS of B is strongly inhibited by carbide formation and matrix effects. Essentials are pyro-coated tubes, high heating rates, background correction, and an addition of alkali-earth salts.

Both losses and contamination are probable during the preparation stage. LTA, dry ashing with ashing aids, and wet digestion under reflux/pressure seem to be suitable for sample mineralization. Preconcentration methods based on extraction or ion exchange which might be adaptable to the ET-AAS have been described.

Chapter 8

CADMIUM

I. DETERMINATION OF CADMIUM IN BIOLOGICAL SAMPLES

The normal Cd concentrations in biological materials range broadly from subnanograms per milliliter to a few nanograms per milliliter in body fluids, and up to micrograms-per-gram levels in some tissue specimens (Table 1). The most sensitive analytical techniques for the determination of nanogram and subnanogram amounts of Cd are[123,389,390] ET-AAS (the method of choice nowadays), ASV or DPASV, NAA, photon activation analysis (PAA), AFS, and SSMS.

The DLs of Cd by ASV or DPASV are as low as 0.001 to 0.01 ng/mℓ; these methods are now the best alternative to ET-AAS in ultratrace Cd assays. The apparatus is relatively inexpensive and a few more elements (Pb, Cu, Zn, etc.) can also be determined in the same digest but the organic matter has to be completely mineralized. Successful applications of ASV/DPASV to analyses of blood,[361,391,392] bone,[392] food,[250] hair,[393] milk,[110,250,394] nails,[133] serum,[321] teeth,[395] tissues,[133] and urine[124,125,396-398] have been described.

NAA has been widely used to determine Cd in various biological specimens: blood,[45] blood platelets,[119] bone,[97] food,[45,116,399] hair,[11,45,48,99,101,104,118,184,399] milk,[110] serum,[111] skin,[112] tissues,[12,45,103,108,372,401] urine,[45,111] etc.[45,121] This technique is attractive for its inherent absence of reagent blanks and high sensitivity, as well as for its instrumental performance (INAA) with some matrices (e.g., hair,[11,99,118] tissues,[12] etc.[45,121]). The in vivo INAA is now performed by a few laboratories especially equipped to detect elevated levels of Cd in kidneys and liver, with DLs of approximately 1 μg/g.[400-402]

PAA may detect down to 0.08 ng Cd in biological samples (hair, diet, etc.) with better selectivity and some other advantages over NAA.[48,184]

The flame AFS may provide DLs as low as 0.01 to 0.1 ng/mℓ, i.e., by two orders of magnitude lower than the flame AAS. This technique has been successfully applied to blood,[384,403,404] tissue digests,[405] and urine.[384,404] All that is needed is simple dilution of body fluids/biological digest. Unfortunately, some modifications of the AAS apparatus are necessary, in order to adapt it to AFS (using uncommon flames such as air hydrogen[403,404] or nitrogen-separated air-acetylene[384], as well as special burners and high-intensity light sources), and this would seem to be a restriction on the routine application of AFS.

The sensitivity of XRF is limited[128] and hence few Cd assays (hair,[293] tissues,[292]) can be performed. The in vivo XRF determination of Cd in kidney cortex was recently claimed to be advantageous vs. the INAA, because of its inherent lower irradiation of the patient;[406] however, the DL was only 20 to 40 μg/g.

The PIXE technique is unique in studying the distribution of Cd and other trace elements in small sample areas (liver biopsies,[407] hair segments[128,408]).

SSMS determinations of Cd in bone,[132] nails,[16] plasma,[55] tissues,[132,275,409] and tooth enamel[25] have been described. Only a few laboratories are able to perform highly accurate and precise isotope-dilution (ID)-SSMS analyses.[410]

The classical AES methods[44] are less sensitive than the AAS, and may be useful in cases requiring simultaneous multielement analyses/screening. The state-of-the-art detection limit of the ICP-ES is 0.07 ng/mℓ,[29] but at least two manufacturers report DLs of about 1 to 2 ng Cd/mℓ. ICP-ES analyses of milk,[35] tissue digests,[348,411,412] urine,[326] etc. (incidentally, after an ion-exchange preconcentration[326,412]) have been described. The arc-source AES, usually after an ashing and/or a preconcentration step, has been widely used in early analyses of blood,[130] saliva,[41,380] serum,[41,42,413] tissues,[43,130] urine,[130,320,342] etc.

Table 1
CADMIUM CONCENTRATIONS IN HUMAN BODY FLUIDS AND TISSUES

Sample	Unit	Selected reference values	Ref.
Blood	μg/ℓ	0.7 ± 0.3	359
		0.79 (0.2—1.4);[a] 3.76 (2.0—6.0)[b]	360
		0.79 (0.3—2.0)	361
		1.0 ± 1.4	244
		1.5 (0.7—3.8)	362
		(<0.25—2.1);[c] (0.3—5.1)[d]	363
Bile	μg/ℓ	2.5 (0.1—11.4)[e]	364
Bone	μg/g	0.06 ± 0.01	268
Feces	μg/g	0.014	365
	μg/day	(41.1—79.4)	366
		50	8
Hair	μg/g	♀, 0.3; ♂, 0.45 (G.D.R.)	367
		x̄, 0.3 (0.33—16.1) (Japan)	11
		0.488 ± 0.406 (0.025—2.072) (U.K.)	368
		♀, 0.62;[f] ♂, 0.96;[f] (0.08—8.73) (New York)	288
		0.79—0.99;[g] (0.08—4.40) (Austria)	99
		0.83—1.10 (U.S.A.)	369
		1.3 (0.34—1.3)[h]	54
		♀1.0; ♂, 2.2	370
		(0.26—2.7);[i] (0.25—3.7)[j] (Canada)	104
		1.38 ± 1.04 (0.34—5) (India)	101
		1.4 ± 2.1 (<0.6—8.2) (Iraq)	118
Kidney	μg/g	x̄, 2.4; 2.7 ± 0.7 (1.67—13.1)	105, 106
		13.9 ± 0.7	3
		16.5 ± 1	268
		17.7 (2.1—22)	371
Kidney cortex	μg/g	14.3 ± 2.9	3
		15	107
		21	369
		♀, 38 ± 20;[h] ♂, 102 ± 66[h]	367
		104 (16—305)[h]	372
		139 ± 88[h]	373
Liver	μg/g	x̄, 0.32; 0.27 ± 0.06 (0.071—1.16)	105, 106
		(0.21—0.40)	109
		0.9 ± 0.1	268
		0.93 ± 0.59	275
		1.1 ± 1.1 (0.26—6.85)	319
		2.03 (1.19—3.71)	371
		2.6 (0.8—7.1)	108
		x̄, 4.8; 6.3 ± 4.9 (0.66—15)	12
Lung	μg/g	x̄, 0.026; 0.019 ± 0.007 (0.001—0.17)	105, 106
		0.04	268
		x̄, 0.12; 0.28 ± 0.31 (0.0088—1.3)	103
Milk	μg/ℓ	<1[k]	374
		<2[h]	110
		5	375
		(2—19)	13
Nails (finger)	μg/g	0.08[l]	133
		(0.99—2.6)[h]	16
Placenta	ng/g	51 ± 20;[c,h] 66 ± 33[d,h]	376
		102 ± 77[h]	377
Plasma or serum	μg/ℓ	1.5 ± 0.7	27
		1.9 ± 0.8 (1.1—3.3)	378
		2.5 ± 1.7	111

Table 1 (continued)
CADMIUM CONCENTRATIONS IN HUMAN BODY FLUIDS AND TISSUES

Sample	Unit	Selected reference values	Ref.
Saliva	μg/ℓ	3.5[m]	379
		(5—11)	380
Skin	μg/g	x̃, 0.09;[h] 0.089 ± 0.05[h]	112
Sweat	μg/ℓ	♂, 24 ± 16; ♀, 90 ± 121	381
Tooth (dentine)	μg/g	0.099	382
Tooth (enamel)	μg/g	0.135	382
		0.51 ± 0.12 (0.03 —2.4)	25
Tooth (whole)	μg/g	0.12[h]	383
Urine	μg/ℓ	0.5 ± 0.3	359
		0.5 ± 0.4 (0.2—1.5)	384
		(<0.14—1.8)	385
		(0.59—0.77)	369
		(0.3—2.8)	386
		x̃, 1.15; 1.59 (<0.5—10.8)	342
		1.6 (0.5—3.2)	362
		1.6 ± 0.9	244
	μg/day	0.98 ± 0.36	387
		(0.15—2)	365
		1.63 ± 1.04	388
		2.1 ± 2.3 (0.24—8.4)	111

Note: Wet weight unless otherwise indicated. Mean ± SD (range in parentheses). x̃, median.

[a] Mainly nonsmokers.
[b] Mainly heavy smokers.
[c] Nonsmokers.
[d] Smokers.
[e] Gall bladder bile from gallstone-operated subjects.
[f] Geometric mean.
[g] Range of medians.
[h] Dry weight.
[i] Rural.
[j] Urban.
[k] Pool.
[l] N = 1.
[m] Parotid.

Finally, the spectrophotometric dithizone procedures may be used in determinations of microgram amounts of Cd (e.g., in tissue digests);[389,414] these procedures give acceptable results in experienced hands, but suffer severely from limited sensitivity and reagent blanks. See also O'Laughlin,[389] Sacchini,[390] Morrison,[123] and Bowen.[45]

II. ATOMIC ABSORPTION SPECTROMETRY OF CADMIUM

The AAS determination of Cd is remarkably selective and sensitive. The characteristic concentration in a lean air-acetylene flame is in the range of 9 to 25 ng/mℓ and the DLs are down to 0.7 to 2 ng/mℓ.[56-59] The recommended light source is EDL, which offers higher intensity (30 ×) and lower DLs (3.5 to 3.7 ×) vs. the HCL.[138] No major interferences should be expected except for background absorption, transport interferences, and nonspecific bulk-matrix effects from an excess of total dissolved solids. All these effects are easily accounted for.

Until recently, two flame accessories — the Delves cup[415] and the Sampling Boat®[416] — with their DLs of 0.05 ng/mℓ (5 pg) and 0.1 to 1 ng/mℓ (0.1 ng), respectively, were popular. Owing to their limitations (poor precision, matrix effects, sensitivity drift, transient background signals, etc.), these accessories have now been almost completely replaced by the more versatile ET-AAS technique.

The classical Fuwa-Vallee long-path absorption tube for flame gases was recently combined with the pulse-nebulization of minute sample volumes;[417] DLs of 0.16 ng/mℓ (8 pg) as well as speed, easy performance, precision, and low cost were claimed.[417]

The ET-AAS of Cd is extremely sensitive but prone to serious matrix effects. The characteristic concentration is from 0.008 to 0.03 ng/mℓ (0.15 to 3 pg) and the DLs are down to 0.0002 to 0.02 ng/mℓ (0.1 to 2 pg).[57,59-61] These figures can easily be obtained with dilute standard solutions, but may be impaired by one order of magnitude in analyses of real matrices. What matters in analyses of biological samples is (1) the background correction efficiency of the system, (2) the temperature programming, and (3) the elimination of matrix interferences. It was only recently that significant progress was made in this field, which will be more conveniently treated under the next heading.

III. AAS METHODS FOR ANALYSIS OF BIOLOGICAL SAMPLES

It should be noted right away that, although AAS is the best analytical technique for trace Cd, the easiness with which Cd is put into solution, chelate extracted, and electrothermally atomized can be misleading in the case of complex biological matrices.

The results of recent interlaboratory comparison studies on Cd determination[389,418,419] as well as the thorough review of the literature reveal that there are numerous sources of systematic errors in all stages of these assays: nonrepresentative sampling; contamination during specimen collection; contamination and/or losses during storage,[420-422] drying,[423] and ashing of samples; sporadic contamination from lab ware,[424] pipet tips,[385,425,426] air particulate matter,[425] etc.; unacceptably high reagent blanks; incomplete extractions; limited and pH-dependent time stability of the Cd(II) dithiocarbamate in organic extracts;[340,427-429] losses due to premature volatization in the Delves cup[430,431] and graphite tube;[432] severe transient background signals; limited linearity of the calibration or standard additions curves; nonreproducible and nonpredictable behavior of some lots of graphite tubes;[360,433] and, finally, serious interferences in the ETA of Cd.[434-438] Most of these problems are duly dealt with in the subsequent text.

There is ample literature on the determination of Cd in biological materials by AAS. Practically all AAS techniques have been applied: flame atomization, microattachments for flame work (pulse nebulization,[417,429,439] the Delves cup, the Sampling Boat®, etc.[440-442]), graphite tube/cup furnaces, Tantalum ribbon,[422,443,444] various noncommercial atomization devices,[445-452] and, recently, the most promising, Zeeman-AAS[363,373,453] and the L'vov platform.[229,454-457] These techniques are classified in Tables 2 and 3, which can serve as a guide to the literature.

Cadmium is a typical pollutant and thus it is nonhomogeneously distributed in some biological materials (e.g., in kidneys[372] and other tissues, food, feces,[365,366] hair,[531] etc.). Clean homogenization procedures have been described,[376,426,532,333] most of which involve grinding of the frozen or freeze-dried tissue in Teflon® at liquid-nitrogen temperature.

The drying of biological specimens is generally safe with respect to Cd.[534,535] There is one report indicating minor Cd losses during the lyophilization or oven drying (>50°C) of tracer-labeled oysters,[423] while Koh[535] safely and rapidly (~15 min) dried liver and carp tissue in a microwave oven, de Goeij et al.[536] lyophilized liver samples without losses of Cd, and Koirtyohann and Hopkins[534] found no losses of the metabolized Cd radioisotope during both oven drying at 100°C and dry ashing at 500 and 600°C of rat blood, kidney,

Table 2
AAS TECHNIQUES FOR Cd DETERMINATION IN BIOLOGICAL MATERIALS

Technique	Biological materials
Flame AAS after digestion	Fish,[458] hair,[288,459-461] milk,[374] tissues[315,319,371,444,462,463]
Flame AAS after TAAH solubilization	Hair,[464] tissues[464,465]
Flame AAS after acid extraction	Hair,[466] tissues[467]
Flame AAS after liquid-liquid extraction	Blood,[7,27,332,468-473] feces,[248,366] fish,[474,475] food,[424,470,476,477] hair,[473,478] milk,[477,479,480] teeth,[481] tissues[7,43,470,472,473,475,477,482] urine[7,27,366,387,469,472,483]
Miscellaneous flame AAS techniques	Serum,[484,485] (solid) tissue,[441] tissues,[442] urine[440,484,485]
Delves cup	Blood,[430,431,486-491] tissues,[491,492] urine[491]
Sampling Boat®	Blood,[493] fish,[494] food,[495] milk,[495] urine[27,496]
Direct ET-AAS	Blood,[57,330,392a,435,497-504] bone,[450] fish (ash),[505] food,[506] hair,[506] serum,[432,507] teeth,[451] tissues,[454,455,505,508,509] urine[330,385,422,434,435,510,511,511a]
Direct Zeeman ET-AAS	Blood,[363] tissues,[373] urine[363,453]
ET-AAS after digestion	Bile,[364] blood,[7,362,388,438,457,512] feces,[365] fish,[513] food,[365] hair,[426,444] liver biopsies,[364] milk,[394,514] nails,[444] placenta,[67,87,376] plasma,[388] saliva,[379] tissues,[7,67,87,256,274,376,457,513,515-520] urine[362,365,388,438,457]
ET-AAS after TAAH solubilization	Blood,[392a,521] hair,[368,521] muscle biopsies,[92] tissues,[521] urine[521]
ET-AAS after liquid-liquid extraction	Blood,[359,392a,444,512,522-526] bone,[527] fish,[527] tissues,[528] urine[27,359,360,386,522,523,525,529,530]
Miscellaneous ET-AAS techniques (and procedures requiring special equipment)	Blood,[447] food,[446] saliva protein fractions,[379] serum,[378] tissues,[445,449,452,509] urine[448]

Table 3
SELECTED AAS PROCEDURES FOR Cd DETERMINATION IN BIOLOGICAL SAMPLES

Sample	Brief procedure outline	Ref.
Blood	1 + 1 dilution with aqueous 0.1% Triton® X-100; Zeeman ET-AAS: 125°C (ramp), 250°C (ramp) for 80 sec, and 2000°C; standards on blood	363
	Fivefold dilution with aqueous 5% $(NH_4)_2HPO_4$—5% Triton® X-100; ET-AAS: 100, 500, and 2200°C (gas stop); Pyro-coated tubes; background correction; standards in diluent	502
	100 $\mu\ell$ of blood + 100 $\mu\ell$ of 0.05% v/v HNO_3 + 100 $\mu\ell$ of 1% w/v $NH_4H_2PO_4$; ET-AAS: 100°C (ramp 40 sec + hold 20 sec), 600°C (60 + 30 sec, O_2 added as alternate gas), 600°C (1 + 25 sec), 400°C (1 + 10 sec), 2100° (1 + 3 sec), and clean-out at 2300°C; background correction; standards on spiked blood	503
	Blood protein pptn with 1 *M* HNO_3 (1 volume of blood + 3 volumes of acid); centrifugation; ET-AAS: 120°C for 20—30 sec, 350 to 500°C (ramp 20 sec + hold 20 sec), and 1600°C (1 + 4 sec, gas stop); background correction; acid-matched standards	360
	Four relevant procedures detailed, compared, and discussed: (1) direct Zeeman ET-AAS after Lumatom® dilution, (2) HNO_3 protein pptn-ET-AAS, m20.9 (3) wet ashing-ion associate extraction-ET-AAS, and (4) wet ashing ASV; standard addition calibration	392a
	TCA protein pptn.; APDC-MIBK extraction at pH 5—6; ET-AAS; background correction; standards extracted similarly	526

Table 3 (continued)
SELECTED AAS PROCEDURES FOR Cd DETERMINATION IN BIOLOGICAL SAMPLES

Sample	Brief procedure outline	Ref.
Blood, urine, tissue digests	TCA pptn of blood proteins; SDDC-MIBK extraction at pH 6.5—7; flame AAS; standards extracted similarly	472
Blood, bone, fish, tissues, urine	Wet digestion with HNO_3 + H_2SO_4 (4 + 1) or HNO_3 + $HClO_4$; saturated $NaHCO_3$ added (pH ~7); APDC-CCl_4 extraction; ET-AAS: 110—130, 390—400, and 1800°C (gas stop); background correction; standards extracted from aqueous standard solutions (pH ~7)	522, 527
Blood	Delves cup procedure: 15-μℓ aliquots of heparinized hemolyzed blood pipetted into cups, and a batch of 100 cups simultaneously oven ashed for 16 hr up to 400°C; background correction; peak-area calibration; standards on spiked low-Cd blood; DL 0.3 ng/mℓ	430
Bone, fish ash, NBS SRMs, teeth	Direct ET-AAS; solid sampling (0.3—3 mg of powdered samples analyzed); 100, 300, and 1000°C; background correction; peak-area mode; standards on solid hydroxyapatite (for bone, teeth) or else standard addition calibration	450, 451, 505
Feces	HNO_3-$HClO_4$ digestion; APDC-MIBK extraction; flame AAS; standards extracted similarly	248, 366
Hair	A 0.4-g sample digested with 10 mℓ of HNO_3 (overnight, at room temperature), evaporated to near dryness, final solutions in 10% v/v HNO_3; flame AAS; standards in 10% v/v HNO_3	460
Hair, nails, tissues	TAAH solubilization: a 0.1-g sample incubated with 1—2 mℓ of the base (10 to 46% solutions in water or toluene or alcohol) at 60—70°C for 2—24 hr; diluted to 1—5 mℓ; flame AAS[464,465] or ET-AAS;[92,368,521] temperature program:[92] 150, 400 (ramp), and 1900°C; background correction; standard addition calibration	
Milk	Protein pptn. with 2 *M* H_2SO_4; centrifugation; ET-AAS; background correction; standard addition calibration	394
Placenta	Thoroughly homogenized (!) samples digested either with HNO_3 in an autoclave or with HNO_3 and $HClO_4$ (Kjeldahl); ET-AAS: 90, 325, and 1900°C; background correction; standard addition calibration	376
Tissues, urine, blood	Kjeldahl flask digestion of 25 mℓ of urine or 0.5 g of tissue with 10 mℓ of HNO_3, evaporated to ca. 5 mℓ, aqueous $NH_4H_2PO_4$ added (to its final concentration of 0.8% w/v), and diluted to 25 mℓ; ET-AAS with tantalized and pyro-coated L'vov platforms: ramp to 750°C, atomization at 2100°C; background correction; standards contained $NH_4H_2PO_4$ and HNO_3; HNO_3-digested blood analyzed similarly	457
Urine	1 + 1 dilution with 5% v/v HNO_3; Zeeman ET-AAS: 110°C for 40 sec, 250°C for 40 sec, and 1500°C for 8 sec; aqueous standard solutions	363
	1 + 2 or 1 + 4 dilution with water; ET-AAS: 120°C for 20—30 sec, ramp to 350°C, atomization at 1200°C (gas stop); background correction; aqueous standard solutions	360
	Direct ET-AAS (5—10 μℓ): 100°C for 30 sec, 400°C for 30—40 sec, and 1100—1500°C (depending on apparatus!); background correction; standard addition calibration	385
	1 + 1 dilution with 0.3 *M* HNO_3; Varian Techtron® CRA-90 furnace: 100°C for 35 sec, 350°C for 30 sec, and 1200°C (ramp rate of 500°C/sec); background correction; peak-area mode; standard addition calibration	434
	1 + 1 dilution with aqueous 2% v/v NH_4NO_3 — 0.026 *M* $NH_4H_2PO_4$ — 0.02% Triton® X-100; ET-AAS: 110°C (20 + 10 sec), 380°C (30 + 15 sec), 800°C (0 + 8 sec, maximum power), and (clean-out) at 2100°C; background correction; standard addition calibration	511a
	SDDC-MIBK extraction at pH 5, centrifugation; ET-AAS: 120, 350, and 1900°C; background correction; standards extracted similarly	360

and liver. Small retention losses on crucibles were established at 600°C.[520,534] The dry ashing techniques have low reagent blanks but should be regarded as potentially prone to volatization and retention losses. There are reports indicating successful dry ashing of food (cereal),[476] hair,[370,459] milk,[374,480] and tissues[462,520,534] at temperatures between 450 and 500°C (preferably in fused silica crucibles), but there are also quite a few conflicting statements.[43,91,430,431,480,537] Even the addition of ashing aids such as $Mg(NO_3)_2$ or HNO_3 has been questioned.[91] The analyte losses obviously depend on many factors and are highly probable above 500°C. Among these factors let us mention the state of binding of Cd in the particular tissue matrix, the heating rate[520] and aeration conditions, the crucible material, age, and surface condition, etc.[91,266,520,534] The pre-ashing of samples with an excess of HNO_3[250,268] or the addition of sulfate or H_2SO_4 as an ashing aid[250,538] should diminish the losses on dry ashing. It should be noted here that the *in situ* ashing of biological samples in graphite furnaces is even more prone to volatization losses (starting from temperatures as low as 250 to 250°C), owing to the reducing conditions and higher heating rates in these furnaces as compared with the oven ashing.

LTA in excited oxygen[315,438,442,462,516,539] as well as oxygen-bomb combustion[273,274,288,540] are very attractive because of their low blank levels and complete recoveries of the analyte.

All wet-digestion procedures can be applied to oxidize the organic matter, provided that high-purity acids (e.g., containing less than 1 ng Cd/mℓ) and a suitable, precleaned lab ware (of quartz, PTFE, glassy carbon) are used. Wet digestions with HNO_3 and $HClO_4$;[248,256,366,426,472,474,494,524,527,541,542] HNO_3 and $HClO_3$;[272] HNO_3 and H_2SO_4;[477,508,527,543] HNO_3, $HClO_4$, and H_2SO_4;[43,543] HNO_3 and H_2O_2;[43,388,444,505] H_2SO_4 and H_2O_2;[67,424] HNO_3, H_2SO_4 and H_2O_2;[170, 475,525] etc. are available to choose from. Under certain conditions, digestions involving $HClO_4$, $HClO_3$, and H_2O_2 are *hazardous*. The choice of the digestion procedure depends, among other factors, on the requirements of the next analytical step. All acids are tolerable in the flame AAS; HNO_3 or H_2SO_4 or H_3PO are preferable in ETA, but HCl or $HClO_4$ should be avoided as signal depressants. Oxidants should be absent if dithiocarbamate or dithizone extractions are to be performed.

There is a tendency now to use simplified preparation procedures (e.g., a single-acid treatment) which release the Cd from its chemical bonds with the matrix and partly destroy the organic matter, while keeping at the same time much lower reagent blanks. Fortunately, Cd is readily acid extracted from biological materials.[7,319,466,467,512] Even extraction with a 1% v/v HNO_3 at room temperature for 24 hr released Cd almost completely from hair[466] and liver.[467] Cadmium is easily put into solution by treatment/boiling of tissue, blood, hair, etc. with (1 + 1) or concentrated HNO_3[257,460,517] or with (1 + 1) HNO_3 + HCl.[319] Pressurized digestions in PTFE-lined bombs[512,518,519,544,545] or in tightly capped tubes are also popular for their low blanks, low reagent (HNO_3) consumption, and speed (Caution! Pressure build-up!).

Deproteinization of blood with TCA[7,472,526] or HNO_3, [7,360,512] or of milk with H_2SO_4[394] can also be applied; Cd is determined in supernatants by direct ET-AAS[7,360,392a,394,512] or after an extraction step.[7,472,526]

Tissue and hair specimens can also be solubilized with TAAH and then analyzed by flame[464,465] or ET-AAS.[92,368,521]

The time stability of Cd in acidic biological digests is quite acceptable:[425,520] Sperling and Bahr[425] stored the dilute biological digests (1:400, containing 0.1 to 8 ng Cd/mℓ in HNO_3 + H_2SO_4) in precleaned bottles of dense polyethylene for more than 8 weeks; similarly, the hair digests (1:1000, containing approximately 0.1 to 2.5 ng Cd/mℓ in HNO_3 + H_2SO_4 or in HNO_3 + $HClO_4$ in precleaned polyethylene bottles) proved to be stable for at least 4 weeks.[426] Tissue homogenates in TAAH-alcohol are stable for at least 2 to 3 days. After enzymatic digestion, tissue digests are stable for at least 1 month at 5°C.[463]

Preconcentration/separation of the trace Cd from the matrix is effected by extraction, coprecipitation (often as a dithiocarbamate[446,540]), ion-exchange chromatography,[326,378,412,546] or controlled-potential electrolysis.[397,440]

The solvent extraction of Cd is widely used;[76,389,547] as a rule, the extractions are made from pH-adjusted and buffered digests or deproteinized blood/serum. Procedures involving direct extractions of Cd from hemolyzed blood/serum or from pH-adjusted urine[359,468,471,486,491,529] have been described but they may entail an underestimation of the analyte. In untreated samples, the Cd may be bound to the matrix,[523] drugs, EDTA,[529] etc. and thus may not be completely chelated and extracted; moreover, persistent emulsions are formed, centrifugation is required, and a standard addition calibration often becomes necessary.

Cd(II) is typically extracted as a dithiocarbamate. The APDC-MIBK extraction system is very popular in the pH range between 2 and 8 but preferably at pH 3.5 to 5.[248,271,366,387,475,481] Organic solvents besides MIBK which offer certain practical advantages may also be used: CCl_4,[439,522,527] $CHCl_3$,[429,439] heptan-2-one,[479] etc. Alternatively, Cd can be extracted from biological digests with each of the following extraction systems: SDDC-MIBK (at pH 5 to 8.5),[7,471,472,494,529,548] diethylammonium diethyldithiocarbamate (DDDC)-MIBK,[477,480] hexamethyleneammonium hexamethylenedithiocarbamate (HHDC)-butylacetate,[43] HHDC-$CHCl_3$,[429] HHDC-diisopropylketone-xylene,[525] dithizone-MIBK,[496,524] dithizone-$CHCl_3$,[474,530] diphenylcarbazone-pyridine-toluene,[473,524] dipivaloylmethane-MIBK-butylamine,[543] 4.25 to 5 *M* H_3PO_4 to 0.7 *M* KI-MIBK,[476] $[CdI_4]^{2-}$-Amberlite® LA-2,[424,478] $[CdCl_4]^{2-}$-triisooctylamine,[332] $CdCl_4^{2-}$-trioctylamine-MIBK.[392a]

The extraction of Cd is traditionally believed to be straightforward, but there is evidence now that the role of some important parameters has to be considered more critically. The main problem is the limited time stability of the Cd(DTC) complex in the extracts.[245,427,428] For example, the Cd pyrolydinedithiocarbamate in MIBK is stable for only 30 min or 2 hr after extractions performed at pH 2 to 3 or 4 to 8, respectively;[427] in a similar solvent, heptan-2-one, the chelate is stable for less than 1 hr after extraction at pH 4.5,[479] whereas in an "inert" solvent, CCl_4, after extraction at pH 7, the time stability of the chelate is at least 16 hr.[527] Hence, such "details" in extraction procedures as pH,[427-429] the organic solvent used,[340,428,429] the completeness of the phase separation,[340,428] the concentration of the APDC, the presence of other coextracted species (Fe, Cu, Zn, excess APDC, etc.),[245,480] and the storage conditions for prepared extracts (temperature, exposure to light, duration, etc.) should be carefully reexamined for every newly adopted procedure. Moreover, the ETA of Cd in extracts, though generally useful and widely applied, is still prone to interferences from chlorinated solvents,[549] coextracted APDC[549] or SDDC,[7] background absorption,[550] etc.

Cd is readily determined by flame AAS at levels above 0.01 μg/mℓ, i.e., in digested tissues, hair, and nails. Any dilute material acid may be present in the final solutions, provided that its concentration is approximately matched in the standard solutions as well. Bulk-matrix effects and background contribution are easily checked by standard additions and background corrector. A three-slot burner and pulse-nebulization technique are to be recommended in analyses of digests/homogenates, while organic extracts should preferably be analyzed by means of pulse nebulization with a single-slot burner and simultaneous background correction.[429,439] Calibration in TAAH-solubilization procedures[464,521] is usually done by standard additions, in order to compensate for the effect of TAAH and its solvent, as well as to ensure stability of the calibration solutions in the organic base.

The Delves cup and the Sampling Boat® accessories have now been almost completely displaced by ET-AAS, but can still be used in analyses of blood,[430,487-490,493] tissue homogenates[492,495] or digests (after neutralization or evaporation of the acids present!),[494] as well as of organic extracts.[27,487,491,494,496] The practical DLs are about 0.2 to 0.4 ng/mℓ. Good accuracy and precision have been obtained only with strict control of all analytical steps.[430,488] The most successful procedures combine some of the following practical approaches: simultaneous background correction (mandatory!), standard additions[431,490] or matrix-matched calibration (advisable),[430,488,490] peak-area measurements,[430,492,493] external ashing at controlled temperature,[430,488,493] and matrix/analyte modification (by adding $(NH_4)_2HPO_4$[489] or K_2SO_4[27]).

The *problems* in the ET-AAS of Cd stem from its high volatility and short wavelength (228.8 nm):

1. Sporadic volatization losses during the ashing step
2. Premature evaporation and losses at the very start of the atomization step
3. Severe background absorption
4. Pronounced matrix effects

In order to eliminate or reduce these effects and to properly account for them, the analyst should apply modern, versatile equipment as well as a rational methodology. Let us consider this matter in the following order: (1) rational sample preparation and matrix/analyte modification; (2) versatile furnace programming; (3) background correction efficiency; (4) choice of calibration mode; and (5) radical solutions and concepts.

The dilute HNO_3 (0.1 to 1 *M*) appears to be a very convenient medium in the ET-AAS of Cd;[436,437] hence, the specimens are often pretreated with HNO_3 (by dilution, digestion, acid extraction, or protein precipitation), which acid is present in the final sample and standard solutions either by itself or in combination with surfactants and/or other matrix/analyte modifiers.[7,360,434,438,500,511a,512,515,517] Partial neutralization of the HNO_3 digests with aqueous NH_3 is also practiced, producing some NH_4NO_3[519] which serves as a suitable matrix modifier.[223,362,513] Other matrix/analyte modifiers are: 0.2 to 2 *M* H_2SO_4,[67,330,394] $(NH_4)_2SO_4$,[223,454,455] aqueous NH_3,[438] $(NH_4)_2HPO_4$,[223,362,502] $NH_4H_2PO_4$,[457] $NH_4H_2PO_4$ + HNO_3,[503] NH_4F,[223,362] HF (in graphite furnaces without quartz windows!),[501] 5% H_2O_2,[511] 1% H_2O_2 + 1% HNO_3,[432] La^{3+},[438] La^{3+} + MoO_4^{2-} + NH_4NO_3,[513] etc.

The analytical solutions should not contain HCl[437,506] and $HClO_4$,[66,437] which are known to be very strong depressants in the ET-AAS of Cd.

Particularly versatile equipment and temperature programming are called for in Cd assays. The analyst should be able to choose between different ramp heating rates in order to ensure smooth drying and ashing. Diluted urine samples were found to spread within the graphite tube more than the aqueous standard solutions,[385,434] thus resulting in both sensitivity and reproducibility impairment; it has therefore been recommended to inject these solutions into a prewarmed graphite tube.[434] The ashing temperature is a critical parameter, as it strongly depends not only on the particular specimen analyzed, the sample pretreatment, and the matrix/analyte modifiers added, but also on such variables as the heating rate, the quality of a particular graphite tube(!),[360,433] and the pretreatment/history of the graphite tube.[438,513] For example, the optimum ashing temperatures ranged between 290 and 400°C[433] or between 350 and 500°C,[360] just due to the individual properties of each of the graphite tubes used. Thus, the maximum ashing temperature should be confined to as low as 250 to 350°C (for untreated biological fluids or tissue homogenates) or 350 to 500°C (for digested or matrix/analyte modified samples).

Delves and Woodward[503] recommended the use of O_2 as an alternate purge gas (internal flow of 50 mℓ O_2/min) to promote in dry ashing of the diluted blood samples; in that case, ramp ashing up to 600°C was applied.

The next critical instrumental parameter to be optimized is the atomization step. The temperature, heating rate, and purge gas flow all strongly affect the sensitivity, extent of matrix effects, precision, and background signal as well as its separation from the analyte signal. It is a difficult task to balance all these requirements. Moreover, the behavior of diluted standard solutions and sample solutions may be quite different;[392a,434,436,502,513] e.g., for undiluted urine the optimum atomization temperature has been as low as 1100 to 1200°C, whereas aqueous standard solutions have been atomized at 1800°C.[385] In presence of organic matter and NaCl, the appearance temperature of Cd is significantly lower, and even a 50-fold dilution of the urine specimens (from exposed sheep)[435] did not eliminate the matrix effect.

There would appear to be two alternatives in the programming of the atomization stage. The first is to apply as high as possible heating rates ("temperature-controlled heating",[497] "maximum power" heating,[60] capacitive-discharge heating",[509] etc.) and relatively low atomization temperatures. For example, at 830[497] or 900°C,[510] the analyte peak appears before the bulk-matrix background. The second way, which may prove practicable with older models of apparatus (with slower heating rates and inefficient background correction ability) is to apply ramp atomization in order to temporally resolve the analyte signal from the background. This way of programming usually suffers from serious interferences, so that the standard addition calibration becomes mandatory.

As a rule, the temperature program for Cd should include a clean-out stage, in order to avoid the accumulation of matrix residues and the associated accuracy and precision impairment.

If an efficient background correction is available, the "gas stop" or, at least, the "miniflow" mode of the purge gas flow during the atomization should be advantageous.

The background absorption is a serious problem in ET-AAS of Cd in biological materials. The deuterium arc background correction system must be carefully aligned and balanced so as to correct background absorbance of at least 0.5 to 0.8 absorbance units. It should be stressed that both the analyte and the background signals are transient, and thus most early AAS equipment cannot accurately correct higher background absorbance signals. Therefore, both reduction and time resolution of the background signal vs. the analyte signal should be favored. This can be done by a clever combination of matrix/analyte modification;[362,388,392a,394,434,501,502] ashing up to the highest tolerable temperatures[330,501,515] or ashing in oxygen[503] or interruption of the Ar flow to allow air access during the ashing stage;[500,501] use of lower atomization temperatures;[451,497,510] applying either ramp atomization[501,513] or — just the opposite — extremely high heating rates;[497,509,510] pretreatment of the graphite tube with Mo(VI)[510] or Mo(VI) + La^{3+},[513] etc. Certain radical solutions (see further on) may also be utilized.

The choice of a proper calibration mode is another major concern. This choice should depend on many factors, such as the particular matrix, the preparation step, the concentration levels of Cd, the versatility of the equipment/programming available, etc. Dilute tissue[444,515] and hair digests[426,444] may be analyzed against simple acid-matched standards; blood samples are quantitated vs. spiked, low-Cd blood or vs. preanalyzed blood specimens employed as standards[362,497,501,503] or, rarer, by standard additions.[330,392a,504,512] The standard addition calibration is often mandatory, in particular with such difficult and variable samples as urine,[385,435,438,453] milk,[394] and TAAH-homogenized specimens.[92,392,521] There are, of course, exceptions to this generalization, both simpler[360,365,388,502,510] and more complicated calibration modes,[376,392a,517,519,520] depending on the above factors. Standard addition calibration, combined with peak-area measurements, is often applied in direct analyses of solid samples (bone,[450] teeth,[451] tissues,[454,455,505] etc.). Standard additions check should always be recommended with newly adopted procedures, and perhaps with new graphite tubes.

The peak-area mode of measurements is now more and more available;[551,552] it can additionally improve the analytical precision, provided the instrument used has a stable baseline and insignificant zero drift.[85,87,245,434,456]

Finally, the matrix effects problems can be settled by radical solutions and concepts such as: use of constant-temperature furnaces or furnaces approximating a constant-temperature performance;[445-448,452,509] atomization off the L'vov platform;[85,229,454-457] application of special background correction schemes,[485,499,500] and, in particular, the Zeeman background correction;[363,373,453] as well as the stabilized-temperature platform furnace (STPF) concept.[553]

In L'vov platform atomization,[229] the appearance time of Cd is delayed by 0.085 to 1.55 sec,[85,229,454] and the analyte is thus volatized when the furnace temperature is substantially higher (by 310 to 570°C) vs. the "off-wall" atomization. Thus, the matrix effects are reduced, the precision and accuracy are improved, and the tube lifetime is extended. The combination of the L'vov platform technique with an analyte/matrix modification[454,455,457] and/or with a standard addition calibration (for solid samples)[454,455] or with peak-area measurements[456] results in very effective ET-AAS procedures for direct analyses of solid specimens[454,455] or simply prepared HNO_3 digest of blood,[457] tissues,[456,457] and urine.[457]

IV. CONCLUSION

ET-AAS is the method of choice at nanograms per milliliter levels of Cd; alternative techniques are ASV/DPASV. Extraction-flame AAS procedures and Delves cup accessory require less expensive equipment, but more preparatory work. Extractions of Cd(II) as dithiocarbamate or halide are useful in both flame and ET-AAS, provided they are preceded by an adequate sample treatment and certain important procedural details (see text) are observed.

ET-AAS involves serious problems with background, premature volatization, and matrix effects. Of great importance are therefore: (1) the efficiency of background correction (high, transient, and incidentally structured background signals are encountered); (2) the analyte thermal stabilization with (composite) matrix/analyte modifiers (often ammonium hydrogen phosphates or dilute HNO_3 plus surfactant); (3) versatile temperature programming (thoroughly optimized ashing conditions, ashing in O_2, maximum heating rates, gas stop, etc.); and (4) adequate calibration (often, matrix matched, standard additions, and/or peak area).

Simple and rapid preparation procedures are available: blood protein precipitation (with HNO_3) or incomplete (incidentally pressurized) single-acid digestion or TAAH homogenation.

Important in direct analyses of biological liquids are Zeeman background correction, L'vov platform atomization (and the STPF concept as a whole), and aerosol deposition technique. There are still serious problems with direct analyses of urine and solid microsamples.

Good quality control schemes are essential in trace Cd assays.

Chapter 9

CHROMIUM

I. DETERMINATION OF CHROMIUM IN BIOLOGICAL SAMPLES

The concentrations of Cr in biological samples range broadly (Table 1), being about or below 1 ng/mℓ level in most body fluids, scores of nanograms per gram in soft tissues, and up to 1 μg/g in other biological specimens (bone, feces, food, hair, etc.). Several analytical techniques are sensitive to detect nanogram amounts of Cr: ET-AAS, SSMS, NAA, and chelate-GC.

NAA has been applied to analyses of blood,[45,573] brewer's yeast,[574,575] diet,[45,116] hair,[9,11,45,48,98,101,118] milk,[182] nails,[560] platelets,[119] serum/plasma,[45,120,562,575] serum protein fractions,[120] skin,[112,565] teeth,[23] tissues,[12,45,106,108,575] urine,[45,165,575] etc.[45,575] This method is attractive because of its inherent protection from extraneous additions of Cr after the irradiation stage of analysis, as well as because of the possibility of instrumental performance (INAA) (e.g., with some specimens such as hair,[11,98,101,102,118] nails,[560] and tissues[12,565]).

SSMS analyses of hair,[54] nails,[16] plasma,[55] teeth,[25] tissues,[132,275] etc. have been described; in some of these procedures, however, the Cr content has been below the DL. A few analytical laboratories are able to perform very accurate and sensitive assays in the subnanograms-per-gram concentration range by ID-GC-MS.[568]

Spectrophotometric[133,342,554,576] and catalytic,[577] methods, and chelate-GC procedures (with electron-capture detector)[575] call for experienced hands and especially purified reagents.

Effective preconcentration is needed in XRF procedures.[578]

The classical AES technique, after an ashing and/or preconcentration step, has been widely applied to analyses of blood,[36,579] erythrocytes,[579] hair,[288,555,556,579] serum/plasma,[413,579] sweat,[21] teeth,[26] tissues,[36,44] urine,[579] and other biological samples;[41,42] in these procedures, one should expect contamination of the sample ash arising from both grinding and homogenation/mixing with spectrochemical buffers.

The ICP-ES should provide better DLs than the flame AAS, e.g., from a few nanograms per milliliter down to 0.08 ng/mℓ in aqueous standard solutions.[29,35] Several successive applications of this technique to biological samples have been described.[30,348,392,580]

II. ATOMIC ABSORPTION SPECTROMETRY OF CHROMIUM

In air-acetylene flame and under properly optimized parameters, the characteristic concentration is 0.04 to 0.1 μg/mℓ, and the DLs are 0.002 to 0.006 μg/mℓ.[56-59] Both the signal-to-noise ratio and the extent of interferences depend greatly on the acetylene flow rate and the observation height. Often, the instrumental parameters chosen for best sensitivity may not be optimal as regards avoidance of matrix effects. Serious interferences from acids and various concomitant ions, e.g., from Fe, Ti, Co, Ni, V, Al, Mo, Mn, etc. are encountered, and their depressive effect is more pronounced in reducing flame and at a lower observation height.

Numerous spectrochemical buffers have been proposed, being effective in different specific cases: 1 or 2% w/v aqueous solutions of NH_4Cl, NH_4ClO_4, NH_4HF_2, Na_2SO_4, $KHSO_4$, $K_2S_2O_7$, $K_2S_2O_8$, NH_4HF_2 + Na_2SO_4, $AlCl_3$, 8-hydroxyquinoline, sodium dodecylsulfonate, $LaCl_3$, etc. Dilute mineral acids, e.g., 1 to 5% v/v of HCl or $HClO_4$ or HNO_3 are all suitable media but their concentration in both sample and standard solutions should be matched.

Chromium is determined more selectively, but at the expense of a somewhat lower sensitivity in a N_2O-acetylene flame. Under thoroughly optimized conditions, the DLs may

Table 1
CHROMIUM CONCENTRATIONS IN HUMAN BODY FLUIDS AND TISSUES

Sample	Unit	Selected reference values	Ref.
Blood	μg/ℓ	2.9	554
		(2—6)	90
Bone	μg/g	0.46 ± 0.05	268
Brain	μg/g	0.01 ± 0.001	3
Feces	μg/day	69 (<10—296)	96
Hair	μg/g	(0.05—0.57)	555
		0.18 ± 0.11 (0.02—0.50) (U.S.A.)	556
		$\bar{x}$, 0.2 (0.15—15.6) (Japan)	11
		♂, 0.192; ♀, 0.265	162
		0.21 ± 0.14 (U.S.A.)	557
		0.46 ± 0.39 (0.05—2.63) (India)	101
		$\bar{x}$, 0.5; 4 (<0.5—65.3) (Italy)	98
		♂, 0.57;[a] ♀, 0.63;[a] (0.06—5.3)[a] (New York)	288
Kidney	μg/g	(<0.003—0.069)	106
		0.03 ± 0.005	3
		0.07 ± 0.02	268
Kidney cortex	μg/g	0.02 ± 0.004	3
		0.033	107
Liver	μg/g	0.0054 (0.002—0.010)	108
		(0.003—0.11)	106
		0.033 ± 0.006	268
		0.045 ± 0.059	275
		0.08 ± 0.06	3
Lung	μg/g	0.052 ± 0.005	268
		$\bar{x}$, 0.082; 0.075 ± 0.045 (0.025—0.72)	105, 106
		0.5 ± 0.07	3
Milk	μg/ℓ	0.45;[b] ~1[b]	558,559
Muscle	ng/g	5 ± 1	3
Nails (finger)	μg/g	(2.4—13)[c,d]; (3.6—11)[c,e]	16
Nails (toe)	μg/g	♂, 4.23 ± 1.25[c] (1.8—6.5)[c]	560
		♀, 6.77 ± 3.99[c] (2.8—12.9)[c]	560
Plasma or serum	μg/ℓ	0.14	561
		0.16 ± 0.083 (0.0382—0.351)	562
		0.20 ± 0.07	563
Saliva	μg/ℓ	0.09[f]	564
Skin	μg/g	$\bar{x}$, 0.9;[c] 1.1 ± 0.51[c]	112
		0.227 ± 0.187;[c,g] 0.985 ± 0.948[c,h]	565
Sweat	μg/ℓ	25	566
Tooth (dentine)	μg/g	1.99 ± 0.84	23
Tooth (enamel)	μg/g	1.02 ± 0.51	23
		3.2 ± 0.8 (<0.1—18)	25
Urine	μg/ℓ	0.27 ± 0.13; 0.60 ± 0.16	567
		0.32[b]	568
		0.2—0.4[i]	563
		0.41 (Max 0.8)	554
		$\bar{x}$, 0.9	561
		1.95 (0.20—10.72)	569
		<9	570
		0.24	559
	μg/day	1.09 (0.45—2.35)	571
		1.6 ± 1.1	572
		<10	96

Table 1 (continued)
CHROMIUM CONCENTRATIONS IN HUMAN BODY FLUIDS AND TISSUES

Note: Wet weight unless otherwise indicated. Mean ± SD (range in parentheses). x̃, median.

[a] Geometric mean.
[b] Pool.
[c] Dry weight.
[d] Adults.
[e] Children.
[f] Whole mixed stimulated.
[g] Dermis.
[h] Epidermis.
[i] Mode.

be as low as 0.005 to 0.01 μg/mℓ. Only some minor interferences should be expected in this flame, in particular, if its red zone is broader than 3 to 5 mm (i.e., under more reducing conditions).

In both these flames, the Cr absorbance depends on its oxidation state, being higher for Cr(III) than for Cr(VI). This effect should always be reckoned with when sample and standard solutions are prepared or treated; it is especially pronounced in reducing flames.

In ETA, the sensitivity and selectivity of Cr assays are quite satisfactory, although this technique poses its own problems. The characteristic concentration ranges from 0.005 to 1 ng/mℓ (0.5 to 20 pg), and the DLs are between 0.0005 and 1 ng/mℓ (from 0.3 to 10 pg).[57,59-61] These broad ranges should not be surprising, as numerous factors affect both the efficiency of atomization and noise levels: volume of furnace and solution injected, pyrolytic coating, heating rate, emission noise, background absorption and noise from the cyanide formed, flow rate of the purge gas, etc.

In order to lower the extent of Cr-carbide formation (and thus improve the sensitivity by a factor of 2 to 4), pyrolytically coated graphite tubes are generally preferred.[60,581,582] Two exceptions from this rule were found in the literature: (1) in direct analyses of serum with *in situ* ashing in oxygen, uncoated tubes proved superior to pyro-coated tubes;[582] and (2) assays of Cr in organic extracts were more precise with uncoated tubes.[583]

The *in situ* pyrolitic coating by adding methane to the argon purge gas has proved useful, as well.[295]

For the best performance, argon (or incidentally argon + 10% CH_4) is recommended; the use of nitrogen results in cyanide formation associated with background absorption, background noise, baseline drift, and a restricted lifetime of the tube.[27,581,584]

Minor interferences[553,554,556,585] should be expected in ET-AAS of Cr, since most volatile matrix components can be selectively vaporized out of the graphite tube before the atomization stage. Interferences from chloride[586] were lowered by adding matrix modifiers (listed in a decreasing order of efficiency[586]): $(NH_4)_4EDTA > H_2SO_4 > (NH_4)_2$ tartarate $> NH_4$ glycinate $> (NH_4)_3$ citrate $> HNO_3$. $Mg(NO_3)_2$ was recently proposed as thermal stabilizer of Cr up to 1650°C.[553]

III. AAS METHODS FOR ANALYSIS OF BIOLOGICAL SAMPLES

The literature on AAS of Cr in biological samples is ample although not consistent. The flame technique has found its early applications to specimens of blood,[587,588] feces,[589] fish,[475,590] food,[588,590,591] hair,[461] milk,[590] tissues,[319,442,475,517,588,590] and urine,[587,588,592] but the sensitivity of most procedures is limited, and an effective preconcentration step is required in analyses of blood,[587,588] milk,[590] urine,[587,588,592] and most food samples.[588,590] Thus, the use of flame

AAS should be confined to the detection of Cr in some tissues and specimens from exposed laboratory animals. The flame AAS is also useful in assays of Cr_2O_3 in feces (in metabolic balance studies).[589] Unless a lean N_2O-acetylene flame (with red zone of approximately 2 to 3 mm) is used, AAS measurements may be prone to serious interferences. It should be mentioned that this flame may be hazardous, especially under the indicated conditions; an automatic gas control is therefore advisable. Acid-matched standards and a check by standard additions should be applied.

At present, Cr is typically determined by ET-AAS with good sensitivity and acceptable selectivity. Numerous applications to samples of blood,[7,457,511,554,593-596] biologically active extracts from SRMs and brewer's yeast,[597] brewer's yeast,[574,575,583,598,599] diet,[567,600] erythrocytes,[7] feces,[567] hair,[78-80,426,556,557,593,601,602] milk,[558,559] mussels,[86] serum/plasma,[7,27,511,559,561,563,575,582,584,594,596,601-608a] serum protein fractions,[603,604] skin,[609] tissues,[7,67,87,457,516,556,561,571,583,594,601,610,611] urine,[7,27,457,511,554,558,559,561,563,567,569-572,575,581,585,594,608,612-618a] and urine fractions[613,618,619] have been described (see Table 2 for selected procedures).

The main problems in subnanograms-per-gram Cr assays are contamination and background absorption. Minor losses during the preparation step (ashing and/or extraction) and minor interferences from matrix constituents should also be reckoned with.[556,585,618a]

Inasmuch as very low concentrations are to be determined in some biological liquids (serum/plasma, CSF, urine, milk), the obtainment of a representative sample without gross extraneous additions of Cr is of vital importance.[49,562,620] It is now recognized that many previous analytical results and published "normal" values have been positively biased by up to one or two orders of magnitude due to severe contamination during sample collection, storage, and/or analysis. To a lesser extent, this statement should also refer to analyses of solid biological specimens.

A very high contamination of blood,[49,562] CSF,[621] liver biopsies,[622] and autopsy tissue specimens[623] from stainless steel needles or knives (stainless steel may contain as much as 10% Cr, i.e., 100,000,000 ng Cr/g!) has been experienced; important extraneous additions of Cr from grinding procedures (agate, corundum[624]), dust, sweat,[162,566] skin, adherent surface dirt, etc., as well as from storage containers and added conservants, are to be expected, checked for, and controlled. Versieck et al.[562] published the lowest normal values for Cr in serum (0.0382 to 0.351 ng/mℓ) after careful sampling prior to RNAA procedure: the samples were taken with a plastic cannula trocar and collected in precleaned, high-purity quartz tubes; the cleaning procedure included 2-hr boiling with a H_2SO_4-HNO_3 mixture, repeated twice, followed by a 3-hr steam-cleaning with quartz-distilled water in a dust-free room; still, the blank levels of the whole procedure were 26 to 70 pg Cr. Other authors reported blank levels higher by as much as one or two orders of magnitude!

Reliable container materials to be used in ultratrace Cr assays proved to be quartz, PTFE, polyethylene, and polystyrene, after proper cleaning involving, e.g., detergents, dilute acids or acidic mixtures, steaming, sonification, etc.[561-563,568,575,623]

A representative sampling and subsequent homogenation are important in analyses of certain specimens, e.g., feces,[96,589] hair,[100,555,556] and tissues.[623]

Urine samples are generally very low in Cr ($\leqslant$1 ng/mℓ) and should be collected in precleaned polyethylene containers; in presence of 1 or 2 *N* HNO_3, the storage time was estimated to be 1 and 20 days, respectively.[569] Cornelis et al.[165] found no losses after a 3-day storage of urine at room temperature, but the precipitate formed in spiked and aged urine specimens was found to contain Cr. For longer storage, urine samples should better be frozen[571] or freeze-dried.[568]

The role of surface contamination of hair by sweat,[162,566] vegetable dye,[566] and environmental factors,[555,566] as well as the washing procedures for hair,[48,100,162,556,566,625,625a] have been discussed in the literature, but there is no consensus on this matter.

Table 2
SELECTED AAS PROCEDURES FOR Cr DETERMINATION IN BIOLOGICAL SAMPLES

Sample	Brief procedure outline	Ref.
Blood	2 μℓ of heparinized hemolyzed (+ Triton® X-100) blood injected into CRA-63 (carbon cup) atomizer, 90° for 40 sec (ramp!), 500°C (ramp), 1200°C, and 2300°C; standards on blood; alternatively, 25 μℓ of blood plus 25 μℓ of HNO_3 pressure decomposed[a] in small Teflon® tubes at 80°C for 2 hr; digests analyzed as above	595
	10-fold dilution with water; ET-AAS; standard addition calibration;	554
Blood, hair	HNO_3 digestion at 110°C overnight in a closed vessel;[a] cooled, opened, and digestion completed with H_2O_2;[a] ET-AAS: 110, 1300, and 2700°C	593
Blood, liver, urine	HNO_3 digestion, $NH_4H_2PO_4$ (0.8% w/v) added as a part of a multielement scheme; ET-AAS with tantalized and pyro-coated grooved L'vov platform, 900°C (ramp!) and 2600°C ("maximum power"); background correction; standard addition calibration (for liver and blood); acid-matched calibration (for blood)	457
Hair	HNO_3 digestion; dilution ET-AAS with HGA-74[b] (Reference 602) or CRA-90[c] (Reference 557) furnaces; background correction; temperature program: 80°C for 10 sec, 130°C for 30 sec, 1200°C (ramp), 2570°C, and clean-out at 2650°C (Reference 602)	557, 602
Hair, liver	0.1-g samples wet ashed overnight at 100°C in tightly capped tubes;[a] cooled, opened, and digestion completed with H_2O_2;[a] ET-AAS: 120°C for 100 sec, 1050°C for 90 sec and 2650°C for 8 sec; standard addition calibration	556
Meat, tissues	1-g sample digested at 80°C with 10 mℓ of (1 + 1) HCl, evaporated to ca. 3 mℓ, and diluted to 25 mℓ; flame AAS (N_2O-C_2H_2); acid-matched standards	517
Milk, serum, urine	1 + 1 dilution with water; ET-AAS; background correction;[d] standard addition calibration	559
Milk, urine	LTA for 3 hr, H_2O_2 (100 μℓ) added and re-ashed, and finally dissolved in 1 *M* HCl; ET-AAS: 150, 300 (ramp), 1100 (ramp) and 2650°C (pyro-coated tubes, miniflow, background correction); see also References 581 and 618a	558, 627
Plasma	1 + 2 dilution with 0.01% Triton® X-100; ET-AAS: 45°C for 40 sec, 105°C for 25 sec, 800°C for 30 sec, 1000°C for 45 sec, 2650°C for 5 sec, and clean-out at 3000°C for 5 sec; background correction by using the U 358.5-nm line and dual-channel spectrophotometer;[e] standards in 0.01% Triton® X-100	608a
Serum, tissues, urine	HNO_3- H_2O_2 digestion overnight at 80°C in fused silica tubes fitted with threaded Teflon® caps;[a] ET-AAS: 110 (ramp), 1100, and 2500°C (gas stop, maximum heating rate); background correction;[d] acid-matched standards	561
Serum	1 + 1 dilution with water; ET-AAS with HGA-500;[b] 120°C (40-sec ramp + 30-sec hold), 600°C (30 + 40 sec, with O_2 added), 600°C (1 + 25 sec), 1000°C (15 + 35 sec), 2700° (0 + 6 sec, "maximum power", miniflow); uncoated tubes! background correction;[d] serum-based standards	582
Serum protein fractions	Gel-permeation chromatography; LTA or bomb[a] decomposition; final solutions in 2% v/v HNO_3; ET-AAS: 1250°C (ramp), atomization at 2300 to 2500°C; background correction;[d] standard addition calibration	603, 604
Tissues	TAAH (Soluene® 350) solubilization and ET-AAS	610
Urine	1 mℓ of urine + 10 μℓ of 1 *M* HCl; ET-AAS: 100°C (15 + 20 sec), 130°C (10 + 20 sec), 1200°C (15 + 60 sec), 2700°C (0 + 4 sec), and (clean-out) at 2700°C; pyrocoated tubes; background correction;[d] standard addition calibration	585
	9 + 1 dilution with water; CRA-63 furnace;[c] 130 (ramp), 900 (ramp), and 2300°C; standard addition calibration	554
	1 + 1 dilution with 1 *M* HNO_3 or 1 *M* HCl; CRA-90 furnace:[c] 100°C for 40 sec, 1000°C for 50 sec, and 2300°C (with a ramp rate of 600°C/sec); H_2 diffusion flame around the atomizer during atomization; background correction; standard addition calibration	571

[a] Safety precautions mandatory.
[b] Perkin-Elmer®.
[c] Varian Techtron®.
[d] With W halide lamp.
[e] Instrumentation Laboratory IL 651.

In principle, both dry- and wet-ashing techniques are suitable for oxidizing the organic matter prior to Cr determination; nevertheless, the choice of preparation procedure should be a serious concern. It is important to use simple procedures, with as little reagent consumption as possible, and to quantitatively transfer the analyte into a proper final acidic solution.

Drying and dry ashing in precleaned quartz crucibles at 450 to 550°C should be safe, provided that a gradual temperature increase and good aeration are secured and airborne contamination is restricted. The drawbacks of dry ashing are the minor retention losses due to a fusion of the matrix with the crucible, as well as the low solubility of chromium oxides;[91,534,599] the ashes thus require a treatment with concentrated mineral acid in order to put Cr into solution.[534,588,599]

LTA proved to be the best method for the digestion of oven-dried or lyophilized biological samples, ensuring both complete recovery and very low blank levels.[601] Successful applications of this technique to blood,[169] hair,[566,601] liver,[275,626] lung,[516] pancreas,[611] milk,[558,627] serum,[167,601,603] and urine samples[558,581,614,627] have been described. The recent report that LTA has given about twice lower results for Cr in milk and urine vs. dry ashing at 500°C in porcelain crucibles[558] was later supplemented by the evidence that the dry-ashing results were positively biased due to contamination from porcelain, and, thus, the LTA results have been correct.[627]

Many wet-digestion procedures can be used,[601] but in practice, digestions with either HNO_3 alone (e.g., samples of blood,[457,593,595] hair,[557,593,602] tissues,[457,517,545] and urine[457]) or with HNO_3 and H_2O_2 (caution!) (e.g., samples of blood,[587] hair,[556] serum,[561] tissues,[556,561] and urine[561,587,612]) are favored since the resulting digests contain only HNO_3 which is a suitable medium (1 to 10% v/v HNO_3) with respect to following ET-AAS assays. In some digestion procedures, decomposition of organic matter has been carried out in closed, pressurized vessels (often with single HNO_3) to ensure more efficient digestion with less acid consumption and lower blanks.[426,545,561,593,595,603] A word of caution should be added here: the digestion in pressurized vessels is potentially dangerous, especially with big/fatty specimens as well as with acidic mixtures containing HNO_3 + H_2O_2[561] or HNO_3 + $HClO_4$.[426]

Oxidation of the organic matrix can also be performed by mixtures of HNO_3 and H_2SO_4;[7,426,590,601] HNO_3, H_2SO_4 and $HClO_4$;[442,583,588,601] or HNO_3, H_2SO_4, and (finally) H_2O_2;[475,583] but these acids will attack the graphite tube during its prolonged heating to high temperatures. Moreover, the residual $HClO_4$ (>0.1 *M*) wil also depress the Cr absorbance,[66] and higher blank corrections should be made as well. Complete ashing of organic matter may be needed in procedures which include the next liquid-liquid extraction step.[583,588,590]

Precipitation of serum proteins with TCA[7] or tissue solubilization with TAAH[610,611] may be useful, as well.

Inasmuch as the ET-AAS of Cr is sufficiently sensitive, the preconcentration step is rarely needed (see, e.g.,[540,583,608,612]). On the contrary, in many early flame AAS procedures an effective preconcentration by extraction[76,547,587,588,590] or coprecipitation[592] was called for. Separation techniques are also useful in Cr speciation studies.

The extraction of Cr is not straightforward: prolonged reaction/extraction times are required, the pH or acidity range is narrow, and the analyte should be released beforehand from its chemical bonds and converted into a definite oxidation state, Cr(III) or Cr(VI). Several extraction systems have been applied to preconcentrate Cr prior to its AAS determinations: Co(III)-APDC-MIBK, at pH 3 to 4 (from digested urine, with R = 87%);[583] Cr(VI)-MIBK[588,619] (from digested blood, food, tissues, and urine[588]); Cr(III)-pentane,2,4-dione-MIBK, at pH 4.5 (from digested food, milk, etc., average R = 93%; for meat, 82%);[590] Cr(VI) — 5% methyltricaprylammonium chloride in MIBK (or in toluene), at pH 2.5 (from tissue digests, average R $\sim$ 90%);[583] and Cr(VI) — 10% tributylphosphate in MIBK (from ashed blood or urine).[587]

Cr(III) can be coprecipitated with $Al(OH)_3$[592] or $Fe(OH)_3$.[612,618] Recently, Arnold et al.[608] found that Cr can be quantitatively precipitated with serum proteins (transferrin) by acetone, and applied this approach to serum and urine analyses.

The ET-AAS assays of Cr are sensitive and relatively selective, and should not be a problem provided that a versatile AAS equipment is available. It is essential to have a high-intensity tungsten-halogen lamp for background correction, pyrolytically coated graphite tubes, multistage temperature programming with very fast, ''maximum power'' heating during the atomization stage, controlling the flow rate of a purge gas,[585] and using alternate gas (oxygen or air) during the ashing stage. It seems that the difficulties experienced by early workers were due to the absence of one or more of these facilities.

The background absorption at the Cr 357.9 nm wavelength is generally small but persistent. The intensity of a deuterium-lamp emission in this spectral region is very low as compared with the HCL; thus, in order to balance intensities of the two sources, the HCL current has to be reduced below its optimum value. As a result, the background emission of incandescent graphite tube may give rise to an additional emission noise.[27,594,615]

There are several ways of dealing with background problems:

1. The use of a high-intensity tungsten-halogen lamp as a continuum source[559,561,582,585,603] should provide a good background correction system. At present, however, few AA spectrometers are equipped with both a deuterium arc and a tungsten-halogen lamp. The use of a special, high-intensity thermal-cathode type of deuterium lamp has also proved to be satisfactory.[612]
2. The Zeeman background correction is effective.[628]
3. If a dual-channel AA spectrometer is available, the second channel can be used for background correction. Spectral lines of the following sources can be utilized (in nanometers): U, 358.5;[608a] Ni, 357.6;[581] Co, 352.6;[617] Ne, 352.0;[302,569] Ti, 364.3;[569] Fe, 358.1;[302] Sn II, 357.5; Au, 358.7; Tl II, 356.1; or else Pb, 357.3.[629] With a single-channel apparatus, these lines can be used to check for background in a sequential approach.
4. The special background-correction scheme (not commercially available), based on a wavelength-modulated, continuum source, echelle AA spectrophotometer, has proved to be very efficient.[485,581]
5. A thorough optimization of the temperature program is called for. Ramp drying and ashing to temperatures as high as tolerable (e.g., between 900 and 1300°C, depending on the matrix) should eliminate most of volatile matrix constituents. Atomization at somewhat lower temperatures (e.g., at 2300 to 2600°C) and at the maximum heating rate available results in lower background absorption and background emission/noise.[572,581,584,594]
6. Introducing oxygen during the ashing stage (at 600°C)[582] or ashing in air[605] were found to decrease both the background absorption and the matrix-residue build-up in direct analyses.
7. In a graphite furnace of an open-type design (Varian Techtron® CRA-90), using a hydrogen diffusion flame around the furnace during the atomization stage led to thrice lower background.[571]
8. Modification of the matrix/analyte by adding 1 or 2% v/v HNO_3 or HCl should lower the background absorption, as well.[571]

Several other problems have been encountered in ET-AAS of Cr in biological samples: (1) significant tube-to-tube variations;[568,581] (2) ''memory'' effects[618a] and matrix-residue accumulation;[594,595,603] (3) somewhat prolonged temperature programs (e.g., up to 5 to 6 min); (4) enhancement of the analyte signal in serum or blood matrix vs. aqueous standards;[595]

(5) poor reproducibility of sample delivery,[594,605] etc. Thus, every newly adopted procedure should be carefully checked; a high-temperature decontamination step should often be included in the temperature program; brushing-out the matrix residues may be necessary after every few samples (e.g., if untreated blood or serum is injected[596,603]); calibration should be typically done by acid-matched standards (in analyses of biological digests) or by matrix-matched standards (in direct assays of blood[595] or serum[582,605]); and a standard additions check is advisable (especially in direct analyses of urine, milk, etc.).[554,559,585,628] The advantages of peak-area measurements[86,567] and L'vov platform[457,553] should be further explored.

Several recent Cr speciation studies with serum,[603,604,608] urine,[613,618,619] and food[597] are worth mentioning. Assays of Cr(VI) in urine should be performed soon after collection;[554,618,619] the hexavalent Cr should be hardly present in normal urine.[554,619]

IV. CONCLUSION

Inasmuch as the Cr concentrations in most biological samples are very low (~ ng/g), extreme care is called for to avoid contamination during specimen collection, storage, and preparation. As simple as possible sample pretreatment and use of versatile ET-AAS apparatus are therefore advisable. Pyro-coated graphite tubes, a tungsten iodide lamp for background correction, high heating rates, and standard addition calibration are required. Suitable diluents are aqueous Triton® X-100, TAAH, dilute HNO_3, and HCl. The best ashing techniques are LTA and (pressurized) HNO_3 digestion. More work is needed on evaluation of the peak-area calibration, L'vov platform atomization, pretreated graphite surfaces, and ashing in O_2 as alternate gas.

Other reliable methods are ID-GC-MS and RNAA.

Chapter 10

COBALT

I. DETERMINATION OF COBALT IN BIOLOGICAL SAMPLES

The normal concentrations of Co in biological liquids are very low — around or below the 1 ng/mℓ level; most tissues, hair, and nail samples should contain less than 0.1 μg Co/g (Table 1). Few analytical techniques are sensitive in this concentration range, and the NAA appears to be the best method available today. Many NAA assays can be performed without any radiochemical separations, e.g., the INAA analyses of CSF,[621] erythrocytes,[631] hair,[11,48,98,102,118,635] milk,[182] nails,[560] platelets,[119] serum/plasma,[120,562,631] teeth,[23,51] tissues,[45,573] etc. Many other NAA procedures for analysis of blood,[45,573] diet,[45,47,116] placenta,[377] serum protein fractions,[120] skin,[112] tissues,[45,108,633] urine,[45,111,165] etc.[45,121] are available to laboratories with an access to nuclear reactor and radiochemical equipment.

The arc-source AES has been applied to analyses of hair,[288] serum/plasma,[36,413] sweat,[21] teeth,[26] tissues,[36,44] urine,[320] etc.

The sensitivity of the XRF[128,636] and ICP-ES[29,35,636] appears to be inadequate for most biological applications.

There are some spectrophotometric procedures[40,133,414,634] which, however, lack sensitivity and require large samples and specially purified reagents.

Incidentally, catalytic procedures,[577,637] capillary GC of volatile dithiocarbamates,[638] SSMS,[25,54,55,132,275,409] and miscellaneous techniques[637] have been used.

II. ATOMIC ABSORPTION SPECTROMETRY OF COBALT

The flame AAS of Co is relatively selective but insufficiently sensitive. The characteristic concentration ranges between 0.04 and 0.2 μg/mℓ, and the DLs are down to 0.003 to 0.01 μg/mℓ.[56-59]

What matters is to isolate the cobalt resonance line (240.7 nm) from its very close neighbors (240.4 and 240.9 nm), as well as to ensure a better intensity ratio of the resonance line to its neighboring lines. Failure to fulfill these requirements is likely to result in both an impaired signal-to-noise ratio and a restricted linearity of the working curve. Hence it is important to (1) employ a high-intensity single-element HCL, (2) optimize the lamp current, and (3) use a narrower bandpass, e.g., only 0.1 to 0.2 nm.

A lean oxidizing air-acetylene flame and a slightly elevated observation height are recommended for better selectivity (i.e., less pronounced condensed-phase interferences), although these instrumental parameters result in a sensitivity loss, as well. Under the recommended conditions, only bulk-matrix effects and transport interferences as well as background absorption are to be expected. Acid-matched standard solutions are usually employed (e.g., in dilute HCl or HNO_3 or $HClO_4$). Under inappropriate conditions, however, i.e., in a reducing flame and/or at a low observation height, one may encounter a marked signal depression!

In ETA, the determination of Co is sensitive and relatively selective. The characteristic concentration is between 0.03 and 1 ng/mℓ (3 to 70 pg), and the DLs are from 0.004 to 1 ng/mℓ (5 to 50 pg), depending on apparatus and conditions.[57,59-61,294] Sensitivity can be improved by a factor of 3 or 4 by using pyrolytically coated graphite tubes[60] or *in situ* pyrocoating.[295] Moreover, lower atomization temperatures can be used with pyro-coated tubes. The "maximum power" heating mode is advantageous, in view of lower atomization temperatures (2200 vs. 2800°C, for the "maximum power" temperature-controlled heating vs.

Table 1
COBALT CONCENTRATIONS IN HUMAN BODY FLUIDS AND TISSUES

Sample	Unit	Selected reference values	Ref.
Blood	μg/ℓ	0.5 (0.1—1.2)	630
		1.1	94
CSF	ng/g	(<0.7—7.3)	621
Erythrocytes	ng/g	0.59 ± 0.23	631
Feces	μg/day	(18—25)	632
		26.3 (6.8—80)	96
Hair	μg/g	x̄, 0.036 (0.0081—2.2) (Japan)	11
		(<0.006—0.017) (U.S.S.R.)	102
		x̄, 0.066; 0.13 (<0.02—1.1) (Italy)	98
		0.07 ± 0.10 (0.01—1.18) (India)	101
		0.030—0.171[a]	100
Kidney	ng/g	x̄, 1.0; 4 ± 3 (0.5—15)	105, 106
		10 ± 30	12
		13	633
Kidney cortex	ng/g	4.8	107
Liver	μg/g	x̄, 0.013 (0.0005—0.15)	106
		0.016	107
		0.034 (0.023—0.039)	108
		0.05 ± 0.02	275
		x̄, 0.06; 0.06 ± 0.03	12
		0.061 ± 0.013 (0.043—0.074)	633
Lung	ng/g	x̄, 1.6; 2 ± 1 (0.4—13)	105, 106
Milk	μg/ℓ	1	8
		2.0 (1.3—3.0)	182
		(1—8.6)	13
Nails (thumb)	μg/g	♂, 0.04 ± 0.03 (0.01—0.15)	560
		♀, 0.07 ± 0.04 (0.02—0.15)	560
Placenta	ng/g	23 ± 15[b]	377
Plasma or serum	μg/ℓ	0.108 ± 0.060 (0.0394—0.271)	562
		0.22 ± 0.14	631
		♂, 0.22 ± 0.10 (0.08—0.33)	633
		♀, 0.37 ± 0.29 (0.1—0.58)	633
		0.30 ± 0.26[c]	119
		0.52 ± 0.43	111
Saliva	μg/ℓ	70	8
Skin	ng/g	x̄, 49;[b,d] 52 ± 21[b,e]	112
		42 ± 25;[b] 21 ± 29[b]	565
Sweat	μg/day	17	21
Tooth (dentine)	μg/g	1.11 ± 0.27	23
Tooth (enamel)	μg/g	<0.02	25
		0.07[f] and 0.09[g]	51
		0.13 ± 0.13	23
		x̄, <0.6; 1.3 (<0.6—30)	26
Urine	μg/ℓ	0.4 (0.1—2.2)	630
		0—3.5	27
		1—7	634
	μg/day	0.73 ± 0.48 (0.23—1.8)	111
		1.8 (1.1—2.8)	320
		(0.26—15.6)	96

Table 1 (continued)
COBALT CONCENTRATIONS IN HUMAN BODY FLUIDS AND TISSUES

Note: Wet weight unless otherwise indicated. Mean ± SD (range in parentheses). x̃, median.

a Variations over the same head.
b Dry weight.
c ng/g.
d Epidermis.
e Dermis.
f European.
g African.

the normal heating rate).[60] A dilute HNO_3 (≤ 1 *M*) appears to be the best medium in ET-AAS determinations of Co. Excess HNO_3,[294] $HClO_4$,[66] HCl, and chlorides all depress the analyte signal; moreover, $HClO_4$ and H_2SO_4 strongly attack the graphite-tube surface during heating to high temperatures.

The temperature programming and the background correction efficiency are important considerations in ET-AAS assays of Co.

III. AAS METHODS FOR ANALYSIS OF BIOLOGICAL SAMPLES

Inasmuch as the concentration levels of Co in biological samples are very low, there are two important problems to be considered: contamination control and efficiency of the preconcentration step.

The flame AAS has been applied to analyses of blood,[7,27,332] feces,[39] fish,[474,590,639] food,[39,590,592] hair,[461] milk,[590] serum,[640] tissues,[7,319,473,482,590,641-643] and urine,[7,27] but in most cases it lacks sensitivity.

The long-path absorption cell has been applied to determine micrograms-per-gram levels of Co in isolated carboxypeptidase, Co-containing alkaline phosphatase, thermolysin, and vitamin B_{12}.[644]

The method of choice is ET-AAS (often combined with a preconcentration step), as described for analyses of blood,[7,27,473,630,645-648] hair,[78-80,473,649,650] milk,[651] serum/plasma,[27,645,652-654] and tissues.[7,446,454,455,473,516,517,630,648,649,655] Unfortunately, it appear that many previous analyses of blood, serum/plasma, urine, etc. have been performed on already-contaminated biological samples, and this may explain the discrepancy in existing "normal values" for Co.[49,620,645]

Severe contamination from needles, stainless steel knives, heparin, air particulate matter, grinding[624] and homogenation procedures, lab ware, and reagents have been established.[49,163,562,622,645,656]

Versieck et al.[622] found marked contamination of small liver biopsies from needles. The heparin solution used as anticoagulant by Kasperek et al.[656] introduced as much as 0.029 ng Co/mℓ of plasma. Under strictly controlled analytical steps and precautions against contamination which in most routine laboratories are difficult to follow, Versieck et al.[562] were able to keep the blank levels down to 27 pg Co (range of 17 to 34 pg Co) per assay of Co in serum. The dust fallout in the same laboratory has been only 0.3 pg/cm^2/day (vs. 3.6 pg/cm^2/day in an ordinary laboratory).[108] During a 3-day contact of polyethylene containers with 8 *M* HNO_3, the amount of the leached Co was as high as 2.6 ng/g.[163]

Although minor retention losses of Co during the preparation step are possible,[91] they are not as pronounced as are the extraneous additions. Oven- or freeze-drying,[166,167,423,536] LTA,

Table 2
EXTRACTION OF Co FROM BIOLOGICAL DIGESTS

Extraction system	Matrix and conditions	Ref.
APDC-MIBK	Blood; pH 2.8—3.2	332
	Fish; pH 1—5 (optimum <2)	639
	Liver	641
DDDC-MIBK	Food, meat, milk, etc.; pH 4.5	590
DPC-Py-toluene[a]	Blood, hair, tissues; pH 5.5—7.5	475, 482
Dithizone-$CHCl_3$	Urine; pH 8—9 (citrate)[b]	27
HHDC-butylacetate	Deproteinized (with TCA and HCl) serum; pH 6 (acetate)	640
	Plant tissue; pH 6 (acetate)	429
HHDC-$CHCl_3$	Food, feces, urine, etc.; pH ~ 6[b]	659
K ethylxanthate-MIBK	Tissues; pH 6.5—11.5 (optimum 8)	643
1-Nitroso-2-naphthol-$CHCl_3$	Blood, serum; pH 4; H_2O_2 added	645
2-Nitroso-1-naphthol-xylene	Plant tissue[c]	658
PAN[d]-$CHCl_3$	Serum; pH 3—4	645
SDDC-MIBK	TCA-deproteinized blood; pH 5.5—7	7
	Urine, tissue; pH 5.5—7	7
SDDC-$CHCl_3$	Blood, urine; pH 6.5 (acetate)[b]	27
5% TOA[e]—10 *M* HCl—CCl_4	Blood, tissue, urine, etc.; 10 *M* HCl	645

[a] 5 m*M* diphenylcarbazone—2.5 *M* pyridine in toluene.
[b] Double extraction.
[c] Four extraction reagents compared.
[d] 1-(2-Pyridylazo)-2-naphthol.
[e] Tri-n-octylamine.

and dry ashing[91,637] are generally safe in view of volatization of the analyte. Dry ashing at 450,[377,442,482,591] 480,[639] 500,[638] and even at 550°C,[646,647] with subsequent conversion of the ashes into nitrates or chlorides, has been applied with apparent success. Still temperatures above 550 to 600°C should be avoided because of the interaction between the sample and the silica crucibles.

Wet-digestion procedures are somewhat less popular, owing to their intolerable blank levels; simple bomb decompositions or acid extractions are therefore recommended, e.g., treatment of tissue samples with (1 + 1) HNO_3,[517] or with (1 + 1) HNO_3 + HCl,[319] or TCA precipitation of blood[7] or serum[640] proteins, etc.

The analyte can be separated and preconcentrated from the digests by chelate extraction,[76,547] or, rarer, by ion exchange[630] or coprecipitation (with APDC[446]).

Some extraction systems for trace Co in biological samples are summarized in Table 2. Dithiocarbamate extractions are widely used, although excess Fe and Mn may interfere with the extraction of Co.[27,427,590,657,658] 1-Nitroso-2-naphthol extractions call for an addition of H_2O_2 in order to oxidize the analyte to Co(III) and prevent the coextraction of the Fe(II) chelate;[645] excess Cu may interfere as well.[642,658] Extractions with dithizone appear to be less convenient, as they involve multiple extractions and masking agents.[27,474]

Organic extracts are easily analyzed by both flame and ET-AAS techniques. The furnace program may include an ashing step between 150 and 600°C, and atomization temperatures from 2400 to 2800°C, depending on the apparatus.[7,27,245,427]

The flame AAS is hardly applicable directly to biological digests (see, e.g., assays of hair[461] and liver[319]), while the direct introducing of biological liquids,[27,651,653] digests,[27,516,517] and solid microsamples[78-80,454,455,649,650,652,655] into a graphite furnace have been described. The calibration is complicated (either by standard additions or matrix-matched standards)

Table 3
SELECTED AAS PROCEDURES FOR Co DETERMINATION IN BIOLOGICAL SAMPLES

Samples	Brief procedure outline	Ref.
Blood, tissues, urine	5-g samples dry ashed at 550°C, and ashes dissolved in 10 mℓ of 10 *M* HCl; extraction with 10 mℓ of 5% trioctylamine in CCl_4; back extraction into 1 mℓ of water; ET-AAS	646—648
Blood, serum	10-mℓ samples wet digested with HNO_3, $HClO_4$, and H_2SO_4 (caution!) and ashes dissolved in 4 *M* HCl and diluted to 5 mℓ; H_2O_2 added (100-volume, 50 μℓ), pH adjusted to 4, and 1% w/v 1-nitroso-2-naphthol (in glacial acetic acid, 0.2 mℓ) added and left for 30 min; extracted twice with $CHCl_3$, combined extract evaporated to dryness and dissolved in 200 μℓ of $CHCl_3$; ET-AAS vs. aqueous standards	645
Blood, hair, tissues	Dry ashing at 350—450°C, HNO_3 added and re-ashed, and ashes dissolved in 0.1 *M* HCl; extraction at pH 6 (acetate buffer) with 5 m*M* diphenylcarbazone—2.5 *M* pyridine in toluene; flame- or ET-AAS	473, 482
Blood, tissues, urine	SDDC-MIBK extraction at pH 5.5—7 from urine or TCA-deproteinized blood or tissue digests (after HNO_3-H_2SO_4 digestion); ET-AAS: 100, 300, and 2400°C; background correction; standards extracted similarly	7
Blood, urine	Wet digestion with HNO_3 (urine) or with HNO_3 containing 1% v/v H_2SO_4 (blood samples), gently boiled to dryness, ashed up to 330—350°C, evaporated to dryness with 9 *M* HCl, and finally dissolved in 9 *M* HCl; ion exchange on Dowex® 1 × 8 resin, washed with 6 *M* HCl, eluted with 30 mℓ of 4 *M* HCl; eluate evaporated to a small volume and analyzed by ET-AAS: 130 (ramp), 900 (ramp), and 2600 °C (gas stop); background correction	630
Hair	1-cm segments directly ashed up to 1100°C and atomized at 2500°C; calibration against spun silk or animal hairs, previously soaked in appropriate standard solutions	79, 80, 649, 650
Fish	Dry ashing at 480°C; APDC-MIBK extraction at pH 1—5; flame AAS; background correction; standard addition calibration	639
Food, milk, tissues	HNO_3-H_2SO_4 digestion DDDC-MIBK extraction at pH 4.5; flame AAS; standards extracted similarly	590
Tissues	1-g sample digested with 10 mℓ of (1 + 1) HNO_3 at 80°C, evaporated to ca. 3 mℓ, and diluted to 25 mℓ; ET-AAS: 150, 800, and 2700°C; background correction; standard addition calibration	517
	Direct ET-AAS (0.2 to 2-mg samples) with L'vov platform furnace: 250, 1150, and 2700°C; background correction; standard addition calibration	454, 455

and the efficiency of background correction must be verified. Ashing temperatures of at least 800 to 900°C should be quite tolerable; "maximum power" heating during the atomization step is preferable. The L'vov platform proved useful in analyses of solid samples (0.2 to 2 mg of NBS SRMs "Bovine Liver" and "Oyster Tissue"); the precision was improved by a factor of 3 (7.3 vs. 2.2% RSD).[454,455]

Selected AAS procedures for the determination of Co in biological materials are briefly outlined in Table 3.

IV. CONCLUSION

NAA is the method of choice at nanograms-per-gram levels of Co in biological samples. ET-AAS often demands a preconcentration of the analyte (chelate extraction). Pyro-coated graphite tubes, high heating rates during atomization, and efficient background correction are required. Direct flame or ET-AAS procedures are prone to (minor) matrix and acid effects and should always be checked by standard additions. Strict contamination control at all stages of specimen collection, storage, and ashing is essential.

Chapter 11

COPPER

I. DETERMINATION OF COPPER IN BIOLOGICAL SAMPLES

The determination of microgram amounts of Cu (Table 1) does not represent a particularly difficult analytical task and is performed routinely in thousands of clinical laboratories. The most convenient analytical techniques are flame AAS, ET-AAS, spectrophotometry (for serum/plasma), and ASV/DPASV.

Well-established spectrophotometric procedures are routinely applied to serum/plasma[678-680] and other biomaterials;[40,133,681] commercially available reagent kits are popular.

Sensitive catalytic procedures[577,682] involving simple and low-cost equipment have been developed.

ASV/DPASV has been applied to analyses of blood,[361,391,683] diet,[250] hair,[393] milk,[250] serum,[321] teeth,[395] tissues,[250] and urine.[124,125,541,684] It is extremely sensitive (DLs down to a few nanograms per liter) as well as relatively simple and inexpensive, but most procedures require a complete release of Cu from its chemical bonds with the organic matrix.[391,683,684]

NAA is a very sensitive method for Cu and has been widely applied to various biological materials: blood,[45,178,685,686] bone,[103] CSF,[621,687] diet,[45,47,116] erythrocytes,[287,665] hair,[9,11,45,48,290,670] milk,[110,182] nails,[15,103,290,670] platelets,[119] saliva,[671,672,687] serum/plasma,[111,287,633,665,680,687] skin,[112] teeth,[50,103,676] tissues,[12,43,45,108,633,680,686] urine,[111,165,416,687] etc.[45] Some specimens can be directly analyzed by INAA, as well (e.g., CSF,[621] hair,[11,290] nails,[15,290] saliva,[672] teeth,[50,676] etc.[45]).

Application of the arc-source AES to various biomaterials has been described; usually, dry-ashed specimens are analyzed for many elements simultaneously, e.g., analyses of blood,[36,37] bone,[38] diet,[37] milk,[14,40] saliva,[41,380] serum/plasma,[36,41,42] sweat,[21] teeth,[26] tissues,[36,43,44] urine,[37] etc.[36,37,41]

ICP-ES, with its DLs from a few down to 0.06 ng/mℓ,[29,35,123] is now becoming an important method for a sensitive and multielement analysis of biological materials. A routine application of this technique to nine elements (Au, Ca, Cu, Fe, K, Li, Mg, Na, and Zn) in 0.5 mℓ of serum has been documented.[688] ICP-ES analyses of blood,[30,32,689] bone,[392] food,[690] milk,[35] serum/plasma,[30,289,689] tissues,[289,348,411,412,518,636] etc.[326,348] have been described.

The XRF,[128] though not very sensitive, can be used to analyze blood,[691] nails,[692] serum,[693,694] tissues,[292,636] and urine.[695]

Few laboratories are able to perform PIXE analyses,[128,408,696-698] which are unique in studying the distribution of Cu in very small samples, e.g., in bone,[97] tissue,[407] single hairs,[195] etc.

MS[53,123] and SSMS[16,25,52,54,55,123,275,409] have been applied in some multielement studies.

II. ATOMIC ABSORPTION SPECTROMETRY OF COPPER

The flame AAS of Cu is remarkably selective and fairly sensitive. The characteristic concentration and DLs are 0.025 to 0.1 and 0.001 to 0.003 μg/mℓ, respectively.[56-59,294] In a lean air-acetylene flame and at a proper (not too low) observation height, one should expect only bulk matrix effects due to, e.g., a high content of total dissolved solids, excess acids, and/or major viscosity and surface-tension discrepancies between the analyzed solutions and the standards. As a rule, acid-matched solutions should be nebulized, dilute mineral acids (e.g., <1 *M* HCl or $HClO_4$) being acceptable media in the flame AAS of Cu. Only under inappropriate measurement conditions (e.g., at a very low observation height

Table 1
COPPER CONCENTRATIONS IN HUMAN BODY FLUIDS AND TISSUES

Sample	Unit	Selected reference values	Ref.
Bile	μg/mℓ	3.29 (0.24—5.38)[a]	660
Blood	μg/mℓ	♂, (0.64—1.22); ♀, (0.82—1.76)	661
		(0.64—1.31)	391
		1.01;[b,c] 0.64—1.28[c,d]	13
Bone	μg/g	0.49 ± 0.03	268
		x̄, 2.94; 4.24 ± 3.43 (0.85—11.8)	103
Brain	μg/g	5.6 (2.7—9.1)	662
		23.9 ± 6.36 (13.1—39.4)[e]	103
		12.4;[e,f] 26.1[e,g]	663
CSF	μg/mℓ	0.06 (0.02—0.11)[h]	660
		0.076 ± 0.017 (0.01—0.28)	664
Erythrocytes	μg/mℓ	0.68 ± 0.06 (0.56—0.86)[i]	665
		0.89 (0.66—1.12)[h]	660
Feces	mg/day	1.96 ± 1.33 (0—4.62)	37
Hair	μg/g	♀, 5.55; ♂, 10.06 (Nepal)	666
		x̄, 10.7 (1.56—97) (Japan)	11
		♀, 14 ± 7; ♂, 15 ± 9 (G.D.R.)	367
		x̄, 14.5—17.5 (4—100) (U.S.A.)	9
		16.2 ± 6.2 (6.8—59.6) (India)	101
		♂, 13.9; ♀, 25.1 (2.22—184) (New York)	288
		19;[b,j] 11—34[d,j]	13
		x̄, 19.1 (7.64—54.5)[e] (U.K.)	103
		♂, 24.2 ♀, 35.5; (9.6—242) (U.K.)	368
Kidney	μg/g	x̄, 1.57; 2.4 ± 0.7 (1.07—4.19)	105, 106
		2.0 (1.2—3.1)	660
		x̄, 3.0; 3.4 ± 1.6 (1.2—8.1)	103
Liver	μg/g	x̄, 3.59; 3.3 ± 0.5 (1.44—6.90)	105, 106
		5.1 (3.0—9.5)	660
		6.0 (3.9—7.7)	108
		x̄, 7.0; 7.1 ± 3.2 (2.6—13.0)	103
Lung	μg/g	x̄, 1.10; 1.2 ± 0.1 (0.62—1.44)	105, 106
		x̄, 2.22; 2.3 ± 0.75 (0.93—3.5)	103
Milk	μg/mℓ	0.34 ± 0.15	667
		0.45 (0.2—1.0)[k]	8
		0.15—0.75[l] (0.091—1.32)	14
		(0.197—0.500)	13
		(0.090—0.630)	668
Nails	μg/g	x̄, 14.9; 18.1 ± 12.1 (3.18—58.2)	103
		(5—50)	669
		13.3 ± 10.9	670
		♀, 4.2 ± 3.4;[e,m] ♂, 4.3 ± 2.8[e,m]	15
Saliva	μg/mℓ	0.026 (0.013—0.053)[n]	671
		0.028 (0.015—0.052)[o,p]	671
		0.088[o]	379
		0.09;[o,p] (0.02—0.25);[p,q] (0—0.1)[n,q]	564
		0.0753 ± 0.0345 (0.0293—0.156)[r]	672
Serum	μg/mℓ	♂, 0.82 ± 0.15;[i] ♀, 0.98 ± 0.11[i]	633
		♂, 1.00 (0.76—1.33)	673
		♀, 1.08 (0.70—1.40)	673
		1.09 ± 0.27 (0.70—1.63)	665
		♂, 1.09 (0.81—1.37)[h]	660
		♀, 1.20 (0.87—1.53)[h]	660
		1.19 ± 0.19 (0.70—1.65)	674

Table 1 (continued)
COPPER CONCENTRATIONS IN HUMAN BODY FLUIDS AND TISSUES

Sample	Unit	Selected reference values	Ref.
Sweat	μg/mℓ	(0.044—0.080)[k]	8
		♂, 0.55 ± 0.35 (0.030—1.44)	675
		♀, 1.48 ± 0.61 (0.59—2.28)	675
		♂, 1.427 ± 0.505; ♀, 1.533 ± 0.115	381
Synovial fluid	μg/mℓ	0.50 ± 0.14	788
Tooth	μg/g	0.21 ± 0.10[e]	676
(dentine)		1.03 ± 0.26;[s] 1.55 ± 0.45[t]	24
Tooth	μg/g	x̃, 0.42; 4.2 ± 3 (0.1—81)	25
(enamel)		1.30 ± 0.54;[t] 1.38[s]	24
		x̃, 7.9[e] 10.1 ± 7.8 (1.59—39.7)[e]	103
Tooth (whole)	μg/g	1.48[e]	383
Urine	μg/ℓ	14 ± 4.5	27
		16.8 ± 7 (6.1—30.3)	674
	μg/day	15 (5—25),[h] normally <60	660
		21.5 ± 8.2 (4.2—63)	677

Note: Wet weight unless otherwise indicated. Mean ± SD (range in parentheses). x̃, median.

[a] Gallbladder post-mortem bile.
[b] Weighed mean.
[c] N = 1941.
[d] Range of published mean values.
[e] Dry weight.
[f] White matter.
[g] Grey matter.
[h] 95% limits.
[i] μg/g (ppm).
[j] N = 3793.
[k] "Reference Man".
[l] Range of mean values.
[m] Toenails.
[n] Rest saliva.
[o] Parotid.
[p] Stimulated.
[q] Whole mixed.
[r] Nondialyzable.
[s] Permanent.
[t] Primary.

and/or in reducing flame), can one encounter a minor signal depression from PO_4^{3-}, CNS^-, salts, etc.

Microdeterminations of Cu in small sample volumes (20 to 200 μℓ) are carried out by means of the pulse-nebulization flame AAS;[417,439,699-703] both the injection and the dipping techniques are now automated[58,704,705] and are becoming more important in biological analyses.

The ET-AAS of Cu is very sensitive although matrix effects are encountered. The characteristic concentration ranges between 0.008 and 0.8 ng/mℓ (0.8 to 30 pg) and the DLs are from 0.001 to 1.4 ng/mℓ (1 to 7 pg),[57,59-61,294] depending on the apparatus.

The use of pyrolytically coated graphite tubes as well as higher heating rates during the atomization state is advantageous. In such furnaces the optimum atomization temperature may be substantially lower (e.g., 2000 to 2300 vs. 2400 to 2700°C) than in the older models.

It has also been found that the sensitivity can be improved by a factor of 4 to 6 vs. normally heated and uncoated graphite tubes.[60]

The ashing stage usually involves temperatures between 600 and 1000°C, practically never higher than 1100°C. Hence, some salts of low or intermediate volatility cannot be completely eliminated from the graphite tube before the start of the atomization stage, and this results in matrix interferences[706-708] as well as in background absorption, which may persist in the atomization stage. The depressive effects of $CaCl_2$, $MgCl_2$, NaCl, $LaCl_3$, KCl,[706,707] as well as of $NaClO_4$ and the salt matrix of sea water,[708] have been thoroughly studied; as has been shown, these interferences — under gas-flow conditions and with a slowly heated furnace — cannot be eliminated by applying peak-area measurements.[706,707] The combined use of a ramp ashing up to 900 to 1100°C and simultaneous background correction and the addition of 0.3% Na_2O_2 has been only partially effective with sea water samples.[708]

The effect of a heavy salt matrix — 1% NaCl — has been eliminated by using the L'vov platform:[229] the evaporation of the Cu off the platform was delayed by 1.25 sec and this resulted in higher furnace temperatures (2300 vs. 1840°C) in the moment of copper evaporation and hence in lower matrix effects.

III. AAS METHODS FOR ANALYSIS OF BIOLOGICAL SAMPLES

AAS is an established analytical technique for the determination of Cu in biomaterials. Most assays can be performed by flame AAS which nowadays should be considered to be the best method for a routine analysis of biological samples. Microprocedures based on pulse nebulization or ETA are available as well.

The AAS techniques for Cu determination are classified in Table 2; selected procedures are given in Table 3.

No major methodological problems should be expected in Cu assays: the sampling, storage, and handling of biological specimens do not pose any particular difficulties or contamination problems; the ashing and chelate extraction steps, when applied, should be straightforward; and, finally, the flame AAS of Cu is selective. ETA of Cu, however, may be prone to matrix effects and thus may require greater attention, a versatile programming, and, often, a complicated calibration.

The sampling step still may introduce systematic errors. The distribution of Cu in liver[108] and hair[161,650] is rather homogeneous, but some microspecimens of nail,[669,780] brain tissue,[776] and small hair segments[101,779] are not truly representative. "Spot" samples of milk,[668] urine, and feces should not be considered as representative, either. The sampling technique for sweat calls for standardization, as the arm-bag collection technique was recently found to give falsely high and variable results as compared with the whole-body wash-down technique.[381] The sampling time of autopsy tissue specimens should also be considered as a source of variability, as post-mortem changes may occur.[789]

The samples of sweat and serum require thorough shaking after storage under refrigeration or deep freeze, since a stratification and enrichment of the upper layer may take place.[790] The same applies to aged urine specimens, the sediment of which contains some Cu.[165,416,771]

For a long-term storage (up to several months[684]) urine specimens should be immediately frozen in precleaned polyethylene containers. Some copper impurities may be released from the plastics if not properly selected and/or cleaned; e.g., Karin et al.[163] leached 0.16 $\mu g/m\ell$ Cu from their polyethylene bottles after a 3-day contact with 8 *M* HNO_3.

Grinding in agate or corundum may also introduce intolerable contamination.[624] The brittle fracture technique[532] and its modifications[533] appear much safer in this respect. In dry-ashing procedures, the porcelain crucibles have often been a source of contamination.[91,733,791] Incidentally, contamination from the Vitreosil® crucible has been experienced.[66]

Table 2
AAS TECHNIQUES FOR Cu DETERMINATION IN BIOLOGICAL MATERIALS

Technique	Biological materials
Direct flame AAS with or without dilution	CSF,[664,709] milk,[681,710-712] plasma,[713,714] serum,[27,57,539,657,664,713,715-723] urine,[539,664,677,715]
Flame AAS after TCA deproteinization	Serum[657,674,680,718,719,722-725]
Flame AAS after digestion	Blood,[726,727] bone,[728] erythroctyes,[727] feces,[248,729] fish,[442,458,475,494,730] food,[537,591,729-731] hair,[31,370,460,461,662,666,732-735] milk,[374,538,591,681,729,730,736] nails,[669,692,737,738] teeth,[22] tissues,[190,268,277,442,462,463,538,541,680,718,729,730,739,740]
Flame AAS after TAAH solubilization	Hair,[464] plasma,[741] tissues,[464,465,741]
Flame AAS after acid extraction	Feces,[729] food,[729] liver,[319,467] meat,[517,729] milk,[729] tissues[517]
Flame AAS after liquid-liquid extraction	Blood,[332,473,482,742] erythrocytes,[743] feces,[39,659,674] fish,[474] food,[39,476,659,690,744] hair,[473,478] milk,[110,745] plasma,[743] serum,[7,473,482,640,657,719] shell fish,[334] sweat,[675] teeth,[481,746] tissues,[7,43,429,473,641,662,674,690] urine[7,27,39,659,674,718,742,747,748]
Pulse nebulization and miscellaneous flame AAS techniques	Blood,[749] bovine milk and intestinal xanthine oxidaze fractions,[750] milk,[749] plasma (speciation),[751-753] serum (speciation),[754,755] serum,[699,700,702-704,721,756,757] synovial fluid,[758] tissues,[417,429,641,759-761] urine[749]
Direct ET-AAS with or without dilution	Blood,[421,500,652,762] CSF,[763] erythrocytes,[764] leukocytes,[158] milk,[651,681] plasma or serum,[27,330,511,652,762-770] saliva,[379] urine[27,330,763,771]
ET-AAS after digestion	Blood,[686,727] bone,[90,97] erythrocytes,[631,727,772] feces,[773] fish,[774] hair,[557] liver biopsies,[515] milk,[681,775] NBS SRMs,[518,686] serum,[766] tissues,[373,515-517,686,766,774] tooth dentine,[24] tooth enamel,[24,84] urine,[27]
ET-AAS after TAAH solubilization	Blood,[521] brain,[776] hair,[521] muscle biopsies,[92] plasma,[610] tissues[92,521,510,776]
ET-AAS after deproteinization	Semen,[7] serum[7]
ET-AAS after liquid-liquid extraction	Tissues,[7] urine[7,27,771]
Direct ET-AAS (solid samples)	Fish (ash),[505] hair,[79,80,649,778,779] mussels,[86] nails,[669,780] tissues[79,80,505,649,778,779,781]
Miscellaneous ET-AAS techniques (and procedures requiring special equipment)	Erythrocytes (speciation),[764] food,[446] milk,[782] plasma protein fractions,[764,783-785] plasma ultrafiltrate,[765] rat liver fractions,[786] saliva protein fractions,[379] serum,[485] serum protein fractions,[786,787] synovial fluid (speciation),[788] tissues,[452] urine[485]

No noticeable contamination of serum or plasma from the commercially available Vacutainer® tubes has been found,[420,725] but some other storage containers (both glass and plastics) have been markedly leached by blood during storage.[421]

Much attention has been paid to the washing procedures for hair[31,290,625,733-735] and nails.[15,290,623,670,692,738] Of the washing procedures studied, the aqueous EDTA[733,735] and the dilute HNO_3[466,735] are to be rejected, as they both partially extract the endogenous Cu; the no-wash approach of Chittleborough[393,625] could be criticized as well. Mild washing with

Table 3
SELECTED AAS PROCEDURES FOR Cu DETERMINATION IN BIOLOGICAL SAMPLES

Sample	Brief procedure outline	Ref.
Blood, serum	1-μℓ sample containing ~1% Triton® X-100 injected. ET-AAS: 600°C for 10 sec, 800°C for 5 sec, 2700°C for 5 sec (miniflow), and 2700°C for 2 sec	1213
Brain	TAAH solubilization: 0.1-g sample + 0.1 mℓ of water + 1 mℓ of Soluene® 350 overnight; dilution with MIBK; ET-AAS	776
Feces, food, meat, milk	1- to 2-g sample boiled for 30 min with 15 mℓ of a (27:3:20, by volume) mixture of HCl-HNO_3-H_2O; diluted to 50 mℓ; flame AAS; standard addition calibration	729
Food, milk, tissues	HNO_3-H_2SO_4 digestion; flame AAS; standards in 5% v/v H_2SO_4	730
Hair	HNO_3 digestion; flame AAS vs. acid-matched standards (e.g., in 10% v/v HNO_3)	460
Hair, nails	HNO_3 digestion, completed with $HClO_4$;[a] flame AAS; standards in 10% v/v $HClO_4$	732, 738
Liver (biopsies)	5-mg sample (dry weight) digested with 1 mℓ of 1 *M* HNO_3 at 80°C for 24 hr, slowly evaporated; 2 mℓ of 0.01 *M* HNO_3 added, and vigorously mixed; ET-AAS: 125, 600, and 1950°C; standards in 0.01 *M* HNO_3	515
Liver (NBS SRM)	2-mg sample (dry weight) digested in a closed vial with 40 μℓ of (5 + 1) HNO_3 + $HClO_4$[a] or else with HNO_3 at 120°C for 3 hr (stepwise TCH essential); pulse-nebulization flame AAS; acid-matched standards	760, 761
Liver, muscle, meat	1-g sample + 10 mℓ of (1 + 1) HNO_3 heated at 80°C, evaporated to ca. 3 mℓ, and diluted to 25 mℓ; flame AAS vs. acid-matched standards; ET-AAS: 150, 800, and 2700°C with background correction and standard addition calibration	517
Milk	1 + 3 dilution with a detergent solution (1% saponine or 0.2% Meriten (nonyl-phenyl-polyglycolether) or 0.2% Na dodecylbenzenesulfonate); flame AAS; standards in detergent; applicable to evaporated milk (Reference 712)	710, 712
Muscle (biopsies)	5- to 20-mg sample solubilized with 200 μℓ of TAAH (Lumatom®) in a sealed vial overnight at 60°C;[a] ET-AAS: 150, 650 (ramp), and 2200°C (miniflow); background correction; standard addition calibration	92
Plasma/serum	1 + 10 dilution with water; pulse-nebulization AAS; standards in 0.1 *M* HCl	703
Plasma	1 + 9 dilution with 0.01 *M* HNO_3; ET-AAS: 120, 800, and 2550°C (gas stop); background correction; standards containing 0.01 *M* HNO_3 and (1 + 9) saline	765
Saliva	1 + 1 dilution with water; ET-AAS: 100, 850, and 2550°C, and clean-out at 2750°C; background correction; standard addition calibration	379
Serum	Pulse-nebulization flame AAS: 100-μℓ injections of 1 + 1 diluted serum; standards containing 0.03% w/v of polyvinilalcohol (Merck-Schuchardt); alternatively, calibration vs. 1 + 1 diluted control sera	700, 701
	1 + 6 dilution with water or 0.01 *M* HNO_3; flameAAS	767
	1 + 9 dilution with 0.1 % Triton® X-100; ET-AAS: 150 (ramp), 700 (ramp), and 2800°C (gas-stop, "maximum power"); background correction; standard addition calibration	766
	100 μℓ serum + 100 μℓ water + 800 μℓ of diluent (containing 0.15 *M* H_3PO_4 — 0.125 *M* NH_4NO_3 — 0.1% w/v Triton® X-100); ET-AAS: 100, 600, and 2700°C (miniflow); background correction; standards in diluent	770
Teeth	$HClO_4$ digestion[a] (Reference 746) or pre-ashing with HNO_3, then dry ashing, and HCl dissolution (Reference 481); APDC-MIBK extraction at pH 1.0—1.4; flame AAS	481, 746
Tissues	TAAH solubilization (e.g., 2 mℓ of base for every 100 mg of dried tissue) at 60°C for 8—24 hr; flame AAS; standard addition calibration	464, 465, 741

Table 3 (continued)
SELECTED AAS PROCEDURES FOR Cu DETERMINATION IN BIOLOGICAL SAMPLES

Sample	Brief procedure outline	Ref.
Urine	24-hr urines acidified with 5 mℓ of H_2SO_4, random urines (~25 mℓ) acidified with 0.5 mℓ of H_2SO_4, and shaken thoroughly before aliquoting; 1 + 1 dilution; ET-AAS: 110 (ramp), 900 (ramp), and 2700°C; background correction; standards containing 0.2 mℓ of 70% w/v HNO_3/100 mℓ; alternatively, APDC-MIBK extraction, centrifugation, and ET-AAS; standards extracted similarly	771
	HCl added to pH ~3, and H_2O_2 added to release Cu^{2+}; APDC-2-heptanone extraction at pH 3 in presence of antifoaming agent, and centrifugation; flame AAS; standards extracted similarly	747
	1 + 1 dilution with 1% v/v HNO_3; ET-AAS: 120, 750, and 2400°C; standard addition calibration	330

[a] Safety precautions mandatory.

organic solvents (acetone) or with nonionic detergents appears to be acceptable, preferably when combined and compared with an analysis of unwashed samples. Nail clippings may even require more vigorous treatment to remove the surface contamination.[15,290,623,692]

Most analyses of biological liquids can be successfully carried out without prior destruction of the organic matter (see Tables 2 and 3). If an ashing step is involved, both wet and dry-ashing techniques can be used. Various wet digestions have been described with few adverse comments. Most of the described procedures are suitable for a successive determination of Cu and other elements in the digests: HNO_3 (blood,[686,726] bone,[90] erythrocytes,[631,772] feces,[729] food,[729] hair,[31,460,557,666,735] liver,[467,515,518,545,686] meat,[729] milk,[712,729] teeth,[22,24,545] tissues,[373,545,718]); HCl (tooth enamel,[84]); $HClO_4$ (teeth[746]); $HClO_3$ (hair and brain[662]); HNO_3 and $HClO_4$ (brain,[680,740] feces,[248] fish,[474,494,792] hair,[732,734] nails,[669,738] and tissues[7,277,541,760,761]); HNO_3 and H_2SO_4 (bone,[97] fish,[458,730] food,[730] milk,[730] nails,[737] tissues,[7,730] and urine[742]); HNO_3 and H_2SO_4, completed with H_2O_2 (fish,[475] food,[690,744] and tissues[170,462,475]); HNO_3, H_2SO_4, and $HClO_4$ (feces,[674] fish,[792] tissues,[43,190,442] and urine[674]); HNO_3 and HCl (feces, food, meat, and milk,[729] liver[319]); H_2SO_4 and H_2O_2 (tissues[774,793]); HNO_3 and H_2O_2;[277] etc. The choice of digestion procedure should be guided, among many factors, by the next analytical step; e.g., the use of $HClO_4$ should be avoided in the ET-AAS procedures, since the final concentration of the $HClO_4$ in digests cannot be exactly controlled and this can lead to irreproducible results[776] as well as to signal depression.[66] The dilute HNO_3 should be very appropriate in ETA but not in procedures involving extraction.

The LTA in excited O_2 has proved to be quite satisfactory for Cu.[442,462,516,773]

Dry ashing is generally safe up to 500 or 550°C (with some specimens even to higher temperatures), but minor losses (5 to 10%) due to retention of the analyte onto the crucible walls or siliceous or other residues are common.[91,729] Thus, dry ashing must often be completed by treatment of the ashes with aqua regia,[370,537,538] or at least by evaporation with HNO_3,[374,473,482,537,731,739,792] in order to effectively put the copper into solution.

Enzymatic digestion was recently applied to tissue analyses.[463]

Homogenization of hair and tissue samples with TAAH (e.g., with Lumatom®, Soluene®, etc.) appears to be a quite convenient and simple procedure prior to flame[464,465,741] or ET-AAS.[92,521,610,776] A standard additions calibration is often required in the TAAH procedures.

The AAS determinations of Cu are typically performed without a preconcentration step. If such a step is needed (e.g., in extraction-flame AAS analyses of urine, or in some older procedure based on nonversatile ET-AAS equipment, etc.), the copper must be released from its chemical bonds with the organic matter. Then Cu(II) is readily extracted, usually as a dithiocarbamate. The most common and effective extraction systems which have been applied to biological digests and/or deproteinized biological liquids are the following: APDC-MIBK (at pH from 1 to 8);[110,337,481,541,662,718,744-746] SDDC-MIBK (at pH from 6.5 to 8);[7,494] and HHDC-butylacetate (at pH 0 to 6).[43,640,657,719] Several other extraction systems have been reportedly applied to analyses of biomaterials, with few or no practical advantages over the above-indicated three systems: APDC-$CHCl_3$;[641] APDC-butylacetate;[743] APDC-heptan-2-one;[747] DDDC-MIBK;[332] DDDC-heptan-2-one;[690] HHDC-$CHCl_3$;[429,439,659] HHDC-8-hydroxyquinoline-$CHCl_3$-i-amyl alcohol;[39] diphenylcarbazone-pyridine-toluene;[473,482] dithizone-$CHCl_3$;[474] TTA-MIBK;[742] 4.25 to 5 *M* H_3PO_4 — 0.7 *M* KI-MIBK;[476] and KI-H_2SO_4 — Amberlite® LA-2-MIBK.[478] It should be recalled here that most of these extractions were aimed at simultaneously extracting several trace elements, including copper, from biological fluids/digests.

The Cu(II) dithiocarbamate complex is very stable in the extracts, provided that a good phase separation is achieved and the extracts are not exposed to excessive light. The $Cu(PDC)_2$ complex in MIBK was found to be stable for 48 and 168 hr, after an extraction at pH 1 and 2 to 8, respectively.[427,794] A few adverse notes on Cu extractions are the interference from excess iron (in erythrocytes[743]); the formation of persistent emulsion (e.g., during extractions from untreated urines); and the restricted time stability of the $Cu(DDC)_2$ complex and of SDDC in presence of acids and oxidants.

Preconcentration techniques such as coprecipitation with APDC,[446] sorption/ion exchange,[326,341,412] etc. may also be used (in multielement schemes).

The flame AAS determinations of Cu in diluted serum, plasma, and milk, and in digested or homogenized tissue, hair, nail, food, and feces are straightforward. Urine specimens from patients with Wilson's disease may also be directly nebulized after a (1 + 1) dilution, whereas normal urines require either preconcentration or ETA. The mode of calibration is the main concern in direct flame AAS assays. There are minor transport interferences and bulk-matrix effects which obviously depend on the apparatus employed (e.g., on the diameter and the length of the nebulizer capillary, the construction of the spray chamber, the observation height, etc.). If serum or plasma samples are diluted more than fourfold, these effects are generally negligible, and the calibration against aqueous solutions is tolerable. Often, in order to obtain higher absorbance signals, lower dilution factors, e.g., 1 + 1, are used. Viscosity-matched standards must then be used for calibration, which may contain, say, 10% v/v of glycerol,[27,539] or 0.03% w/v of polyvinylalcohol,[700] or 3% w/v of bovine albumin,[718] or else (1:12) ethyleneglycol plus 0.62 m*M* of L-histidine monohydrochloride.[716] Including NaCl and KCl in standard solutions[701,721,722] does not necessarily improve accuracy. The calibration may also involve standard additions, spiked pooled serum which has been stripped off the copper by treating with Chelex®-100 resin,[153] preanalyzed sera, internal standardization, etc.

The injection method of Berndt and Jackwerth[700,701] and some other procedures based on pulse-nebulization flame AAS,[702,703,756] as outlined in Table 3, appear to be the best technique for automated, direct, micro, flame AAS analyses of serum samples. Inasmuch as only 10 or 25 $\mu\ell$ of serum is needed, the use of standard sera or preanalyzed serum specimens for calibration is justified.

The flame AAS of Cu should always go with a thorough optimization of the signal-to-noise ratio, in order to improve the baseline stability and thus to be able to use higher scale expansion factors and higher dilution ratios. This can be accomplished by a careful alignment of the optical system, by using a somewhat higher lamp current, broader slitwidth, as well as proper observation height and an acetylene flow rate.

The ET-AAS of Cu is more prone to interferences. A simultaneous background correction is called for, and its efficiency must be checked. The intensity of the deuterium lamp at 325 nm is low and, in order to balance the intensities of the HCL and the D_2 lamp, one will have to lower the current of the HCL and thus to impair the baseline stability. It is preferable to substitute a tungsten halide lamp for the D_2 lamp, or to apply a Zeeman background correction.

In presence of chlorides, the copper may be partially lost during the ashing stage at temperatures around 600 to 800°C.[766] The temperature program should be carefully optimized, and typically involves a ramp ashing to 600 to 900°C and a "maximum power" atomization of 2000 to 2700°C, preferably with a restricted flow rate or just stopped purge gas.

Air or O_2 can be used as an alternate gas during the charring stage of the graphite-furnace program,[582,781,795] provided a versatile furnace is available and uncoated graphite tubes are used.

In view of recent experience with the L'vov platform,[229,454,455,553] one might expect new developments in the methodology of the ET-AAS of Cu in difficult biological matrices (hopefully, in urine). Chakrabarti et al.[454,455] demonstrated the use of the L'vov platform in direct analysis of powdered biological samples to improve the precision by a factor of 2.

Efforts have been made to directly determine Cu in small pieces of tissue, hair, and nail (from 0.1 to 5 mg) by direct ET-AAS, but serious problems have been encountered: complicated calibration, background absorption, accumulation of matrix residues, memory effects, etc. Less sensitive analytical lines vs. the resonance line at 324.7 nm are involved in these direct assays: 327.5 (2.5 ×), 222.6 (20 ×), 249.2 (100 ×), and 244.2 nm (400 ×); unfortunately, these lines exhibit not only lower sensitivity, but also impaired baseline stability and higher background contribution.

Atomization off the L'vov platform,[454,455] as well as the peak-area calibration with standard additions[86,505] and the Zeeman background correction, should be further explored in the direct ET-AAS of Cu.

And, finally, several interesting speciation studies on copper binding in erythrocytes, milk, plasma, serum, saliva, and synovial fluid have been realized (Table 2); most of these procedures involved chromatographic separations and ET-AAS.

IV. CONCLUSION

AAS is the best current method for the determination of Cu in biological samples. Several other instrumental techniques can also be used: spectrophotometry, ASV, ICP-ES, NAA, XRF, etc., the first of which is well established in many clinical labs. The flame AAS with diluted, solubilized, digested, or (dithiocarbamate) extracted biological samples is straightforward. Only minor transport interferences should be reckoned with. The pulse-nebulization technique (in its injection or, still better, dipping versions) is applicable to sample volumes of less than 0.1 to 0.2 mℓ. Extractions are hardly needed except in urinalysis.

The drawbacks of ET-AAS are marked matrix effects, minor but persistent background, complicated calibration, low sample throughput, and higher cost. Versatile equipment is required (high heating rates, L'vov platform atomization, gas stop, and perhaps peak-area calibration). Most of the assays can actually be effected by automated flame/microflame procedures.

Chapter 12

GALLIUM

I. DETERMINATION OF GALLIUM IN BIOLOGICAL SAMPLES

The Ga content of most biological samples is as low as subnanograms per gram (Table 1); few analytical techniques — RNAA, ET-AAS, and SSMS — are sensitive and useful for these ultratrace assays.

RNAA, with a DL down to 0.05 ng/g,[797] should be the best method; successful analyses of biological tissues,[45,796,797] plasma,[796] and serum[797] have been described.

Gallium has been determined in tissues,[3,132] and hair[54] by SSMS; the DL was as low as 0.9 ng/g in soft tissues.[3] In two other reported SSMS procedures for analyses of plasma[55] and teeth,[25] the concentration of Ga in samples was below the DL — 6 and 20 ng/g, respectively.

The fluorimetric assay with 2,2′,4-trihydroxy-5-chloro-1,1′-azobenzene-3-sulfonic acid (Lumogallion) is not sufficiently sensitive (DL, 15 ng/g), and requires a tedious separation scheme.[798]

AES with hollow-cathode discharge[799,800] or arc-source AES[36,44,800] are relatively simple though not as sensitive (DL, 10 ng/g[800]).

II. ATOMIC ABSORPTION SPECTROMETRY OF GALLIUM

There are several Ga line options, depending on the HCL quality/age and the optical parameters of the equipment. Gallium is mostly determined at the 287.4-nm line. Two other lines — 403.3 or 417.2 nm — provide a better emission intensity and are less prone to background absorption but also less sensitive (three or two times, respectively) and require background check or correction by means of a tungsten iodide lamp. The Ga doublet at 294.36/294.42 nm can also be used but at the expense of slightly bent calibration curves.

Gallium determination in a lean N_2O-C_2H_2 flame is relatively selective though insufficiently sensitive. The characteristic concentration is about 1 μg/mℓ, and the DLs are down to 0.04 to 0.08 μg/mℓ.[56-59] Acid-matched standard and sample solutions, in any dilute mineral acid, should be useful; slight ionization interference as well as transport interferences have to be expected.

If another flame (stoichiometric air acetylene) is used, the sensitivity will be impaired by a factor of 2 or 3, and the determination will be prone to serious matrix effects. Matrix-matched standards and standard addition checks are mandatory with this flame.

In ETA, the characteristic concentration is within the range of 0.05 to 1 ng/mℓ (5 to 20 pg), and the DLs are down to 0.01 to 0.1 ng/mℓ (1 to 10 pg).[59,61,294] The volatility of Ga halides and other species may give rise to some problems: losses of the analyte during the ashing stage, premature evaporation of molecular species at the very onset of atomization, low tolerable ashing temperatures (e.g., 800°C[800]), and background and matrix effects.[777]

Several reagents have proved to be useful as analyte/matrix modifiers in ET-AAS of Ga: 1 or 2% v/v HNO_3;[57,64,223,800] $(NH_4)_2$EDTA;[777] 0.5 m*M* Na_4EDTA;[801] 1 or 2% v/v HNO_3 + 1 or 2% v/v H_2O_2;[223] aqueous NH_3;[294] and 1% v/v ascorbic acid.[802] Reportedly, the latter reagent has improved sensitivity by a factor of 20.[802]

The temperature program must be thoroughly optimized. The use of L'vov platform and "maximum power" heating should be advantageous.

Table 1
GALLIUM CONCENTRATIONS IN HUMAN BODY FLUIDS AND TISSUES

Sample	Unit	Selected reference values	Ref.
Brain	ng/g	0.6 ± 0.03	3
Hair	μg/g	0.05[a,b]	54
Kidney	ng/g	0.9 ± 0.3	3
Kidney cortex	ng/g	0.7 ± 0.2	3
Liver	ng/g	0.7 ± 0.1	3
		~1	796
Lung	ng/g	5 ± 2	3
Muscle	ng/g	0.3 ± 0.04	3
Plasma or serum	μg/ℓ	0.074	797
		0.22	796
Tooth (enamel)	μg/g	<0.02	25

Note: Wet weight unless otherwise indicated. Mean ± SD (range in parentheses).

[a] Dry weight.
[b] N = 1.

III. AAS METHODS FOR ANALYSIS OF BIOLOGICAL SAMPLES

ET-AAS can be applied to determine subnanogram amounts of Ga in biological materials only after an efficient preconcentration step. The literature is very scant.[64,777,800,801] Newman[801] studied the pharmacokinetics of Ga administered as an anticancer drug to animals and humans. Severe depression of Ga absorbance by $CaCl_2$ was observed, and an addition of 0.5 m*M* Na_4EDTA was recommended. The samples (blood, tissues, and urine) were digested with HNO_3 and analyzed against matrix-matched standards. The temperature program included drying at 125°C (ramp), ashing at 900°C and atomitization at 2500°C (gas stop). Similarly, Caroli et al.[800] found pronounced tissue-specific interferences in HNO_3 digests; pyro-coated tubes [100, 800 (ramp), and 2500°C (gas stop)] and standard additions were used; and the DL was 5 ng Ga/g tissue.

Ranisteano-Bourdon et al.[64] determined therapeutic levels of Ga in blood, plasma, and serum (1 + 19) diluted with 0.25% v/v Triton X-100® and 0.5% v/v HNO_3. The Varian Techtron® CRA-63 furnace was programmed at 100°C for 30 sec, 450°C for 30 sec, and 2050°C for 1.5 sec; calibration by standard additions was applied. Tissue samples were digested with HNO_3, diluted, and analyzed analogously. The same authors analyzed urine samples by ET-AAS after an extraction of Ga with 8-hydroxyquinoline in MIBK at pH 4.5 (acetate buffer).[64]

In fluorimetric[798] and RNAA[797] procedures, gallium was preconcentrated by more tedious ion-exchange techniques.

Numerous extraction procedures for Ga(III) have been compiled and discussed by Cresser.[76] Popova et al.[803] studied in detail the Ga(III) extraction as cupferronate in butanol or i-pentanol (at pH 4 to 6), or as a chloride ion-association complex in MIBK (from 4.5 to 5.5 *M* HCl), or else in butylacetate (from 5.5 to 7.5 *M* HCl).

Pelosi and Attolini[804] introduced extracts of gallium cupferronate into a graphite-tube furnace, ramp ashed to 630°C (butanol extracts) or 1020°C ($CHCl_3$ extracts), and atomized at 2635°C; the sensitivity dramatically depended on the organic solvent, being much better (25-fold) in butanol than in $CHCl_3$.

There is no reason to expect pronounced losses of Ga during the ashing of biomaterials. Treatment with HNO_3 alone[64,777,799-801] or with HNO_3 and H_2SO_4[797] would seem simple and satisfactory. The use of $HClO_4$, however, will involve a signal depression and irreproducible ET-AAS results. See also the recent study by Nakamura et al.[777]

Freeze drying[3,797] and LTA[3,626] proved to be safe for sample preparation.

Finally, in ultratrace determinations of Ga, precautions should be taken against contamination due to, e.g., grinding procedures: from corundum (1 μg Ga/min of grinding!),[624] agate,[624] alumina, dust fallout, etc.

IV. CONCLUSION

The ET-AAS of Ga is inadequately sensitive and prone to interferences; it should therefore be preceded by an efficient concentration step (extraction). The determination of higher-than-normal levels involves HNO_3 digestion and ET-AAS with simultaneous background correction and matrix-matched calibration; the STPF concept may be relevant as well. Other reliable methods are AES (with HCL or ICP excitation source) and RNAA.

Chapter 13

GERMANIUM

I. DETERMINATION OF GERMANIUM IN BIOLOGICAL SAMPLES

The concentrations of Ge in most biological specimens should be well below 0.1 μg/g (Table 1);[3,13,805] the scant early values appear to be unreliably high. A few SSMS applications are known,[3,16,25,54,55,132] but some of them seem either insufficiently sensitive[25,55] or positively biased.[16,54] DL of 7 ng Ge/g of soft tissue was obtained by Hamilton et al.[3]

Nixon,[192] by ICP-ES after preconcentration of Ge as volatile $GeBr_4$, and Robbins et al.,[131] by hydride-generation MIP-AES, determined Ge in digested biological samples. The DLs of ICP-ES are down to a few nanograms per milliliter (in aqueous standard solutions), and the state-of-the-art figure appears to be 0.5 ng/mℓ.[29]

II. ATOMIC ABSORPTION SPECTROMETRY OF GERMANIUM

Ge determination in a reducing N_2O-C_2H_2 flame is selective but not sensitive; the characteristic concentration and DLs are 0.8 to 2.5 μg/mℓ and 0.02 to 0.2 μg/mℓ, respectively, depending on apparatus and light source.[56-59 294] EDL provides a significantly higher emission intensity (up to 100 ×)[138] and lower DLs (2 to 3 ×) vs. the HCL.[138] Organic solvents (acetone, ethanol, hexane, MIBK, etc.) strongly enhance sensitivity (by up to one order of magnitude).[806,807]

In ETA, the characteristic concentration ranges between 0.4 and 15 ng/mℓ (0.04 to 0.4 ng), and the DLs are between 0.1 and 3 ng/mℓ (0.01 to 0.3 ng).[59,808,809] The main problem with ET-AAS is volatility of some Ge species (halides, sulfides, GeO, etc.) and thermal stability of others (GeO_2, carbides), so that both losses and inefficient atomization may be expected.

Ediger[223] established the following series of matrix-modifying reagents, in order of their increasing effect on sensitivity:

1% H_3PO_4 < aqueous solutions < 1% HCl < 1% aqueous NH_3 < 1% H_2SO_4 < 1% HNO_3 < 1% $HClO_4$.

Recently, Studnicki[809] confirmed the strong effects of acids on Ge absorbance and found that $HClO_4$, HNO_3, or H_3PO_4 (0.1 to 1%) increased the sensitivity, whereas HCl and H_2SO_4 (approximately 1%) were signal depressants. Effective in eliminating chloride interference (e.g., due to $MgCl_2$, NaCl, KCl, $CaCl_2$, etc.) proved to be additions of HNO_3 (e.g., 0.6 *M*) or NH_4NO_3 or H_2SO_4.[809] The temperature program of the Instrumentation Laboratory® 455 graphite furnace has been 75°C for 20 sec, 600°C for 15 sec, 900°C for 15 sec, and 3500°C (?) for 10 sec.[809]

Aqueous 0.3 *M* NaOH and 0.01 *M* $Ca(NO_3)_2$ have also been applied as matrix/analyte modifiers (with pyro- rather than TaC-coated tubes).[810]

Hydride-generation AAS has not been particularly successful until recently. The sensitivity is generally poor (DLs of 3 to 10 ng/mℓ or 0.05 to 0.5 μg absolute),[140,155,232,242,586,587,806] and the assay is prone to interferences (e.g., from excess (500 ×) of As, Au, Bi, Cd, Co, Cu, Fe, Ir, Ni, Pd, Pt, Rh, Ru, Sb, Se, Sn, Te[155]). Recently, Castillo et al.[806] found depression from Fe(II), Fe(III), Te(IV), Se(IV), Sb(III), Sn(II), oxalate, and tartrate at pH 4.75 (acetate buffer); iron interference was prevented by using a more concentrated $NaBH_4$ solution (e.g., 3.5% w/v).

Table 1
GERMANIUM CONCENTRATIONS IN HUMAN BODY FLUIDS AND TISSUES

Sample	Unit	Selected reference values	Ref.
Blood	μg/g	0.2?[a]	3
Feces	mg/day	0.1?[b]	8
Hair	μg/g	2.2?[c,d]	54
Kidney	μg/g	9?[a]	3
Liver	μg/g	0.04[a]	3
Lung	μg/g	0.09[a]	3
Muscle	μg/g	0.03[a]	3
Nails (finger)	μg/g	(0.48—11)?[c]	16
Plasma	μg/ℓ	<30[d]	55
Tooth (enamel)	μg/g	<0.02	25

Note: Wet weight unless otherwise indicated. Mean ± SD (range in parentheses).

[a] Preliminary data.
[b] "Reference Man".
[c] Dry weight.
[d] N = 1.

Some other instrumental techniques for quantitation of the evolved germane (GeH_4) have provided better DLs: GC (30 ng or 0.3 ng/mℓ in 100 mℓ);[147] GC-MS (5 ng or 1 ng/mℓ in 5 mℓ);[146] and MIP-AES (3 ng or 0.15 ng/mℓ in 20 mℓ).[131]

There seem to be several reasons for the lower sensitivity of HG-AAS: (1) low GeH_4 evolution rate; (2) low chemical yield of reaction; (3) low atomization efficiency of Ge in both cool flames (H_2-inert gas-entrained air)[155,811,812] and heated silica tubes;[140] and (4) use of HCL in most early studies instead of EDL,[138] a better light source.

GeH_4 is evolved in a very broad range of acidity/pH: from 4 *M* HCl to a slightly alkaline medium. Acidity should determine the rate of germane generation, the completeness of the reaction, and the extent of interferences. Many workers prefer slightly acidic media, e.g., pH between 0 and 2,[140,155] adjusted by dilute HCl or H_2SO_4 or oxalic or tartaric acid.

The recent work by Andreae and Froelich[813] deserves special attention: after a reduction with $NaBH_4$, the germane was stripped from solution by a helium gas stream, and was collected in a liquid-nitrogen cooled trap; it was then released by rapid heating of the trap; a slightly modified graphite-tube furnace was finally entered, with its heating cycle synchronized with the arrival of GeH_4 to the heated (2600°C) tube; and peak-area measurements were used for quantitation. Several points in this study should be stressed: (1) a maximum reaction yield in the near-neutral pH range; (2) the importance of proper buffering of the reaction medium for good sensitivity and precision; (3) collection of germane in a cooled trap to eliminate potential interferences; (4) rapid release of the trapped GeH_4 for sensitivity improvement; (5) very high atomization temperatures (>2500°C) for effective GeH_4 pyrolysis; and (6) presilanization of all glass surfaces. Impressive DLs — 0.14 ng Ge absolute or 0.00056 ng/mℓ (in a 250-mℓ sample of water) — were obtained by those authors.[813] With another atomizer — a hydrogen-rich flame burning inside a quartz tube — the DL was much worse (3 ng).[813] The need for effective atomization of GeH_4 was also recently demonstrated by Castillo et al.:[806] the evolved germane was introduced into a N_2O-C_2H_2 flame without using a carrier gas, providing characteristic concentration and DL of 0.012 and 0.0038 μg/mℓ, respectively.[806]

III. AAS METHODS FOR ANALYSIS OF BIOLOGICAL SAMPLES

No reports seem to have been published on Ge determination in human biological materials. Such analyses call for highly effective preconcentration.

Germanium can be extracted as a chloride complex from a strongly acidic HCl medium (e.g., 7 to 9 *M*),[76,807] and back extracted with water[808] or another suitable reagent prior to its ET-AAS assay. Coprecipitation with $Fe(OH)_3$ at pH 9; distillation as a volatile bromide (from biological digests[192]); or germane evolution (from wet-digested blood, with recovery of 87%[131]) can be used for preconcentration/separation from the matrix.

Both wet- and dry-ashing techniques are prone to volatization and retention losses of the analyte. In dry ashing of coal, slow heating rates as well as additions of basic ashing aids (CaO or $Ca(NO_3)_2$) proved to be important. Oxygen-flask combustion with trapping solution of 0.3 *M* NaOH + H_2O_2 (one drop per milliliter) appears very appropriate.[810] Wet ashing should be started with HNO_3 at room temperature.

IV. CONCLUSION

The normal levels of Ge may prove much lower than hitherto assumed. Both flameless AAS techniques — ETA and HG — are insufficiently sensitive and prone to acid and matrix effects. Germane (GeH_4) generation, trapping, and atomization at high temperatures (e.g., >2600°C) or else $GeCl_4$ extraction, back extraction, and ET-AAS should be further explored and adapted to biological samples. Some ashing techniques may involve analyte losses. Plasma-source AES is a promising alternative to AAS.

Chapter 14

GOLD

I. DETERMINATION OF GOLD IN BIOLOGICAL SAMPLES

Normal gold levels in human biological materials are exceedingly low: generally less than 0.1 ng/mℓ in body fluids and ≤1 ng/g in most tissues. Somewhat higher (while strongly variable and still below 0.1 μg/g) is the gold content of hair and nails (Table 1).

Few analytical methods are sensitive below the 1 ng/mℓ concentration level: NAA, ET-AAS, and ASV. Among these, NAA has been widely used in assays of normal concentrations of Au in human biological specimens:[45] blood,[45] diet,[45,116] hair,[9,11,45,48,101,118,184,290,670,815] nails,[290,560,670] plasma or serum,[45,111] plasma protein fractions,[816] platelets,[119] saliva,[672] skin,[112] teeth,[23,45,51,676] tissues,[45,105,106] urine,[111] etc.[45,121] Some of these assays have been performed by INAA.[11,23,51,101,118,121,290,672,676]

SSMS analyses of biological tissues have been incidentally attempted,[3,25,55,325] but most reported DLs (e.g., 3 ng/g in soft tissue;[3] 40 ng/g in dried plasma;[55] 20 ng/g in teeth;[25] and 0.3 ng/g in blood[3]) appear to have been above the normal levels encountered in these samples.

Gold salts are widely used in the treatment of rheumatoid arthritis. Thus, sensitive, rapid, and reliable methods for monitoring gold in body fluids at *therapeutic levels* are needed. These levels are much higher than normal (e.g, up to 0.1 to 10 μg/mℓ), and several more techniques can be used; the flame AAS (the method of choice), ICP-ES,[688] ASV,[817] XRF, spectrophotometry,[818] etc.

II. ATOMIC ABSORPTION SPECTROMETRY OF GOLD

The determination of Au in a lean air-acetylene flame is sensitive and selective. The characteristic concentration and DLs are 0.08 to 0.25 μg/mℓ and 0.006 to 0.02 μg/mℓ, respectively.[56-59,294] It would be preferable to use a somewhat higher lamp current and a broader bandpass, to improve the signal-to-noise ratio of the resonance line (242.8 nm). Occasionally, the 267.6-nm line, although twice less sensitive, may prove useful (owing to better intensity and lower background). Only minor transport interferences and bulk matrix effects as well as minor ionization[57] and background are to be expected. Acid-matched standard solutions and standard addition checks are advisable with high-salt solutions. The observation height should not be too low.

In ETA, the characteristic concentration is between 0.1 and 2 ng/mℓ (often ~10 ng absolute), and DLs are in the same range,[57,59-61,819] and even down to an impressive 0.013 ng/mℓ.[59]

The temperature program in ET-AAS of Au should be carefully optimized so as to avoid losses of volatile gold species. Ashing temperatures have to be restricted to 500 to 600°C; accordingly, most salts of moderate and low volatility cannot be eliminated from the furnace prior to the atomization stage. Moreover, the analyte is readily reduced to its elemental state which sublimes from the graphite-tube surface at temperatures as low as 1000°C.[819] Hence, serious matrix effects from many concomitants are observed: chloride,[816,820] excess of HNO_3 or H_2SO_4 (>2 or 3 *M*),[294,821] phosphate,[822] TAAH,[823] organic matter,[816,822] etc. Simultaneous background correction with a properly aligned and balanced deuterium lamp is always required.

Versatility of the graphite furnace design and programming is very important; advisable are ramp-ashing mode, as fast as possible atomization heating rates, gas-stop mode of the purge gas, and atomization off the L'vov platform.[229]

Table 1
GOLD CONCENTRATIONS IN HUMAN BODY FLUIDS AND TISSUES

Sample	Unit	Selected reference values	Ref.
Blood	μg/ℓ	0.055 ± 0.030	178
Hair	μg/g	x̃, 0.0016—0.0055[a] (U.S.A.)	9
		x̃, 0.0099 (0.0006—1.36) (Japan)	11
		x̃, 0.014 (<0.001—0.175) (Canada)	10
		♂, 0.05—0.1[b] ♀, 0.15—0.20[b] (U.S.A.)	814
Kidney	ng/g	0.2 ± 0.1 (<0.01—1.1)	105, 106
Kidney cortex	ng/g	0.39	107
Liver	ng/g	0.09 ± 0.03 (0.01—0.46)	105, 106
Lung	ng/g	x̃, 0.62; 0.45 ± 0.19 (0.080—7.9)	105, 106
Nails (finger)	μg/g	0.171 ± 0.27	670
Nails (thumb)	μg/g	♂, 0.42 ± 0.32 (0.1—0.7)[c]	560
		♀, 2.64 ± 3.81 (0.1—6.4)[c]	560
Plasma or serum	μg/ℓ	0.009 ± 0.010	111
Saliva	μg/ℓ	0.6 ± 0.2 (0.1—1.3)[d]	672
Skin	ng/g	x̃, 2; 2.8 ± 1.8	112
Tooth (dentine)	μg/g	0.03 ± 0.01[c]	676
		0.07 ± 0.04	23
Tooth (enamel)	μg/g	<0.02	25
		0.015[e] and 0.025[f]	51
Urine	ng/day	11 ± 25 (0.1—83)	111

Note: Wet weight unless otherwise indicated. Mean ± SD (range in parentheses). x̃, median.

[a] Range of medians.
[b] Mode.
[c] Dry weight.
[d] Nondialyzable.
[e] African.
[f] European.

Calibration against matrix-matched standards appears inevitable with most of the current graphite-tube furnaces.

III. AAS METHODS FOR ANALYSIS OF BIOLOGICAL SAMPLES

Both flame and ET-AAS are routinely applied in many clinical laboratories for the determination of Au in biological liquids at therapeutic levels. If normal levels (subnanograms per gram!) are to be determined, then the ETA has to be preceded by efficient preconcentration.

Flame AAS procedures for analysis of blood,[824] serum/plasma,[7,821,824-828] synovial fluid,[825,827] tissues,[7] and urine[7,298,821,824,825,827,829] have been described. ET-AAS has been applied in microprocedures for blood,[820,830,831] feces,[832] hair,[823] nails,[823] packed blood cells,[820] serum or plasma,[7,330,770,816,820,821,830,833,834] serum/plasma protein fractions,[816,821,822,835] synovial fluid,[788] tissues,[832] and urine.[7,330,821,828,834,836]

No major problems are encountered in AAS assays of therapeutic gold levels in blood, serum/plasma, and urine, and simple direct procedures are available (Table 2).

In the sampling step, an uneven distribution of Au in certain specimens such as hair,[100,161] nails, and tissues (kidney) is encountered. So is surface contamination of nails, hair, and skin from gold jewelry,[118,560] and of saliva from dental alloys.[672] This may account for the observed higher gold concentrations in thumbnails vs. toenails (by a factor of 4[560]), higher Au levels in thumbnails of women vs. men (by a factor of 6[560]), as well as for some peak Au values in saliva[672] and hair[9,118] assays.

Table 2
SELECTED AAS PROCEDURES FOR DETERMINING THERAPEUTIC CONCENTRATIONS OF Au IN BIOLOGICAL MATERIALS

Sample	Brief procedure outline	Ref.
Blood	1 + 19 dilution with water; ET-AAS: 125 (ramp), 800 (ramp 10 + 80 sec), and 2200°C; background correction; standards on pooled blood	831
Hair, nails, tissues	H_2SO_4-HNO_3 digestion; ET-AAS: 130, 480, and 2200°C; background correction; standards in 5% v/v H_2SO_4	823
Plasma/serum	1 + 1 dilution with water; a drop of CH_3OH added; a lean air-C_2H_2 flame; standards on pooled serum	821
Plasma/serum, urine	Plasma/serum diluted 1 + 4 with 0.1% Triton® X-100, and urine diluted 1 + 9 with 0.01 *M* HCl; ET-AAS: 100, 500, and 2400°C; standard addition calibration; 267.6 nm preferable	821
Plasma, plasma fractions	1 + 9 dilution or gel-permeation chromatography on Sephadex® G-200; ET-AAS: 100, 500, and 3000°C (CRA-63 graphite cup); background correction; matrix-matched standards	816
Serum	1 + 3 dilution with water; flame AAS; standards containing albumin (30 g/ℓ)	826, 828
	100 μℓ of serum + 100 μℓ of water + 800 μℓ of a diluent (0.15 *M* H_3PO_4—0.125 *M* NH_4NO_3—0.1% w/v Triton® X-100); ET-AAS: 100, 600, and 2700°C (miniflow); background correction	770
Serum protein fractions	Separation scheme involved electrophoresis, gel chromatography, and (for albumin) polyethyleneglycol precipitation; ET-AAS: 130, 780 (critical!), and 2200°C (gas stop); background correction; matrix-matched calibration	835
Synovial fluid	Total content and distribution of Au studied; gel filtration on Sephadex® G-75 superfine gel; dilution (3= to 17-fold); ET-AAS: 120, 575, 2650, and (clean-out) 2700°C; background correction; calibration vs. preanalyzed (by standard additions) specimens	788
Urine	1 + 3 dilution with water, and a drop of CH_3OH (antifoam) added; a lean air-C_2H_2 flame; background correction; standard addition calibration	821

Ashing step should be avoided as it is both time consuming and prone to analyte losses. Gold is generally difficult to put into and keep in solution. Both retention and volatization losses have been observed in dry ashing[91,169] and LTA[169] of organic samples. Certain wet-digestion procedures may also involve losses of the analyte, owing to its reduction to an elemental state and pronounced adherence on glass and plastic surfaces. TAAH solubilization of tissues would also seem iffy due to Au precipitation during the preparatory step, as well as severe depression of Au signal in the ET-AAS.[823] (Recently, Turkall and Bianchine[832] managed to obtain good results with TAAH in toluene (Soluene® 100) and next dilution with MIBK.)

A few wet-digestion techniques have reportedly been successful: HNO_3 and H_2SO_4 (hair, nails, plasma, and tissues);[7,823] HNO_3 and $HClO_4$, completed with dissolution in 3 *M* HCl (urine);[829] $KMnO_4$ and HCl (caution!) (serum, synovial fluid, and urine);[827] $HClO_4$ alone (caution!) (serum and urine);[834] and HNO_3, completed with $HClO_3$ (serum; ASV procedure).[817]

Trace gold can be preconcentrated by liquid-liquid extraction;[76] coprecipitation with Te,[828,836] sulfides, or else with sulfur-containing organic ligands; sorption on activated charcoal, etc.

Gold is mostly extracted as a halide complex (chloride and, more rarely, as a bromide or iodide); this extraction is very effective (distribution coefficients $>10^4$) and relatively selective (from 0.5 to 3 *M* HCl). It has been widely applied to analyses of biological samples.[298,330,827-829,836] Several other extraction systems have also been found useful for gold preconcentration from biological digests: APDC-MIBK (blood, serum, urine),[824] SDDC-MIBK (plasma, tissues, urine),[7] and dimorpholinethiuramdisulfide-MIBK (serum, urine).[834]

MIBK is the best solvent owing to its good burning properties, but $CHCl_3$ in pulse-nebulization flame AAS is 3.6 times as sensitive.[429]

In flame AAS determinations, serum and urine are diluted (1 + 1 to 1 + 3) with water and directly sprayed into the flame;[821,825,826,828] a drop of methanol[821] or another surfactant may be added to prevent excessive foaming. Calibration standards contain albumin[826] or, better yet, diluted pooled serum spiked with known gold additions.[821] Standard additions calibration and simultaneous background correction are preferred in urine analyses.[821,825] Pulse nebulization of small sample aliquots (50 to 100 μℓ) which is now automated should be favored.[58,705]

In ET-AAS, samples can be diluted 5 to 20 times with water[7,816,831] or surfactant (Triton® X-100)[821] or an aqueous matrix/analyte modifier.[770,822] Phosphate buffer[822] or H_3PO_4 + NH_4NO_3 + Triton® X-100[770] have been used as diluents. The temperature program often involves a ramp ashing to, say, 500°C, high-speed heating and ''gas stop'' in the atomization step, and simultaneous background correction. The L'vov platform technique[229] is promising.

Considerable tube-to-tube variations have been observed.[822] Any newly adopted procedure should be adapted to a particular graphite furnace. Urine analyses call for standard addition calibration. Biological digests should preferably contain a dilute H_2SO_4 or H_3PO_4, and acid-matched calibration, with frequent standard addition checks, being performed.

IV. CONCLUSION

NAA is the best technique at normal (nanograms per gram) gold levels, while AAS is the method of choice at therapeutic (~micrograms per gram) levels. Flame AAS, preferable in its pulse-nebulization version and with matrix-matched standards, is fast and reliable. ET-AAS calls for thorough temperature programming, background correction, and matrix-matched calibration. L'vov platform (STPF) atomization may be expected to be most useful.

Gold may be nonhomogeneously distributed in certain specimens as well as difficult to put and keep into solution. It is readily extracted (e.g., as aurichloride complex).

Chapter 15

INDIUM

I. DETERMINATION OF INDIUM IN BIOLOGICAL SAMPLES

Very little has been published on the determination of In in biological materials. The concentrations to be determined should generally be in the lower nanograms-per-gram range (Table 1). ET-AAS, ASV or DPASV,[125] NAA,[12,45] and SSMS[55,132,409] are the most sensitive analytical methods for trace indium.

II. ATOMIC ABSORPTION SPECTROMETRY OF INDIUM

The flame AAS determination of In is selective but not adequately sensitive. At the resonance line of 303.9 nm, the characteristic concentration is between 0.15 and 1 μg/mℓ, and the DLs are 0.02 to 0.04 μg/mℓ.[56-59,294] No major interferences should be expected provided a lean air-acetylene flame and a not too low observation height are used. Acid-matched standard solutions, preferably in dilute HNO_3 or H_2SO_4 or $HClO_4$, should be useful for calibration. Standard addition checks should be incidentally applied, and the absence of background absorption should be verified, when high-salt solutions are analyzed.

In ETA, the characteristic concentration ranges between 0.1 and 1.3 ng/mℓ (11 to 45 pg), and the DLs are 0.02 to 0.3 ng/mℓ (10 pg).[59-61,294]

Pyrolytically coated tubes offer better sensitivity (2 ×) and lower atomization temperature (by about 200°C) as compared to uncoated tubes.[60] A suitable acidic medium is a dilute HNO_3 or H_2SO_4 (1 to 2% v/v), while HCl or chlorides depress the analyte signal. The charring temperatures are relatively low (e.g., 500°C) and many salts of intermediate volatility cannot be expelled from the graphite tube prior to the atomization stage; therefore, matrix interferences and severe background absorption persist in the ETA of In. The versatility of furnace programming is very important in analyses of complex biological samples and digests: the introduction of oxygen as alternate gas during the charring stage;[582] ramp-ashing mode; high-speed heating during the atomization stage (this should significantly decrease the optimum atomization temperatures, e.g., down to 2000 to 2200 instead of 2600 to 2800°C[60]); interruption of purge-gas flow at the atomization stage; etc. The L'vov platform[229] (STPF[553]) should be useful in ET-AAS of In as well.

III. AAS METHODS FOR ANALYSIS OF BIOLOGICAL SAMPLES

Only two applications of flame AAS to analyses of urine have been found so far. Torres[837] determined microgram amounts of In in urine after an ion-exchange preconcentration. The same author extracted In from urine by dithizone-MIBK and atomized the extracts in Sampling Boat® flame accessory.[496]

Numerous extraction procedures for In have been compiled and discussed by Cresser;[76] extractions such as bromide, iodide, dithiocarbamate, or oxinate are most popular.

IV. CONCLUSION

Major preparatory steps (ashing, preconcentration) may be needed in ultratrace In assays. ET-AAS is sensitive but prone to matrix effects. Other relevant techniques are ASV and NAA.

Table 1
INDIUM CONCENTRATIONS IN HUMAN BODY FLUIDS AND TISSUES

Sample	Unit	Selected reference values	Ref.
Kidney	μg/g	0.03 ± 0.05 (≤0.14)	12
Liver	μg/g	x̃, 0.002; 0.02 ± 0.04 (≤0.13)	12
Lung	μg/g	0.04 ± 0.07 (≤0.14)	12
Plasma	μg/ℓ	<4[a]	55

Note: Wet weight unless otherwise indicated. Mean ± SD (range in parentheses). x̃, median.

[a] N = 1.

Chapter 16

IRON

I. DETERMINATION OF IRON IN BIOLOGICAL SAMPLES

The normal levels of Fe in most human biological materials are above 1 μg/mℓ; somewhat lower in iron are only CSF, milk, saliva, and urine (Table 1). The relatively high Fe concentrations can be determined by many analytical methods, among which spectrophotometry and flame AAS are the most widely applied and convenient.

Spectrophotometric procedures for serum iron are simple, inexpensive, and reliable; commercially available reagent kits are often used.[188,679,846-850] Spectrophotometry has also been applied in analyses of milk,[40,110] nails,[133,737] sweat,[843] tissues,[133] etc., as well as in assessing the serum iron-binding capacity.[849,850] Some catalytic procedures which do not require expensive equipment are also useful.[577]

ICP-ES offers the advantages of a simultaneous multielement assay with DLs around or below 1 ng Fe/mℓ,[29,35,123,289] and a minimum sample preparation.[30,32,688] This method has been applied to blood,[32,689] bone,[392] diet,[690] milk,[35] serum/plasma,[30,289,580,688,689] tissues,[289,412,518,580,636] and other biological samples.[35,348]

Many early applications of the arc-source AES are documented.[26,36,38,41,42,44,380] XRF[128,636,691-695,851] and SSMS[16,25,52,54,55,132,275,409] require expensive equipment, and their use is justified in simultaneous multielement assays.

The PIXE technique[128] is unique in studying iron distribution in a small sample area, (e.g., in small liver biopsies,[407] along[128] or across[195] a single hair, in bone, teeth, etc.), as well as with small samples.[128,407,696-698,852]

There are abundant literature data on NAA application:[45,121] blood,[45,573] CSF,[621] diet,[45,47,116] erythrocytes,[631,840] hair,[11,45,48,98,102,118] platelets,[119] skin,[112] serum/plasma,[111,120,631,633,840] teeth,[23,51] tissues,[12,45,108,631,633,853] urine,[111] etc.[45,121]

Serum iron and total iron-binding capacity (TIBC) can be measured radiometrically in 0.4-mℓ samples.[854]

Exogenous magnetite in lungs has been evaluated by magnetic methods.[855,856]

II. ATOMIC ABSORPTION SPECTROMETRY OF IRON

Flame AAS is the best method for the determination of microgram amounts of Fe. Its characteristic concentration and DLs are in the range of 0.03 to 0.2 μg/mℓ and 0.003 to 0.01 μg/mℓ, respectively.[56-59, 294] A lean air-acetylene flame is recommended for better selectivity; in a more reducing flame the sensitivity is better but at the expense of more pronounced interferences (condensed phase effects[857]). Under inappropriate conditions — fuel-rich flame and low observation height — one may encounter a depression from Ca, Si, acids, SO_4^{2-}, PO_4^{3-}, etc., as well as bent calibration curves (even with inflections and maxima[857]). Hence, a thorough optimization of both observation height and acetylene flow and calibration against viscosity-matched and/or acid-matched standards are essential. Simultaneous background correction is often needed. A narrow bandpass (say, 0.2 nm) is required to separate the resonance line at 248.3 nm from its adjacent line, 248.8 nm; bent calibration curves are still observed. A slightly higher lamp current is advisable. Incidental standard additions checks should not be considered as overcautiousness.

In ETA, the characteristic concentration is between 0.003 and 0.6 ng/mℓ, and DLs down to 0.0007 ng/mℓ[59] (0.3 to 20 pg) are claimed.[57,59-61,294] Due to inevitable blank levels and random contamination, the practical DLs are worsened by orders of magnitude.

Table 1
IRON CONCENTRATIONS IN HUMAN BODY FLUIDS AND TISSUES

Sample	Unit	Selected reference values	Ref.
Bile	μg/mℓ	(0.3—3.8)	838
Blood	μg/mℓ	430 ± 38[a]	178
		434 ± 11	573
		(200—680)	96
		♀, (291—523); ♂, (450—630)	661
		447[b,c]	13
Bone	μg/g	90.6 ± 9.1	268
Brain	μg/g	52 (26—164)	662
CSF	μg/mℓ	(0.030—0.100)	763
		(0.20—0.80)	839
		0.47	52
Erythrocytes	mg/g	♀, 0.973 ± 0.134; ♂, 1.072 ± 0.122	840
		3.1 ± 0.09 (3.0—3.2)[d]	631
Feces	mg/day	♀, 11;[e] ♂, 15[e]	8
		22 (5—38)	96
Hair	μg/g	x̄, <10; (<10—2400) (Italy)	98
		x̄, 10; (11.1—300) (Japan)	11
		13 (7—28)	662
		15.3 (9.2—38.6)[d]	732
		♂, 18.0; ♀, 33.9 (Nepal)	666
		10.4—19[f]	100
		♂, 19.8;[g] ♀, 24.0;[g] (3.6—177) (New York)	288
		21.96 ± 11.47 (5.2—45) (U.K.)	368
		37 ± 13 (22—77)	841
		(<10—180) (U.S.S.R.)	102
		60 ± 38 (8.4—334) (India)	101
Kidney	μg/g	73 ± 10	268
		x̄, 79; 79 ± 7 (51—148)	105, 106
		88 ± 48	739
Kidney cortex	μg/g	41.6	107
		97.8 ± 10.2	3
Liver	μg/g	52	268
		120	462
		166	107
		176 ± 29	3
		183 ± 86 (42—252)	633
		205 (21—450)	108
		237 ± 95	275
		x̄, 246; 271 ± 56 (81—381)	105, 106
Lung	μg/g	128 ± 12	268
		174 ± 106	739
		293 ± 47	3
		x̄, 303; 390 ± 58 (122—1448)	105, 106
Milk	μg/mℓ	0.46	667
		0.72[h]	374
		0.84	736
		0.20—0.52;[i] (0.016—1.10)	14
		(0.26—0.73)	842
		(<0.1—1.6)	668
		(0.150—1.50)	13
		(0.016—2.54)	40
Nails	μg/g	27 ± 4	737
Nails (finger)	μg/g	♀, 38 (14—90);[d] ♂, 41 (28—109)[d]	738
		(16—200)[d]	16
Placenta	μg/g	731 ± 231[d]	377

Table 1 (continued)
IRON CONCENTRATIONS IN HUMAN BODY FLUIDS AND TISSUES

Sample	Unit	Selected reference values	Ref.
Plasma	μg/mℓ	1.2 ± 0.3 (0.37—1.69)	631
		1.1[b,j]	13
Saliva	μg/mℓ	(0.11—0.19)	380
		(0.05—0.10);[k] 0.2;[l,m] (0.—0.60)[l,n]	564
Serum	μg/mℓ	(0.50—1.50)	27
		1.09[b,o]	13
		(0.65—1.70)	673
		1.21 ± 0.78	111
Sweat	μg/mℓ	♀, 0.163 ± 0.029; ♂, 0.630 ± 0.587	381
		0.298 ± 0.085;[p] 0.412 ± 0.125[q]	843
		(0.020—0.450)	844
Tooth (dentine)	μg/g	2.18 ± 0.94;[r] 3.53 ± 3.16[s]	24
		42.8[d]	22
		70[d]	845
Tooth (enamel)	μg/g	2.77;[r] 5.82 ± 3.18[s]	24
		4.4 ± 0.95 (0.8—21)	25
		(8—40)[d]	845
		39.5[d]	22
		51 (11—759)	26
		34.2;[t] 86.3[u]	51
Urine	mg/day	0.131 ± 0.084 (0.051—0.350)	111
		♀, 0.20;[e] ♂, 0.25[e]	8
		(<0.1—0.447)	96

Note: Wet weight unless otherwise indicated. Mean ± SD (range in parentheses). x̃, median.

[a] μg/g (ppm).
[b] Weighed mean.
[c] N = 971.
[d] Dry weight.
[e] "Reference Man".
[f] Variations over the same head.
[g] Geometric mean.
[h] Pool.
[i] Range of mean values.
[j] N = 890.
[k] Parotid.
[l] Whole mixed saliva.
[m] Stimulated.
[n] Unstimulated.
[o] N = 2662.
[p] Cell free.
[q] Whole sweat.
[r] Permanent.
[s] Primary.
[t] African.
[u] European.

Some analytical lines of lower sensitivity are also used in ETA of Fe, when dilution is undesirable: 252.3 (2 ×), 372.0 (3 ×), 302.1 (3 ×), 305.9, and 386.0 nm, etc.[56,57,294,455,858]

III. AAS METHODS FOR ANALYSIS OF BIOLOGICAL SAMPLES

The flame AAS is currently a routine technique for analyses of serum and urine in many

Table 2
AAS TECHNIQUES FOR Fe DETERMINATION IN BIOLOGICAL MATERIALS

Technique	Biological materials
Flame AAS with or without dilution	Blood,[859,860] hemoglobin,[860-862] milk,[710,711] milk (evaporated),[712] plasma,[863] urine[864,865]
Flame AAS after deproteinization	Serum,[27,539,719,723,847,865-869] TIBC,[27,863,867,869] urine[27,869]
Flame AAS after digestion	Blood,[726,727] bone,[268,728] erythrocytes,[727] feces,[248,729] fish,[442,458,494,792,853] food,[591,729-731] hair,[368,662,732,733] meat,[517,730] milk,[110,374,538,729,730,750,842] milk (fractions),[842] nails,[692,738] tissues[190,268,442,462,517,520,538,541,662,729,730,739,740]
Flame AAS after TAAH solubilization	Plasma,[741] tissues[741,776]
Flame AAS after liquid-liquid extraction	Blood,[332] feces,[39] food,[39,659,690] milk,[745] serum,[640,657,719,870] tissues,[690] urine[39,871]
Microflame AAS	Serum,[700,701,872-874] tissues[760,761]
Direct ET-AAS with or without dilution	Blood,[762] CSF,[763] milk,[651] plasma,[763] serum,[769,866,875] urine[7,763]
ET-AAS after deproteinization	Milk,[775] serum,[7,27,57,858,866,875] serum TIBC[7,875]
ET-AAS after digestion	Hair,[876] tissues,[67,516] tooth (enamel/dentine),[24] tooth enamel (surface biopsies),[84] urine[773]
Direct ET-AAS (solid samples)	Hair,[78,79] tissues (NBS SRMs)[454,455]
Miscellaneous ET-AAS techniques (and procedures requiring special equipment)	Liver biopsies (iron proteins speciation),[877] serum proteins[786,787]

clinical laboratories. The use of ET-AAS seems to be rarely justified, perhaps with some iron-poor samples (CSF, saliva) or with small samples (plasma, serum). AAS applications are classified in Table 2, and selected procedures are briefly outlined in Table 3.

While iron determinations are generally straightforward, there are still some sources of errors which have to be considered: (1) the sampling step may be nonrepresentative with some specimens (e.g., small samples of brain,[740,776] lung,[856] nail,[16] etc., "spot" samples of feces,[96] food, sweat,[96] etc.); (2) minor contamination during specimen collection (due to steel needles and knives as well as storage containers) and more pronounced extraneous additions of Fe (due to airborne particles, reagent blanks, leaching from lab ware) during sample preparation step; (3) retention losses in dry-ashing techniques;[91,520] (4) minor inaccuracies in serum protein precipitation procedures; and (5) problems intrinsic with ETA: matrix effects, background absorption, random contamination (e.g., from disposable pipet tips).

Serum analyses can hardly do without a minimum of hemolysis; consequently, a few tenths of micrograms per milliliter of hemoglobin iron are typically counted up to the serum iron. The results of direct AAS assays ("total iron in serum"[875]) are thus persistently higher vs. the "serum iron"[875] obtained after a TCA precipitation of serum proteins.[657,719,866-868] All but 5% of the hemoglobin iron can probably be removed from the supernatant through this precipitation step, even with serum samples containing as much as 30 mg hemoglobin per 100 mℓ of serum.[867]

Different variants of protein precipitation are currently used: with[657,700,701,719,847,865] or without[27,657,719,858,867,869] previous incubation with a dilute HCl; with[27,657,719,858,867,869] or without[7,700,701,865,873] heating at 90°C for 15 to 20 min; and with[860,873,874] or without adding

Table 3
SELECTED AAS PROCEDURES FOR Fe DETERMINATION IN BIOLOGICAL SAMPLES

Sample	Brief procedure outline	Ref.
Blood	1 + 50 dilution with 0.5% Triton® X-100; flame AAS	860
Bone, meat, tissues	HNO_3 digestion; flame AAS;[a] standards in HNO_3	268, 517
CSF, plasma, urine	CSF or urine (~0.02 *M* in HCl), and 1 + 10 diluted plasma; ET-AAS: slow ramp up to 1200°C, atomization at 2500°C;[a] standard additions advisable	763
Feces, food, meat, milk	1- to 2-g sample heated with 15 mℓ of HCl-HNO_3-H_2O_2 (54:6:20, by volume) for 30 min at 100°C, and diluted; flame AAS;[a]s36standard addition calibration	729
Food, meat, milk, etc.	HNO_3-H_2SO_4 digestion; flame AAS;[a] standards in 5% v/v H_2SO_4	730
Hair, nails	20- to 100-mg sample ashed with 1 mℓ of HNO_3, finally[b] completed with 0.5 mℓ of $HClO_4$, and diluted to 5 mℓ; flame AAS;[a] standards in 10% v/v $HClO_4$	732, 738
Liver (small samples)	2-mg sample (dry weight) pressure decomposed with 40 μℓ of (5 + 1) HNO_3 + $HClO_4$, and stepwise (!)[b] heated for 3 hr at 90°C and for 3 hr at 120°C; cooled; diluted to 1 mℓ 75-μℓ aliquots pulse nebulized (dipping microflame AAS)[a]	760, 761
Milk	1 + 3 diluted with aqueous detergent (1% saponin or 0.4% Meriten (notyl-phenylpolyglycolether or 0.4% Na dodecylbenzenesulfonate); flame AAS; standards in detergent solution	710, 712
Serum	500 μℓ of serum + 250 μℓ of 1 *M* HCl incubated at 30°C for 45 min, 250 μℓ of 1.5 *M* TCA added, mixed, and centrifuged; 50-μℓ aliquots of the supernatant injected into an air-acetylene flame;[a] standards containing 0.25 *M* HCl, 0.375 *M* TCA, and serum electrolytes (K^+, Na^+, Ca^{2+}, Mg^{2+} from a Merck Titrisol®); likewise References 865, 873, and 874	700, 701
	50 μℓ of serum + 50 μℓ of 10% w/v TCA heated in an air-tight plastic tube at 90°C for 15 min, mixed, and centrifuged; ET-AAS: 100, 900, and 2400°C;[a] λ 302.1 nm; standards in 5% w/v TCA	858
Tissues	0.1-g sample solubilized with TAAH (1 mℓ of Soluene® 100 containing 2% w/v APDC) at 60°C for 24 hr; diluted with toluene; flame AAS;[a] standard addition calibration; see also Reference 777 (Soluene® 350 — MIBK procedure)	741
Urine	Flame AAS;[a] standard addition calibration	864, 865
	1:5 dilution with water; ET-AAS: 100°C for 40 sec, 800°C for 40 sec, and 2400°C for 6 sec;[a] standard addition calibration advisable	7

[a] Background correction.
[b] Safety precautions mandatory.

a reducing agent. Certain sources of minor inaccuracy ensue from TCA precipitation procedures, e.g., changes of the supernatant volume caused by evaporation[869] as well as protein exclusion from solution ("volume displacement error"[878]); minor changes of TCA concentration in the supernatant;[869] adsorption of Fe(III) anionic chloride complex on the positively charged precipitate;[27,878] depression of iron absorbance by TCA; changes in background absorption due to the simultaneous presence of TCA + NaCl + HCl;[879] and minor matrix interferences in the final AAS measurement step. Several practical hints may not be out of place when faced with these problems: using a lean air-acetylene flame and higher observation height; simultaneous background correction;[879] lowering the TCA concentration (e.g., <7%[27]); introducing serum electrolytes and TCA in calibration standards;[27,700,701] using an internal standard (e.g., 150 μg Mn/mℓ, in a lean air-acetylene flame, and with a dual-channel AA spectrometer[27,869]); and frequent standard addition checks.

The ashing step can be avoided in most analyses of biological liquids. Wet-digestion techniques would seem to give good results with most biological samples but need pure reagents and certain safety precautions. Decompositions with many common acidic mixtures have been described: HNO_3 and $HClO_4$ (feces,[248] hair,[732] nails,[738] tissues,[442,541]); HNO_3 and H_2SO_4 (tissues[7,190,730]); HNO_3 and H_2SO_4, completed by H_2O_2 (tissues[170]); H_2SO_4 and H_2O_2 (hair,[368] tissues,[341,541] urine[871]); HNO_3 and (finally) H_2O_2;[853] HNO_3, $HClO_4$ and H_2SO_4 (fish[792]); $HClO_3$ (brain and hair; gradual temperature increase essential!);[662] and HNO_3 vapor attack, completed with $HClO_4$ (blood).[726] Pressure decompositions in PTFE-lined bombs with HNO_3[760,761] or with HNO_3 and H_2SO_4[458,792] or with HNO_3 and $HClO_4$[494,760,761] have also been described. (*Caution!* Gradual temperature increase, limited simple size, and small amounts of $HClO_4$ have been used by original authors!).

Single-acid extraction and incomplete digestion are preferable owing to their simplicity, lower blanks, and "in-batch" performance with many samples simultaneously. Heating with HNO_3 (bone,[90,268,728] feces,[729] food,[729] hair,[666,876] meat,[517,729] milk,[729] tissues,[268,517,520] tooth dentine[24] and enamel,[24,50] etc.) or with HCl (tooth enamel[84]) or with HNO_3 + HCl (feces, food, meat, milk, etc.[729]) seems to have given satisfactory results.

In milk analyses, dissolution of evaporated milk or dilution of liquid milk with an aqueous detergent,[710,712] or else precipitation of milk proteins with a dilute H_2SO_4[775] proved to be suitable preparation steps prior to flame[710,712] or ET-AAS[775] assay of Fe.

So far, only two preparation techniques seem to be unsuitable for Fe: acid extraction of hair[466] and the NBS SRM "Bovine Liver"[467] with 1% HNO_3, as well as the decomposition of the "Bovine Liver" with HNO_3 + NH_4Br, both of which have given low results for Fe.

LTA is a clean ashing technique for serum,[167] tissues,[462,516,626] and urine;[773] still, Behne and Matamba[167] were able to prove small (−5%) but significant (P = 0.02) differences between LTA-INAA and INAA results for total iron in serum. Common dry-ashing procedures may give even lower results, due to retention of iron in silica, pyrophosphates, and other insoluble residue, as well as on crucible walls.[91,520] Volatization losses are improbable, at least up to 500 or 600°C.[91,534] Analyte recovery is improved by a thorough treatment of the ashes with acids (e.g., with H_2SO_4 and H_2O_2[462] or HCl, HNO_3, and H_2O_2[842]). Dry ashing is therefore recommended only with large and/or fatty samples.

TAAH solubilization provides a simple and unattended preparatory step and an additional sensitivity enhancement owing to the organic solvent diluent (MIBK, toluene, propanol); a standard addition calibration is usually involved.[741,776]

Preconcentration is rarely, if ever, needed in AAS assays of Fe. The extraction procedures cited below are usually steps of multielement extraction/determination analytical schemes: APDC-MIBK at pH 2.5 from dry-ashed milk;[745] DDDC-heptane-2-one at pH 4.5 ± 0.2 from ashed food;[690] HHDC-n-butylacetate at pH 6 from TCA supernatant of serum;[640,657,719] HHDC-$CHCl_3$ at pH 6 from ashed biological samples (food, etc.);[659] HHDC-8-hydroxyquinoline-$CHCl_3$-pentanol at pH 4.8 to 5.0 from ashed feces, food, and urine;[39] bathophenantroline-MIBK from TCA supernatant of urine;[870] cupferron-MIBK from wet-ashed blood (2 *M* HCl);[332] and TTA-MIBK at pH 2 from wet-ashed urine.[871] Extraction times for iron may be somewhat prolonged (e.g., 2 min),[427] and nonchlorinated organic solvents should be used.[429,880]

More involved (chromatographic) separations are resorted in speciation studies of Fe in milk,[750,842] liver (biopsy) protein fractions,[877] etc.

In ET-AAS, dilute mineral acids — HNO_3 or H_2SO_4 or HCl — are tolerable, while $HClO_4$ should better be avoided as a severe depressant.[66] Depending on the particular acid and matrix involved, as well as on heating rates and pyro-coating of the graphite tube, the ashing temperatures are up to 800 to 900°C (with TCA or HCl present)[7,773,858] or even to 1100°C, and atomization temperatures are generally above 2300°C. Pyro-coated tubes are advantageous in view of lower atomization temperatures and higher (2 to 3 $\times$) sensitivity.[60] Ramp drying and ashing, ''maximum power'' heating, simultaneous background correction, and matrix-matched standards are typical and recommended. Autosampling devices and peak-area calibration[866] are highly praised as well.

IV. CONCLUSION

Most iron assays can be performed by simple and reliable flame AAS procedures. However, TCA precipitation of serum proteins is required. Microscale pulse-nebulization procedures are preferable. Spectrophotometry is an established alternate technique in many clinical labs. Possible source of inaccuracy: random contamination and reagent blanks, inadequate calibration standards (e.g., viscosity discrepancies, etc.), as well as retention losses during dry ashing. There are problems with the ET-AAS of Fe.

Chapter 17

LEAD

I. DETERMINATION OF LEAD IN BIOLOGICAL SAMPLES

Lead levels in different biological specimens vary broadly, being as low as subnanograms per milliliter in CSF and plasma and up to scores of micrograms per gram in bone, hair, nails, and teeth (Table 1).

Few analytical methods are useful in the low nanograms-per-gram range (ET-AAS, ASV/DPASV, IDMS, and SSMS), while some other, less sensitive methods can also be applied to higher, micrograms per gram, concentration levels (AAS, spectrophotometry, AES, XRF, etc.).

ET-AAS and ASV/DPASV, with DLs down to 0.01 to 0.1 ng/mℓ, are the best techniques for routine determinations of trace lead in biological samples (liquids). ASV or DPASV has been applied to blood,[361,391,683,897] bone,[392] food,[250] hair,[393] milk,[250,394,898] nails,[133] teeth,[395] tissues,[133,250,899] and urine.[124,125,396,398,684] These procedures are quick and suitable for screening,[124,125] but the analyte should be released from its chemical bonds with the matrix[683] and chelating agents,[398] usually done by complete oxidation of the organic matter.

The flame AAS, often coupled with preconcentration, is routinely used in many laboratories.

Dithizone spectrophotometric procedures require purified reagents and several steps of digestion, masking, extraction, etc.[414,882,900-902]

The arc-source AES is a classical method for Pb in biomaterials: blood,[36,37,130] bone,[38,902] CSF,[903] erythrocytes,[37] food,[659] milk,[14,40] saliva,[41] sweat,[21] teeth,[26] tissues,[43,44,130,904] urine,[37,130,320] and other biomaterials.[41,42]

Due to inadequate sensitivity, XRF[128] is of limited use, e.g., in some multielement analyses of dried or ashed bone,[851] blood,[691] and tissues.[128,636] The PIXE technique[128] is unique in studying the Pb profiles in limited areas of bone,[97] tooth enamel,[852] single hairs,[128,195] etc., as well as small samples of blood,[697] erythrocytes,[698] tissue,[407] etc.[128]

A few analytical laboratories are able to perform accurate and sensitive IDMS determinations of Pb in blood,[410,888,905] CSF,[881] hair,[410,905] nails,[410] plasma,[881,888] etc. Another powerful technique, SSMS, has been applied to multielement analyses of blood,[3,52] bone,[3] hair,[54] nails,[16] plasma (ash),[55] teeth,[25] and tissues.[3,132,275,409]

Few applications of PAA,[48,104,184,902,906] the flame AFS,[907] and ICP-ES[30,32,326,348,392,411,636,908] have been described (likewise the recent reviews[909-911]).

II. ATOMIC ABSORPTION SPECTROMETRY OF LEAD

Flame AAS is remarkably selective and fairly sensitive. The characteristic concentration is between 0.08 and 0.7 μg/mℓ, and DLs are down to 0.01 to 0.04 μg/mℓ.[56-59,294]

EDL is the favored light source, owing to its higher intensity (60 ×), lower DLs (7 ×), and longer lifetime vs. the HCL.[138] Two resonance lines can be used — 217.0 and 283.3 nm; the second line, although less sensitive (~2.5 ×), is usually preferred for its (1) higher intensity, (2) better signal-to-noise ratio, (3) broader linear working range, (4) lesser background absorption, and (5) (often) lower DLs in real matrices.

No chemical interferences in an air-acetylene flame seem to have been reported; only bulk-matrix effects are to be expected, e.g., transport interferences, effects of excess dissolved solids, nonvolatile acids, etc., which are easily checked and accounted for. Background absorption, especially at the 217-nm line, should be expected. Acid-matched standards are usually appropriate, although in presence of excess nonvolatile constituents (sulfates,

Table 1
LEAD CONCENTRATIONS IN HUMAN BODY FLUIDS AND TISSUES

Sample	Unit	Selected reference values	Ref.
Blood	μg/ℓ	♀, 96 ± 25 (76—142)	675
		♂, 160 ± 49 (111—295)	675
		117 ± 55[a]	881
		125 (70—190)	361
		♀, (67—132); ♂, (77—179)	661
		158 ± 40	359
		♀, 160 ± 100 (40—410)	882
		♂, 200 ± 120 (30—790)	882
Bone	μg/g	2.85 (2.4—4.9)[b]	97
		2.8 ± 0.3	268
		x̄, 25;[c,d] 38 (9—253)[c,d]	851
		♀, 6.77 ± 5.08 (0.85—22.6)[e]	882
		♂, 8.85 ± 5.81 (0.90—28.5)[e]	882
		♀, 15.99 ± 12.49 (1.50—48.0)[f]	882
		♂, 23.40 ± 15.96 (3.00—73.0)[f]	882
		♀, 16.46 ± 13.08 (6.00—54.2)[g]	882
		♂, 20.17 ± 16.00 (3.90—79.0)[g]	882
Brain	μg/g	0.3 ± 0.1	3
CSF	ng/g	0.53 ± 0.44	881
Erythrocytes	μg/ℓ	371 (116—700)	883
Feces	μg/g	0.28	365
Hair	μg/g	x̄, 4.5 (0.1—30) (Denmark)	884
		x̄, 3.7,[h] 5.0,[i] 7.0,[j] (Denmark)	884
		1.5, 6.1, and 8.9[k] (U.S.A.)	180
		10.1 (0.50—30) (Canada, rural)	104
		16.9 (0.50—40) (Canada, urban)	104
		12.8 (1.2—110.9) (New Zealand)	885
		♀, 10.97;[l] ♂, 13.95;[l] (1.96—155) (New York)	288
Hair (beard)	μg/g	2.09 ± 0.99	886
Kidney	μg/g	x̄, 0.23; 0.31 ± 0.12 (0.10—0.60)	105, 106
		0.47 ± 0.03	268
		♀, 0.46 ± 0.29 (0.11—1.31)	882
		♂, 0.64 ± 0.32 (0.14—1.40)	882
Kidney cortex	μg/g	♀, 0.55 ± 0.39 (0.10—2.20)	882
		♂, 0.78 ± 0.38 (0.15—1.85)	882
		2.25 ± 1.20[m]	373
Liver	μg/g	x̄, 0.28; 0.37 ± 0.12 (0.21—0.55)	105, 106
		0.60 ± 0.05	268
		0.67 ± 0.39	275
		♀, 0.66 ± 0.38 (0.19—1.72)	882
		♂, 1.03 ± 0.62 (0.18—3.13)	882
		1.06 ± 0.57 (0.41—2.5)	319
Lung	μg/g	0.009	268
		x̄, 0.072; 0.09 ± 0.05 (0.0045—0.17)	105, 106
		♀, 0.22 ± 0.12 (0.04—0.55)	882
		♂, 0.22 ± 0.11 (0.05—0.59)	882
		x̄, 0.68;[m] 0.87 ± 0.56 (0.10—3.04)[m]	887
		♀, 0.57;[m] ♂, 1.04[m]	887
Milk	μg/ℓ	<10[n]	374
		12 (6—22)	736
		(12—30)	13
Muscle	μg/g	0.05 ± 0.03 (0.01—0.23)	882
Nails	μg/g	(1.7—24);[m,o] (5.3—64)[m,p]	16
Placenta	μg/g	1.83 ± 2.56[m]	377

Table 1 (continued)
LEAD CONCENTRATIONS IN HUMAN BODY FLUIDS AND TISSUES

Sample	Unit	Selected reference values	Ref.
Plasma	μg/ℓ	0.02[q]	888
		1.40 ± 0.68[a]	881
		2.8 ± 1.1 (1.4—5.7)	889
Saliva	μg/ℓ	31[r]	890
Skin	μg/g	♀, 0.15 ± 0.11 (0.03—0.42)	882
		♂, 0.19 ± 0.14 (0.01—0.60)	882
Sweat	μg/ℓ	♀, 53 ± 12; ♂, 62 ± 40	381
		♂, 51 ± 42 (8—184)	675
		♀, 118 ± 72 (49—283)	675
Tooth (dentine)	μg/g	0.46 ± 0.25; 0.71 ± 0.82	891
Tooth (enamel)	μg/g	3.6 ± 0.24 (1.3—6.5)	25
		8.5 (<2—1000)	26
		19.6	892
		69.9 ± 68.4; 81.4 ± 44.6	891
Tooth (whole)	μg/g	7.29[m]	383
		11.1 ± 14.8 (<2—80)[s]	893
		(1—110)[c]	894
Urine	μg/ℓ	11 (up to 30—40)	895
		16 ± 9.4	359
		16 (8—40)	362
		(1.5—29.4)	896
	μg/day	(7—31)	365
		18 (9—29)	320

Note: Wet weight unless otherwise indicated. Mean ± SD (range in parentheses). x̃, median.

[a] ng/g (ppb).
[b] Femur.
[c] Ash.
[d] Vertebral.
[e] Rib.
[f] Tibia.
[g] Calvaria.
[h] Grey.
[i] Blond.
[j] Dark.
[k] Mean values of three towns with low environmental exposure.
[l] Geometric mean.
[m] Dry weight.
[n] Pool (New Zealand).
[o] Adults.
[p] Children.
[q] N = 1.
[r] Parotid.
[s] Deciduous.

phosphates, Si, Fe) minor depression may persist and hence may involve a more complicated calibration.

The microflame accessories, the Delves cup[415,912] and the Sampling Boat®,[913] provide DLs down to 1 ng Pb/mℓ (1 and 0.1 ng absolute, in a 100-μℓ and 1-mℓ sample aliquots, respectively). These devices have been widely used in the last decade, in spite of drawbacks (matrix effects, background problems, sensitivity drift, involved calibration, etc.). See also the techniques of loop-flame AAS,[914] pulse-nebulization long-path tube,[417] and slotted quartz tube.[442]

ET-AAS determinations are very sensitive; the characteristic concentration is between 0.15 and 1 ng/mℓ (3 to 40 pg), and DLs are down to 0.007 to 0.05 ng/mℓ (a few picograms absolute).[57,59-61,294] This technique is, however, not as straightforward as generally expected, especially with older models of graphite furnaces.

ETA gives rise to several problems: (1) matrix effects; (2) premature volatization; (3) background absorption; (4) accumulation of nonvolatile residues; (5) sensitivity drift; (6) bending of calibration curves; and (7) prolonged temperature programs. Most of these problems are due to the volatility of lead species (Pb, PbO, PbX, organoleads, etc.).[915] Thus, the ashing temperatures (which are strongly matrix dependent) are often kept as low as 350 to 550°C. Few matrix constituents can be expelled from the graphite tube at these relatively low temperatures; most concomitants will thus evaporate together with the analyte during the next atomization stage giving rise to severe background absorption and interferences (occlusion, fractional volatization, vapor-phase reactions, etc.), as well as to sensitivity and precision impairment, and other adverse effects. Among the strongest suppressants are the alkali and alkali-earth halides: $MgCl_2$,[916,917] $CaCl_2$,[917,918] NaCl,[915-917,919] etc. The extent of signal depression is less pronounced in presence of NO_3,[223,330,432,915,917,920] PO_4^{3-},[457,787,921-923] EDTA,[924] organic acids[916] (ascorbic,[802] tartartic, oxalic, citric, maleic), NH_4^+,[223,362] MoO_4^{2-},[513,922,923] La^{3+};[438,513] accordingly, some of these salts are used as constituents of the spectrochemical buffers for Pb.

Inasmuch as most acids interfere with ETA, their concentration should preferably be kept low (<5% v/v) and be matched in sample and standard solutions. Generally, the best acidic medium is a dilute HNO_3 or H_3PO_4. These acids as well as their ammonium salts — either by themselves or in combination with surfactants and/or other reagents — are often used as analyte/matrix modifiers, aimed at thermally stabilizing the analyte and partly eliminating interferences. Wherever possible, H_2SO_4, $HClO_4$, and HCl should be avoided. Acid- and matrix-matched standards, as well as frequent standard additions and/or peak-area calibrations are typical.

The background correction efficiency is a major concern in ET-AAS of Pb. It is essential to thoroughly align and balance the deuterium arc corrector, and to verify whether it is capable of correcting the background, which may be high (0.5 to 1 A), transient, and structured; otherwise recourse should be had to a Zeeman correction (preferable at nanograms-per-milliliter levels) or other approaches. Often (though not always[919]), the background contribution is lower at the 283-nm line. The background may prove intolerable if the temperature program and/or the purge-gas flow are not properly optimized. This may happen, e.g., under gas-stop conditions; at higher atomization temperatures; at an improper temperature ramp during the atomization stage; after an unsuitable dry/ash stage; etc.

Background absorption can be effectively lowered by adding matrix/analyte modifiers; HNO_3 or NH_4NO_3, in particular, depress the NaCl background.[223,915,916]

The problems of the lead ET-AAS are best solved by radical developments in instrumentation: temperature-controlled heating with as fast heating rates as possible during the atomization step;[60,509,925] use of the L'vov platform;[229,454-457,553,919] use of a Zeeman correction;[373,453,917,926-929] pretreated graphite tubes or platforms[457] (Ta + pyro-coating,[457] Mo,[916,929] Mo + La,[513] Mo + H_3PO_4,[922,923] W,[930] etc.); and, finally, with especially designed furnaces which approximate the constant-temperature conditions of the classical L'vov cuvette.[452,509,553,928,931] Clever combinations of these approaches with matrix modification and peak-area measurements, e.g., the STPF concept of Perkin-Elmer®,[553] are commendable. Eventually, the application of ETA in speciation studies of organoleads is worth mentioning.[284,932]

HG-AAS is not a popular technique for trace lead, because of its limited sensitivity and pronounced matrix effects. The experimental conditions: the concentrations of $NaBH_4$[140] and acids,[140,150,933] and the presence of certain matrix or buffer components [Cu, Fe, Ni, tartrate, acetate (depressants) and H_2O_2, HNO_3, Cr(VI) (enhancement agents)][145,150,933,934] are critical. Three reportedly optimal media for plumbanne (PbH_4) generation are as follows: 0.5% HNO_3 + 3% H_2O_2,[150] 0.5 *M* HCl to 0.8 *M* H_2O_2,[934] and the mixture of 1% v/v HCl (10 mℓ) + 3% w/v $K_2Cr_2O_7$ (4 mℓ) + 7.5% w/v tartaric acid (6 mℓ).[933] Standard additions calibration is advisable.

III. AAS METHODS FOR ANALYSIS OF BIOLOGICAL SAMPLES

AAS is currently the most popular method for routine determinations of Pb in biological samples. The literature on this subject is ample, and thousands of laboratories throughout the world are able to perform these assays. All the same, much attention has to be paid to all steps: specimen collection, storage, handling, ashing, preconcentration, instrumental parameters, calibration, etc., in order to obtain reliable results. It seems quite probable that at least part of the staff engaged in one or another step of these assays does not always recognize that this routine is in fact trace analysis, with its problems and demands.

In the specimen-collection step, there are at least two major points: (1) to obtain a representative sample and (2) to avoid exogenous contamination.

The analyte is quite nonhomogeneously distributed within certain specimens, e.g., between blood fractions, along single hairs,[779,884,885,935,936] between dark and grey hairs,[937] and within teeth and bone.[22,891,893] "Spot" samples of capillary blood, urine, feces, food, tooth enamel, etc. may well prove nonrepresentative.

Samples of capillary blood can be contaminated by lead derived from skin surface (in exposed workers);[938-940] therefore, a thorough washing of the sampling site with soap and brush, and with dilute aqueous solutions of HNO_3 and EDTA,[938-940] as well as coating of skin surface with collodion,[940,941] are practiced. The collection devices: evacuated tubes,[420,942] rubber stoppers,[420,888] anticoagulants,[888] urine conservants,[924] etc. should always be checked for their Pb blank contribution. Everson and Patterson[888] recently described an ultraclean sampling and analysis of plasma lead which seems difficult to follow in most ordinary labs; the contribution of heparin lead was 3 to 6 ng Pb per 1 to 6 g of plasma, i.e., an estimated error of $\pm 300\%$ at endogenous lead-in-plasma levels. Moreover, as erythrocyte lead is much higher (by two or three orders of magnitude) than plasma lead, it should be very difficult, if at all possible, to obtain plasma or serum samples with no contribution from erythrocyte lead, owing to inevitable hemolysis.[888,889] Serious contamination may also be involved in sampling and storage of CSF, saliva, and milk specimens.

Hair samples usually need a mild washing with nonionic detergent and/or organic solvent. The other two extremes, the "no-wash" approach[393,625] and rigorous washing with an aqueous EDTA[180,943] or with dilute acids,[466] cannot be justified, except in comparative studies (washed vs. unwashed specimens). The first 1-cm segment close to the hair root is preferably analyzed.[884,935]

Certain ordinary grinding procedures (in agate, corundum, etc.[624]) may introduce intolerable extraneous additions; therefore, clean homogenation techniques (often, in PTFE-coated mills at a liquid nitrogen temperature, variants of the "brittle fracture" technique[532]) are adopted.[426,533,893,944]

During storage of biological liquids, both contamination and losses of lead may occur. Conservants used (heparin, EDTA, citrate, acids, etc.) have to be prechecked for their lead content, as well as for their possible interference with the following analytical steps.

Lead is readily absorbed on glass surfaces[420,421,945-947] and, to a lesser extent, on plastics.[421,945,946]

A simple cleaning and silanization procedure for glass tubes has been described by de Haas and de Wolff:[947] the tubes have been cleaned with detergent, distilled water, and APDC-MIBK extraction, dried, and sprayed with 10% v/v dichlorodimethylsilane in toluene.

In aged urines, lead tends to coprecipitate with calcium phosphate;[396,911] hence, such samples should be vigorously shaken, acidified, or briefly heated to 60°C,[948] in order to recover the lead before analysis.

Blood samples, when stored at room temperature, should be analyzed within 1 or 2 days;[949] otherwise the accuracy and precision of the assay are significantly impaired.[950] For longer storage (e.g., for a few days to 1 or 2 weeks), blood specimens are kept at 4°C. The time stability of lead has been shown to depend on such variables as the individual characteristics of the specimen (!),[897] the anticoagulant added,[531,951] the container material,[421,946] the pretreatment applied (e.g., sonification,[531,952] γ-irradiation[531]), as well as on whether the behavior of the endogenous lead or a lead spike[421] is concerned.[949,953] Pozzoli et al.[27] stored blood samples at 5°C for 1 or 2 days. Matsumoto et al.[951] stored heparinized blood for 15 days, while samples containing heparin + K_2 EDTA proved stable for 30 days. Méranger et al.[421] found important losses of lead from all container types within 50 hr; heparinized, spiked porcine blood was stored at four temperatures (−70, −10, 4, and 22°C) and in six different container types (Pyrex®, soda glass, polyethylene, polypropylene, polystyrene, and polycarbonate).[421] Sabet et al.[953] found spiked blood to be stable for less than 2 weeks, while samples containing only endogenous lead were stable for many weeks. Stoeppler et al.[531] stored and incidentally analyzed a sample of citrate-stabilized blood at 5°C for up to 3 months; the results of 12 successive assays gave 88.9 ± 7.14 (SD) ng Pb/g. These authors were able to store blood samples at 4°C for up to 12 months, provided the blood specimens had been pretreated by EDTA stabilization and sonification and γ-irradiation. Eventually, Morrell and Giridhar[897] found some blood specimens to be stable for 6 weeks, while others spoiled within a week under the same storage conditions (4°C)!

For long-term storage, samples of blood,[531,954,955] urine,[684] and tissue[106] should be frozen soon after collection, and then stored at $\leqslant -20$°C. An even better approach is to lyophilize specimens (blood,[955] urine,[396] tissue,[106]). Blood samples can also be dropped onto a piece of filter paper, dried, and stored thereon until analyzed.[419,954,956-958] Acidic digests in clean (plastic) containers are stable (e.g., for weeks[426,959]).

AAS procedures for the determination of Pb in biological materials are classfied in Table 2. Typically, the analyte is quantitatively put into solution (although there are a few direct applications of ET-AAS to solid microsamples).

The classical approach is to completely oxidize the organic matter by dry or wet ashing.[91,266] Dry ashing is attractive for its simplicity, unattended performance, low blanks, and capability of handling large and fatty samples. Both volatization and retention losses are to be expected, unless ashing temperatures are kept below 500°C and the ashes are adequately treated (e.g., evaporated with HNO_3 or even with aqua regia).[370,537,538]

Dry ashing appears to be a convenient technique for bone,[976] milk,[374,394,479,537,898,977] and teeth.[481,894,976] Retention losses are more pronounced with silica and porcelain crucibles, depending on the crucible material and age, and the particular matrix, as well as on the presence of ashing aids.[91] Several ashing aids have been applied: $Mg(NO_3)_2$ (cocoa,[91] fish,[271,442] food,[271] milk[1032]); $Mg(CH_3COO)_2$ (tissues[190]); HNO_3 (blood,[971] feces,[39] fish,[442] food,[39,537] milk,[537,1033] teeth,[50] tissues,[442] urine[39]); H_2SO_4 (food and tissues[537]); K_2SO_4 in HNO_3 (blood,[971] feces,[971] food,[971] milk,[538] tissues,[538,971] urine[971]). During dry ashing, the samples may yet be contaminated by extraneous lead derived from (porcelain) crucibles, airborne dust, ashing aids, and acids.[971a]

After LTA in excited oxygen,[169] lead has been quantitatively recovered in samples of blood,[169,438,493,1037] liver,[275] lung,[516] and NBS SRMs ("Coal", "Orchard Leaves"),[626] but not in mussels.[442] This technique is slow and inefficient with samples of bone and teeth.[3]

Oxygen flask combustion has been applied to samples of hair,[288] nails,[133] skin,[133] etc.; possible alloying of Pb with the Pt sample support has been noted.[91] Combustion of large

Table 2
AAS TECHNIQUES FOR Pb DETERMINATION IN BIOLOGICAL MATERIALS

Technique	Biological materials
Flame AAS after digestion	Blood,[901] fish,[334] food,[537] hair,[370,460,461,885,935] milk,[374,736] teeth,[894] tissues,[268,463,538,739] tooth enamel[50]
Flame AAS after TAAH solubilization	Hair,[464] tissues[464]
Flame AAS after acid extraction	Liver[319]
Flame AAS after liquid-liquid extraction	Blood,[7,27,57,190,332,468,471,901,949,960-975] bone,[902,971,976] erythrocytes,[886,965] feces,[39,248,971] feed,[248] fish,[271,474,475,477,548] food,[39,476,477,480,970] hair,[478] milk,[477,479,480,898,977] placenta,[377] serum,[27,640,657] sweat,[675] teeth,[481,976] tissues,[7,43,190,271,429,477,541,971,972,978] urine[27,39,339,479,483,960,962,964,965,971,975,979-981]
Miscellaneous flame AAS techniques	Blood,[546,935,937,952,982] fish,[442] food,[591] mussels,[983] tissues,[442,937,984,985] urine[27,546,914,935,937,986-988]
Delves cup	Blood,[27,486,487,912,940,941,953,956,958,989-1001] hair,[886] milk,[1002-1004] saliva,[890,1005] tissues (homogenates),[1006] urine[948,996,1000,1007]
Sampling Boat®	Blood,[27,493,901,1008,1009] fish,[494] food,[495] milk,[495,1004] mussels,[1010] urine[27,913,1008,1009]
Direct ET-AAS with or without dilution	Blood,[27,57,330,421,438,500-502,511,762,883,920,925,938,947,950,953,954,957,973,1011-1028] bone,[450,1029] erythrocytes,[883,1017,1023] fish meal (ash),[505] hair,[78,79,649,778,779,884,936] liver,[505,781] milk,[514,1003] mussels,[86] serum,[432] teeth,[451] urine[27,330,438,511,911,920,922-924,929,1011]
Zeeman ET-AAS	Blood,[453,686,917,921] kidney cortex,[373] liver,[453,686,928] plasma,[926] urine[917,921]
ET-AAS after digestion	Blood,[362,438,686,901,1031] bone,[97,959] feces,[365,959] food,[365] hair,[426,885,954] milk,[514,1032,1033] tartar,[1034] tissues,[67,256,513,516,518,519,686] tooth enamel,[84] urine[362,365]
ET-AAS after acid extraction	Blood,[7,512,1035-1037] hair,[426,1035] meat,[517] milk,[394] tissue[517]
ET-AAS after TAAH solubilization	Blood,[521,1017-1019] erythrocytes,[1017] hair,[368,521] liver,[1038] placenta,[1038] tissues[521,1038]
ET-AAS after liquid-liquid extraction	Blood,[7,359,419,528—530,897,911,947,1014,1039-1042] hair,[426] milk,[333] plasma,[889,1043] tissues,[7,528] urine[7,359,419,530,531,1020,1039]
ET-AAS with L'vov platform	Blood,[457] tissues,[454-457] urine[457]
Miscellaneous ET-AAS techniques (and procedures requiring special equipment)	Blood,[931,1012,1044,1045] erythrocytes (speciation),[1046] fish (tetraalkyl lead),[1047-1049] liver,[446,452,928] "Oyster Tissue" (NBS SRM),[509] serum protein fractions (of marine crabs),[787] tissue (tetraalkyl lead),[899] tooth enamel (biopsies),[1050] urine[931]

organic samples (up to 20 g of food, meat, milk, tissues, etc.) in the Siemens BIOKLAV® has been described by Scheubeck et al.[273,274,540]

Wet-digestion techniques are more useful and universally applicable to trace lead analyses. There are several points to consider: (1) reagent blanks[971a] should be kept low by using high-purity reagents and reflux conditions; (2) clean lab ware materials (quartz, PTFE, etc.) are preferable to glass or porcelain; (3) partial losses due to coprecipitation (with $CaSO_4$[91]) or adsorption of the analyte are to be expected with H_2SO_4-containing mixtures, as well as with certain Ca-rich (bone, milk, teeth) samples; (4) safety precautions should be taken with procedures involving $HClO_4$, H_2O_2, or pressure; and (5) the final acidic digest medium should be compatible with the following analytical step. Some important parameters of wet-digestion techniques — speed, completeness of digestion, reagents consumption, blank contribution, etc. — can be improved by using microwave-assisted digestion,[277,446] closed pressurized vessels,[391,518] or reflux conditions,[684,959] as well as close temperature control.[256,541] Freeze-drying urine samples prior to HNO_3-$HClO_4$ decomposition, Golimowski et al.[396] found the acids consumption and digestion time were lower by a factor of 5 and 3, respectively, vs. the digestions of intact urines; blank levels also dropped substantially.

Organic matter can be completely decomposed by using the acidic mixtures: HNO_3 + $HClO_4$ (blood,[901,935] feces,[248] fish,[442,474,494] hair,[935,954] tissues,[256,442,535,541] urine,[396,1051] etc.);

HNO_3 + $HClO_4$ + H_2SO_4 (blood,[332,391,542] food,[591] tissue,[43,190,984,985] urine,[684] etc.); and HNO_3 + H_2O_2 (blood,[528,530] hair,[885] tissues,[505,528] urine,[530] etc.).

The trend now is to use simpler and faster procedures, based on single-acid treatment of samples. Although the organic matter is not completely destroyed with most of such treatments, the analyte is extracted into an acidic solution (often HNO_3) which is suitable for either direct injection into a graphite tube or, alternatively, for liquid-liquid extraction of the released lead (from HCl or TCA solutions).[27,893,963,964] A few typical examples are worth giving:

1. Protein precipitation techniques: precipitation of blood[27,962,968,969,1008] and urine[960,962,964] proteins with TCA[27,958,962,968] or $HClO_4$[27,963,964] or TCA + $HClO_4$.[963,1052,1053] There are conflicting reports on the efficiency of lead extraction by means of these treatments,[27,1053] and it appears that some analyte losses (e.g., 5 to 30%) may occur, depending on some experimental factors: acid concentration, incubation time, heating, additional washing of the precipitate with one or two portions of dilute acid or water, etc. Stoeppler et al.[1036,1937] and Jönsson[394] were able to quantitatively release lead in protein-free filtrates of blood[1036,1037] or milk[394] by treatment with 2 *M* HNO_3 [1036,1037] or 2 *M* HCl,[394] respectively.
2. Extraction with dilute acids or single-acid digestion: HNO_3 (blood,[726,901,1031] bone,[90,959] feces,[959] food,[250] hair,[466] meat,[515] teeth,[22,50,395] tissues,[373,446,456,515,518,519,545,1054] urine,[363] etc.); $HClO_4$ (blood,[391] teeth,[893] and urine[398]); HCl (tooth enamel[84]); and HCl-HNO_3 (liver[319]). While these procedures usually leave some organic residue (fat, etc.) not quite oxidized, the analyte is almost completely transferred into an acidic solution. The efficiency of HNO_3 digestion can be improved and the blank contribution lowered by using PTFE-lined bombs[250,395,518,545,686] or tightly capped tubes/vessels[983,1030,1031] (caution!).

Biological samples can also be solubilized in TAAH solutions (aqueous,[464,1017] alcohol,[521] toluene,[368,1038] Lumatom®, Soluene®, etc.), and then diluted and directly sprayed into a flame (hair, tissues)[464] or injected into a graphite-tube furnace (blood,[521,1017-1019] erythrocytes,[1017] hair,[368,521] tissues[521,1038]). Homogenates can be stored for 2 or 3 days. The preparatory time, however, is somewhat long (2 to 24 hr), and a standard additions calibration is called for.

Enzymatic digestion was recently applied to liver and kidney tissue.[463]

Preconcentration of the analyte is a usual analytical step in many flame AAS procedures. Liquid-liquid extraction is a relatively simple, fast, and convenient procedure; hence, many small laboratories prefer extraction-flame AAS procedures to graphite furnace analyses.

Lead is readily extracted from hemolyzed blood;[7,27,57,190,359,468,471,486,965-967,972,974,989,1041,1042] pH-adjusted urine;[7,339,359,483,529,972,975,979,981,987,1020] protein-free supernatants of blood,[960-962,964,968,969] serum,[640,657] and urine;[960,962,964] and from pH-adjusted biological digests. Extracts can be directly sprayed in a very fuel-lean air-acetylene flame or injected into a graphite furnace or Delves cup[27,494,496,901] or Sampling Boat®[27,494,496,901] (Table 2).

The extraction techniques for lead in biological samples are summarized in Table 3. Lead is mostly extracted as dithiocarbamate, dithizonate, or as an ion-association complex. In dithiocarbamate extractions, the limited time stability of the extracts[480,965] should not be neglected, nor the potential interferences from excess Fe(III),[377,427,480,541,971] Cu^{2+},[480,971] Cd^{2+},[971] Zn^{2+},[480,971] NO_3^-, and PO_4^{3-}.[949] Hence, addition of masking agents is required in some extraction procedures, e.g., ascorbic acid (→ Fe(II)[377,480]), citrate,[947,949] and citrate + KCN.[971,1042]

There are several common problems encountered in extraction techniques: (1) chelating agents such as EDTA[7,27,529,911,960,965,972,979,980] and penicillamine[965] depress the extraction

Table 3
EXTRACTION SYSTEMS FOR Pb FROM BIOLOGICAL LIQUIDS/DIGESTS

Extraction system	Matrix and conditions[a]	Ref.
APDC-MIBK	Hemolyzed blood; Triton® X-100 often added; no pH adjustment	27,190,232, 359,468,486, 911,966,967, 970,974,989, 1041
	Hemolyzed blood; pH 5.0 (citrate)	947
	Hemolyzed blood (KCN + NH_4 citrate, pH 8.5, 30 min at 50°C)	1042
	Plasma; no pH adjustment	27,889,1043
	Blood and urine; pH 2—2.5[b]	10,39
	Urine; pH 5	359
	Blood, erythrocytes,[c] urine (formamid added)	965
	TCA-deproteinized blood (second washing the ppt with aqueous TCA[962,975] or water);[968] pH 2.2—3.0[b]	960,962,968, 969,975
	TCA-deproteinized urine; pH 2.2—2.8[b]	960, 962
	Deproteinized blood (by TCA, thrice; or by 6 *M* $HClO_4$, with two subsequent washings the ppt with water); pH 3.0[b]	27
	$HClO_4$-deproteinized blood/urine; pH 1.0—2.8[b]	964
	Ashed or pH-adjusted urine; pH 2.2—3.5[b]	27,483,975
	Wet-ashed tissues; preextraction of Fe(III)	541
	Wet-ashed fish/liver; 4% v/v H_2SO_4[b]	475
	Ashed placenta; pH 3.0; ascorbic acid added	377
	Ashed tissues; pH 7.0	190
	Ashed food, fish, tissues, etc.; pH 3—4 (acetate)	271
	Wet-ashed blood; pH 2.6—3.0[b]	901
	Wet-ashed blood; pH 8.5 (citrate + KCN)	901
	Wet-ashed urine	980
	Wet-ashed feces/feed; pH 4.0 (citrate)[b]	248
	Ashed blood, feces, food, tissues, urine, etc.; pH 8.5 (citrate + KCN)	971
APDC-butylacetate	Dry-ashed milk; pH 5.4	898
APDC-2,4-dimethyl-6-heptanone	$HClO_4$-digested teeth; pH 2.2—2.8[b]	893
APDC-2-heptanone	Wet-ashed or pH-adjusted urine; pH 3—4[b]	339,981
	Dry-ashed milk or urine; pH 4.5	479
APDC-xylene	Deproteinized urine (2 *M* $HClO_4$)[b]	963
APDC + DDDC-MIBK	Wet-ashed blood, fish, milk, etc.; 5% v/v H_2SO_4[b]	477
DDDC-MIBK	Dry-ashed food, milk, etc.; ca. 0.5 *M* HCl + ascorbic acid[b,c]	480
Dithizone-$CHCl_3$	Wet-ashed blood[d]	1040
	Urine; $CaCl_2$, NH_4 citrate, NH_3, KCN, and Na_2SO_3 added	979
	Wet-ashed blood, urine, and food; pH 9 (citrate)[d]	530
	Fish digests; pH 8—9.5 (citrate + NH_3); 3×[e]	474
	Urine (containing 10% v/v aqueous NH_3) extracted twice[d]	1020
Dithizone-MIBK	Urine; pH 8—8.5	27,496
K ethylxanthate-MIBK	Tissue digests; pH 3.5—9.5 (Cu masked with thiourea)	978
HHDC-$CHCl_3$	Dry-ashed food; pH 6[c]	659
	Wet-ashed liver/plant; ca. 0.5 *M* $HClO_4$[b]	429
HHDC-butylacetate	Deproteinized (with TCA and HCl) serum; pH 6	640,657
	Tissues digests; dilute H_2SO_4[b]	43

Table 3 (continued)
EXTRACTION SYSTEMS FOR Pb FROM BIOLOGICAL LIQUIDS/DIGESTS

Extraction system	Matrix and conditions[a]	Ref.
HHDC + 8-hydroxyquinoline (4 + 1) in $CHCl_3$ + i-pentanol (3 + 1)	Dry-ashed food, feces, and urine; pH > 1[e]	39
SDDC-MIBK	Hemolyzed blood (citrate/heparin and Triton® X-100) added	471,973
	TCA-deproteinized blood (twice); pH 6.5—7	961
	pH-adjusted urine; pH 7 (tris buffer + $CaCl_2$)	27,529
	Dry-ashed milk	977
	Fish digests; pH 5.5—7.5	548
	Wet-ashed fish/liver; pH 8.0	494
SDDC-CCl_4	Wet-ashed blood/tissues; pH 4—5 (EDTA + citrate)	528
15% Aliquat® 336 bromide in toluene	Dry-ashed bone/teeth; HBr medium	976
0.2 *M* KI—1.8 *M* H_2SO_4—5% Amberlite® LA-2	Biological digests (hair, etc.)	478
(4.25—5) *M* H_3PO_4—0.7 *M* KI-MIBK	Dry-ashed food (e.g., cereal)	476
HCl-triisooctylamine-MIBK	Wet-ashed blood (part of a scheme)	332

[a] Centrifugation often needed.
[b] Limited time stability of extracts.
[c] Interference from excess iron.
[d] Back extraction.
[e] Triple extraction.

from undigested blood and urine samples; (2) the formation of stable emulsions should be prevented by adding Triton X-100® (or, rarer, formamide,[965,1041] Rohnagel 12 n®,[967] Triton X-100 + saponin[57]); (3) lead tends to adsorb on glass surfaces (unless silanization is performed);[947] and (4) the calibration standards should pass through the same extraction procedure.

Two less popular techniques, coprecipitation[27,446,533,540,591,935,984-988,1051] and ion exchange,[326,419,546,897,937] can also be used for separation/preconcentration of Pb from biological liquids or digests. These techniques, however, are more tedious and time consuming, and their use is rarely justified. There are obsolete lead coprecipitations with $Th(OH)_4$ (from urine, at pH 5 to 6, and in presence of Cu^{2+});[27,988] $Bi(OH)_3$ (from ashed blood, hair, and urine, at pH 8.5 to 9);[935,986,1053] $SrSO_4$ (from ashed food[591,984] and tissues[984,985]); $Ca_3(PO_4)_2$ (from urine, at pH 9);[987] APDC (from tissue digests, at pH 3);[446,533] and with DDDC (from tissue digests, at pH 5.0 to 5.5).

Lead has been effectively preconcentrated (125 ×) from acidified (pH ~2), filtered urine by ion exchange on poly(dithiocarbamate) resin.[326] Anion-exchange separations of lead in acidified (~1.5 *M* HCl) blood,[419,546,937] urine,[546,937] and tissue homogenates[937] have been described.

Certain separation techniques have found their application in lead speciation studies: extraction of tetraalkyl leads from biological and environmental samples;[899,1047-1049] electrophoretic[787] or gel-chromatographic[1046] separation of protein fractions; LC[284] or HPLC[932] coupled with ET-AAS detectors; etc.[1049]

The preconcentration step can be skipped in the flame AAS analyses of some specimens containing micrograms-per-gram levels of Pb: bone, hair, nails, teeth, and some tissues. Digests/homogenates are relatively high in salts; background correction and standard addition checks are therefore advisable. The burner height should be adjusted to obtain lesser background and matrix effects, i.e., a somewhat higher position of the light beam above the burner. A three-slot burner is recommended. The same goes for analyses of undiluted urine samples from patients on chelation therapy (pulse-nebulization advisable).

Thompson and Godden[952] described a pulse-nebulization technique for blood lead: the sample was diluted (1 + 1) with an aqueous 0.2% w/v Triton X-100® and sonificated for 10 min; 200-μℓ aliquots were then pulse nebulized into the flame using a slightly modified nebulizer. Silicon antifoaming agent was sprayed between the samples. Standard additions calibration was required. The DL was 0.05 μg Pb/mℓ of blood, and the RSD was 4.9% at 0.65 μg/mℓ.[952]

Analysis of organic extracts does not pose any special problems. Pulse nebulization can be applied and hence many early macroprocedures can be modified and simplified. Incidentally, $CHCl_3$ and CCl_4 can also be used as organic solvents, thus offering some practical advantages;[429,439] several elements can be successively determined in that extract.

The Delves cup technique was first introduced by H. T. Delves[912] for the determination of Pb in small (10 μℓ) portions of (capillary) blood: the 10-μℓ aliquot was pipetted into a small nickel crucible, dried, partially oxidized with 20 μℓ of H_2O_2, redried at 140°C, and finally inserted into the flame, just a few millimeters below the orifice of a Ni absorption tube aligned with the light beam. This technique was subsequently applied to other biological samples (Table 2), as well as to blood spotted onto a filter and punched-out (''disc-in-a-cup technique''), organic extracts,[27,486,989] and to some other volatile elements. It is relatively inexpensive, rapid (20 to 60 assays per hour), and fairly sensitive.

Numerous modifications of the original Delves technique are described, aimed at mechanically stabilizing the burner mount[948,1000] and cup holder;[948,992] at extending the useful lifetime of cups[990,1001,1002] and absorption tubes;[948,953] and at optimizing certain critical parameters.[953,990,1000] Different cup materials have been compared: Ni,[912] Ta,[912,983] and fused quartz;[1001] Ni cups appear to be the best solution, although the quality of an individual cup (thickness, microcracks, etc.) is important, and sensitivity-matched cups should be used. The lifetime of these cups is between 20 and 100 assays; a gradual sensitivity loss is observed. Certain matrices (urine, paper discs, acidic solutions) markedly destroy Ni cups. A uniform distribution of the sample drop within the cup is essential,[1002] and this can be facilitated by first introducing a drop of aqueous NH_3[1002] or aqueous albumin.[1000] The quality of the cup holder is also important; wire loops made of Ni, Pt, Pt-Ir, Nichrome, or Inconel® have been tried out, the latter two alloys being thermally and mechanically most stable. Absorption tubes are made of alumina (perhaps the best),[953,984] silica,[992,1000,1005] and Ni.[912]

The Delves cup technique is prone to serious matrix effects. Attention should be paid to numerous details: alignment of the cup and the absorption tube (the distance of about 2 to 3 mm between the cup and the tube orifice being critical in view of sensitivity, precision, and time separation of the lead and smoke peaks);[953,1038] maintaining the same geometry of the system; reproducible injection and drying of samples; etc. Therefore; matrix-matched calibration and peak-area measurements,[872,948,990] simultaneous background correction, frequent recalibration, running samples in duplicate, and good quality control are essential.

ETA of Pb is attractive as it offers a high sensitivity and an opportunity to avoid the lengthy, contamination/losses-prone preparatory steps of ashing, extraction, etc. Most applications involve graphite-tube furnaces (Table 2), but other ET-AAS apparatus — the Carbon Rod Atomizer (Varian Techtron®),[762,973,1011,1028] Tantalum Ribbon (Instrumentation Laboratory® Model 355 Flameless Sampler),[1012,1013,1032,1041,1055] as well as some homemade equipment[1035] — have also been used. Many laboratories are now able to do accurate, precise, and, recently, automated[512,950,953,1015,1024,1037,1044] microassays of blood lead after simple preparatory steps (Table 4). Nevertheless, as a routine practice, the ET-AAS determinations of Pb in biological materials (particularly, in urine and in biological digests) cannot be called straightforward. Numerous important details of specimen preparation, furnace

Table 4
SELECTED AAS PROCEDURES FOR Pb DETERMINATION IN BIOLOGICAL SAMPLES

Sample	Brief procedure outline[a]	Ref.
Blood	1 + 20 dilution with 0.1% v/v Triton® X-100; ET-AAS: 110 (ramp), 550 (ramp, critical step!), 1800 (miniflow), and 2500°C (clean-out); pyrolytic tubes. Standard addition calibration; see also References 950, 953, and 1022	1015
	1 + 4 dilution with 0.5% Triton® X-100 — 0.5% $(NH_4)_2HPO_4$; ET-AAS: 100, 700, and 2300°C (gas stop); standards in diluent	502
	1 vol of blood (e.g., 20 to 200 μℓ) + 1.5 vol of water + 1.5 vol of 2 *M* HNO_3 vigorously vortexed for 30 min, and centrifuged; supernatant analyzed by automated ET-AAS: 100°C for 30—60 sec, 600° for 30 sec, 2200°C (gas stop), and clean-out at 2600°C; standards on blood; likewise References 512 and 1031	1036,1037
	10-μℓ aliquot pipetted into a Delves cup, dried, and partially oxidized with 20 μℓ of a 100-vol H_2O_2; standards on low-Pb blood; later modifications see text; alternatively, blood is spotted onto filter paper and a punched disc introduced into the cup[487,940,956,958,998,1000]	912
	APDC-MIBK extraction from hemolyzed blood (+ Triton® X-100) at pH 5.0 (citrate); ET-AAS: 100, 200, 600, and 2600°C; peak-area, standard addition calibration; see also Reference 359 and flame AAS[468,966,972,974] or Delves cup versions[486,989]	947
	0.1 mℓ of blood + 5 mℓ of water + 0.2 mℓ of 12 *M* HNO_3 vigorously vortexed, and centrifuged; supernatant analyzed by IL 254 FASTAC® Autosampler and IL 555 CTF Atomizer (Instrumentation Laboratory); aerosol mist deposited onto a graphite tube at 125°C for 10 sec, then ramp ashed to 500°C and atomized at 1800°C; calibration by standard additions or, better, preanalyzed blood specimens	1044
Blood, erythrocytes	1 + 0.1 + 4 dilution with aqueous 10% TAAH and aqueous 0.1% Triton® X-100, and vortexed; ET-AAS: 100°C for 80 sec, 500°C for 50 sec, and 2300 (blood) or 2100°C (erythrocytes); standards on blood	1017
Blood, tissues, urine	HNO_3 digestion (reflux), dilution, and $NH_4H_2PO_4$ added (0.8% v/v); ET-AAS with tantalized and pyro-coated L'vov platform, ramp to 950°C, 2100 and 2700°C (clean-out); standards containing $NH_4H_2PO_4$ and HNO_3	457
Blood, urine	1 + 9 dilution with water (for blood) or 1 + 1 dilution with 5% v/v HNO_3 (for urine); Zeeman ET-AAS with Hitachi® 170-70 cup-type cuvette: 110, 360 (ramp), and 2400°C; standards on blood or standard additions (for urine)	917
Bone, feces	HNO_3 digestion (reflux); ET-AAS: 150, 550, and 2200°C; standard addition calibration	959
Erythrocytes (speciation)	Hemolyzates fractionated on Sephadex® G-75 column; ET-AAS	1046
Food, urine, tissues	HNO_3 digestion; ashing completed at 430°C; ET-AAS: 100, 250 (?), and 2200°C; standard addition calibration	365
Food	10-g samples dry ashed at 450°C in presence of 10 mℓ ethanolic $Mg(NO_3)_2$ (10% w/v), and ashes dissolved in HCl-HNO_3-H_2O (20 + 60 + 15 by vol). APDC-MIBK extraction at pH 3—4 (acetate); flame AAS	271
Hair	1-cm segment directly analyzed by ET-AAS; ramp to 750°C, and atomization at 2500°C; calibration vs. spun silk or animal hairs; see also References 778, 884, and 936	78,79,649
	TAAH solubilization: 50-mg sample with 2 mℓ of Soluene® 350 at 55°C for 1 hr, cooled, diluted to 10 mℓ with toluene; ET-AAS: 60, 450, and 2100°C; standard addition calibration	368
	HNO_3 digestion; flame AAS; standards in 10% v/v HNO_3	460
	Dry ashing at 450°C (ramp!), and ashes taken up in 2 *M* HNO_3 ET-AAS (ramp); standards in 2 *M* HNO_3, and peak-area mode	885

Table 4 (continued)
SELECTED AAS PROCEDURES FOR Pb DETERMINATION IN BIOLOGICAL SAMPLES

Sample	Brief procedure outline[a]	Ref.
Hair, tissues	TAAH solubilization; flame[464] or ET-AAS;[521] standard addition calibration	
Kidney cortex	HNO_3 digestion; Zeeman ET-AAS: 100, 360, and 2400°C; cup version of a Hitachi® 170-70 instrument; standard addition calibration	373
Liver	HCl-HNO_3 (1 + 1) digestion; flame AAS; standard addition check	319
Liver, placenta	TAAH solubilization of a 0.5-g sample with 5 mℓ of Soluene® 350 at 55°C for 4 (placenta) or 24 hr (liver), and dilute to 10 mℓ with toluene; ET-AAS: 60, 450, and 3100°C; standard addition check	1038
Meat, tissues, viscera	1-g sample boiled with 10 mℓ of (1 + 1) HNO_3, evaporated to ca. 3 mℓ, and diluted to 25 mℓ; ET-AAS: 150, 600, and 2300°C; standard addition calibration	517
Milk	Protein pptn with an eq vol of 2 *M* HCl, and centrifugation; ET-AAS (ashing ≤400°C); standard addition calibration	394
Milk (evaporated)	Ashing at 500°C; APDC-butylacetate extraction at pH 5.4; flame AAS	898
Plasma	1 + 2 dilution with water; Zeeman ET-AAS: 100 (ramp!), 500, and 2400°C; standard addition calibration	926
	APDC-MIBK extraction; ET-AAS; standards extracted similarly	889,1043
Saliva (parotid)	100-μℓ aliquot injected into a Delves cup; dried at 140°C for 30 sec, partially oxidized with 20 μℓ of 30% H_2O_2; standard addition calibration	1005
Teeth	$HClO_4$ digestion, and centrifugation; extraction with APDC-2,4-dimethyl-6-heptanone at pH 2.2—2.8; flame AAS; limited time stability of extracts	893
Tooth enamel (biopsies)	Acid etching with 10μℓ of 1.6 *M* HCl in 70% glycerol for 35 sec, repeated with 10 μℓ of 70% glycerol, and diluted to 1 mℓ with water; ET-AAS: 100, 550, and 2000°C; Ca determined by flame AAS so as to calculate the amount of enamel taken (say ~0.1 mg)	1050
Urine	Undiluted urine directly nebulized into a flame; background correction; standard addition calibration; applicable to exposed or EDTA-treated subjects; DL 0.05 μg/mℓ	935,1056
	10 + 1 dilution with 20% H_3PO_4; ET-AAS with Mo-treated tubes: 130 (ramp), 850 (ramp, critical setting!), 1100 (gas stop), and 2500°C (clean-out); standard addition calibration	929
	500 μℓ of urine + 20 μℓ of iodine solution (1 *N* I_2 in 1 *N* KI) heated at 40°C for 5—10 min; 1 mℓ of a diluent added (0.05% w/v $(NH_4)_2MoO_4$—2% v/v H_3PO_4—0.25% ascorbic acid); ET-AAS: 100°C for 60 sec, 900°C for 40 sec, and 2200° C (ramp); standards on low-Pb urine	922,923
	APDC-MIBK extraction at pH 5.0, and centrifugation. ET-AAS: 150, 350, and 2400°C; EDTA interference	359
	SDDC-MIBK extraction at pH 7.0 (Tris buffer) in presence of 0.01% $CaCl_2$, and centrifugation; ET-AAS	529

[a] Background correction mandatory.

programming, and calibration have to be considered. To a certain extent, the problems experienced are due to using nonversatile equipment/methodology: inadequate background correction capability, low heating rates during the atomization step, unsuitable diluents, low dilution ratios, high purge-gas flow rates, etc.

The general problems of the ET-AAS, as already discussed, completely refer to lead determinations in biomaterials. Let us now consider the most popular ET-AAS assay, lead in blood. There are over 60 variants of procedures (see Table 2), and the most reliable of them seems to fall into three main groups.

1. Direct procedures which require only dilution of blood samples. Dilution ratios are between 1:1 and 1:20, and lesser adverse effects are observed with samples diluted at least 5- to 10-fold. Various diluents have been studied: water;[57,511,917,953,1016] dilute acids (0.01 *M* HNO_3,[883] 0.05 *M* HCl[1037]); aqueous detergents which serve as hemolyzing, antifoaming, and ashing aids [Triton X-100®,[27,362,921,950,953,954,1015,1022,1055] silicon antifoaming agents,[1024] aqueous NH_3,[918,1020] TAAH (e.g., Lumatom®,[1018,1019] etc.[1017])]; as well as matrix/analyte modifiers (HNO_3,[883,1023] NH_4NO_3,[1018] La^{3+},[438] etc.). While most of these diluents are useful by themselves, even better are their rational combinations, e.g., detergent + matrix/analyte modifier[284,438,502,925,1017] (Table 4). Serious drawbacks are intrinsic to many of the reported direct procedures, being most pronounced with insufficiently diluted, not-completely hemolyzed, and/or aged specimens: poor pipetting reproducibility and carryover, residue buildup, prolonged and critical temperature programs, sensitivity drift, matrix-matched calibration (standards on pooled, spiked low-lead blood or secondary standards, i.e., preanalyzed blood specimens), and marked background. Anticoagulants seem to affect the slope sensitivity ($\Delta A/\Delta C$): EDTA > citrate > oxalate > heparin > aqueous.[1015,1016]
2. Procedures involving simple pretreatment of blood specimens (protein precipitation or incomplete digestion). Samples are treated with a dilute[7,1036,1037,1044] or concentrated HNO_3[512,1031] or with $HClO_4$ + TCA[1026,1052] in small, tightly capped test tubes (with or without heating), centrifuged, and supernatants are analyzed by ET-AAS. The final concentration of acids is preferably kept ≤ 1 *M;* the sensitivity, precision, accuracy, sample throughput, and the useful lifetime of graphite tubes are significantly improved. Many samples can be simultaneously treated (aged specimens are tolerable as well). Automated sample injection and quantitation[512,1037] (280 to 370 firings per 10 hr[1037]) and aerosol mist deposition[1044] (95 samples in 90 min![1044]) have been documented.
3. Microprocedures for blood lead, involving chelate extraction from as small as 50 to 100-$\mu\ell$ blood aliquots, are available[7,359,911,1041] (Tables 2 to 4).

Urine samples are very troublesome, even with versatile apparatus. The tolerable ashing temperatures are as low as 300 to 400 (intact samples) or 850 to 900°C (after matrix/analyte modification[922,923,929]); the persisting salts cause severe background and depression of lead absorbance. The performance is improved somewhat by adding HNO_3,[330,365,438,911,917,920] H_3PO_4,[922,923,929] NH_4NO_3,[362] $NH_4H_2PO_4$ or $(NH_4)_2HPO_4$,[457,921] $HClO_4$ + NH_4NO_3, etc.; if alkyl leads are present in urine samples, a pretreatment with I_2 + KI[923] or HNO_3[365,457,929] is required. Ramp atomization may prove advantageous with certain graphite furnaces.[362,911,923] Standard addition calibration is typical, but *not always* efficient! The most reliable direct ET-AAS procedures appear to be those reported by Hodges,[922,923] Hinderberger et al.,[457] and Knutti;[929] see Table 4. The key to success is a combination of (1) simple pretreatment, (2) effective thermal stabilization of the analyte (as phosphate), (3) pretreatment of the graphite surface (molybdate[922,923,929]), and (4) tantalized and pyro-coated L'vov platform.[457]

IV. CONCLUSION

AAS is the best routine method for the determination of Pb in biological samples. Other sensitive and reliable methods are DPASV and IDMS.

Flame AAS assays generally involve a preparatory step (often, dithiocarbamate extraction); its main drawbacks are a limited (pH dependent) time stability of extracts and an analyte masking by certain concomitants.

While ET-AAS is currently the technique of choice, it still faces serious problems which are especially pronounced with older models of graphite furnaces: matrix effects, premature volatization, background, critical ash/atomization settings; prolonged heating cycles, complicated calibration (often, matrix matched or standard additions), etc.

Essentials: matrix/analyte modification, preferably with composite diluents (e.g., detergent + $NH_4H_2PO_4$); carefully optimized dry/ash stages (ramp); temperature-controlled rapid atomization heating; efficient (critically checked) background correction; and effective quality control scheme. Useful approaches to the ETA of Pb are the Zeeman background correction, L'vov platform atomization, aerosol deposition technique, pretreated graphite surfaces (pyro-/Mo/La/W/Ta, etc.), and gas-stop conditions. Radical approaches should be STPF or one including a simple prefurnace treatment (such as the HNO_3 protein precipitation).

Other sources of analytical inaccuracies may be contamination and losses on storage or dry ashing. Trace lead analyses are still to be perfected.

Chapter 18

LITHIUM

I. DETERMINATION OF LITHIUM IN BIOLOGICAL SAMPLES

Normal concentrations of Li are around or less than 0.01 μg/mℓ in blood, serum, and urine, and well below 1 μg/g in most other biomaterials (Table 1). Only a few spectroscopic techniques are sensitive down to nanograms-per-milliliter levels of Li: AAS (flame and ETA), AES with various atomization/excitation sources (flames: air-propane, air-C_2H_2,N_2O-C_2H_2,[1060,1061] etc.), ICP,[29,35,688,1057] and MS.[3,25,123,688,1062]

The use of ICP-ES and SSMS, which are essentially multielement techniques and require expensive equipment, is justified only when many elements in blood,[3] serum,[688] tooth enamel,[25] tissues,[3] etc. have to be determined simultaneously. Highly accurate results are obtained by IDMS.

Lithium-based drugs are used in the treatment of manic psychosis; hence, many clinical laboratories need a rapid, simple, and reliable technique for monitoring *therapeutic concentrations* of Li in blood, erythrocytes, saliva, serum, and urine. These levels (e.g., 1.4 to 7 μg/mℓ in serum; 7 to 200 μg/mℓ in urine) are substantially higher (100 to 1000 ×) than the normal Li levels, and their determination involves no particular difficulties. Flame-emission photometry,[188,288,1057,1058,1063-1066] often with added K^+ as an internal standard and ionization buffer,[1065] is the technique of choice in routine practice, with flame AAS as runner-up. Flame photometry is simpler, cheaper, somewhat more sensitive, easier to use, as well as linear in a broader concentration range vs. flame AAS. The latter, however, is slightly more precise,[1062,1066] in particular with urine samples which exhibit variable salts content.[1067]

II. ATOMIC ABSORPTION SPECTROMETRY OF LITHIUM

In flame atomization (air-acetylene flame, 670.8 nm), the characteristic concentration ranges between 0.016 and 0.05 μg/mℓ, and DLs are down to 0.0003 to 0.003 μg/mℓ.[56-59,294] Minor ionization interference does exist at low Li concentrations, its extent depending on air-to-acetylene ratio, burner type/design, observation height, etc. This effect may produce slightly convex calibration curves in their lower part and hence involves a minor inaccuracy. It can be suppressed by adding an ionization buffer (e.g., 0.1% KCl), as well as lowered or eliminated by working with a three-slot burner or a slightly fuel-rich flame or else with lesser dilution factors. Minor background absorption due to SrOH may be accidentally encountered. If acids are present, their concentration should be approximately matched in both sample and standard solutions. In one study, 10% TCA, 0.1 *M* HCl, 0.1 *M* $HClO_4$, and 0.1 *M* HNO_3 were found to depress Li absorbance by 57, 12, 10, and 4%, respectively.[1068] Transport interferences and bulk-matrix effects, due to excess salts, surfactants, viscous matrix constituents, etc., are also to be expected and accounted for.

In ETA, the characteristic concentration and DLs are between 0.005 and 1 ng/mℓ (1 to 10 pg).[57,59,61,294] Pyro-coated tubes, fast heating during the atomization step, as well as the presence of dilute H_2SO_4, are all advantageous in ET-AAS of this element. Somewhat higher lamp currents are advisable, being useful in decreasing the background emission noise due to the incandescent graphite tube. Background absorption should always be checked/corrected by means of a tungsten halide lamp or by Co 670.8-nm[629] or Ne 671.7-nm[302] lines.

Severe depressions of the analyte signal by chlorides, organic matter, and excess acids have been observed.[294,769] Ashing temperatures between 500 and 800°C, depending on the

Table 1
LITHIUM CONCENTRATIONS IN HUMAN BODY FLUID AND TISSUES

Sample	Unit	Selected reference values	Ref.
Blood	μg/ℓ	3	332
		2;[a,b] 6 ± 2[b]	3
		8.8 (3—290)[c]	906
Brain	ng/g	4 ± 1	3
Diet	mg/day	<0.1	731
Feces	mg/day	1.2 ?	8
Hair	μg/g	♂, 0.04;[d] ♀, 0.05;[d] (0.009—0.228)	288
Kidney	μg/g	0.01 ± 0.003	3
Liver	ng/g	7 ± 3	3
Lung	μg/g	0.06 ± 0.01	3
Muscle	ng/g	5 ± 2	3
Plasma or serum	μg/ℓ	8	1057
		9 (<3—44)	1058
		11 (6—22)	1059
		12[c]	906
Tooth (enamel)	μg/g	1.13 ± 0.13 (0.23—3.4)	25
Urine	μg/ℓ	4 (2—10)	1059

Note: Wet weight unless otherwise indicated. Mean ± SD (range in parentheses).

[a] Pool.
[b] ng/g.
[c] Review data.
[d] Geometric mean.

particular matrix, are typical.[57,294,1069] Welz[294,769] found that the organic as well as the salt matrix of serum samples thermally stabilized the analyte, probably as a lithium carbide, up to 1100 to 1200°C, while depressing the absorbance signal by a factor of 2.

III. AAS METHODS FOR ANALYSIS OF BIOLOGICAL SAMPLES

Flame AAS is now routinely used in many clinical laboratories for the determination of therapeutic Li levels; applications to blood,[332,1067,1070,1071] CSF,[1072] erythrocytes,[1067,1070,1071] saliva,[1074,1075] serum or plasma,[7,57,188,298,539,700,1064,1066-1084] and urine[7,1067,1072,1074,1079,1082] have been described (Table 2).

Microflame AAS procedures based on large dilutions,[1067,1070,1079] flow injection,[1084] or pulse nebulization of small sample aliquots (e.g., 10-[704] or 50-fold[700] diluted serum) are available. ET-AAS is rarely used; microscale procedures for (rat) brain tissue (<10- to 100-mg samples)[1085] and serum (μℓ volumes)[769,1069] have been documented.

No major analytical problems are encountered in these analyses. Flame AAS of Li is straightforward, and the preparatory step is often as simple as dilution with water. Sampling and storage steps do not pose any specific problems, either. Plasma and serum samples appear to be equivalent in their Li content,[1068,1083] and can, respectively, be stored for 24 hr, 1 week, and for months at room temperature, 4°C, and frozen.[1068,1071,1083] Immediate separation of the serum from the clot is not necessary, and no changes in Li concentration have been observed after a 24-hr aging of clotted blood.[1083] Only gross hemolysis might alter the results.[1071] Blood and urine samples are said to be stable for 1 week at room temperature,[1072] but urine samples should preferably be acidified with HCl (e.g., 5 mℓ of

Table 2
SELECTED AAS PROCEDURES FOR DETERMINING THERAPEUTIC CONCENTRATIONS OF Li IN BIOLOGICAL SAMPLES

Sample	Brief procedure outline	Ref.
Blood, plasma erythrocytes	1 + 99 dilution with water; flame AAS; standards on blood/plasma	1067
	1 + 5 dilution with water; flame AAS; standards on blood/plasma; erythrocyte Li calculated (see text)	1071
Saliva, serum	1:50 dilution with aqueous 200 μg/mℓ K^+ (as chloride); flame AAS; standards containing K^+	1073
Serum/plasma	1 + 10 dilution with water; flame AAS; standards containing K^+ (15 μg/mℓ) and Na^+ (350 μg/mℓ) as chlorides; alternatively, standards on pooled serum;[1068,1077,1083] pulse nebulization — injection[700] or dipping[704] — advisable	1066, 1082
Serum	Direct flow-injection flame AAS (10-μℓ aliquot); viscosity-adjusted standards containing NaCl and KCl	1084
	1:25 dilution with water; ET-AAS with CRA-63 (Varian Techtron®): ca. 750 and 2100°C	1069
Tissues	HNO_3-H_2SO_4 digestion; flame AAS	7
Urine	1:100 dilution with water; flame AAS; standards containing 75 μg NaCl/mℓ	1082
	50 mℓ of urine + 5 mℓ of 2 *M* HCl, and diluted 50- to 250-fold; standards on acidified diluted pooled urine (~0.2 *M* HCl)	1067

2 *M* HCl per 50 mℓ of urine), in order to prevent coprecipitation of Li_3PO_4 with the $Ca_3(PO_4)_2$ precipitate formed on storage.[1067]

Precipitation of blood or serum proteins with TCA[57,298,1075] or ethanol[1064] is used occasionally, although not superior to simpler direct procedures; moreover, TCA strongly depresses Li absorbance.[1064,1068,1072]

Minor ionization and transport interferences are common and well documented in flame AAS of Li in diluted serum and urine.[1068,1072,1077,1082] In serum analyses, dilutions from none or 1 + 1[1081] to 1 + 99[1067] have been used, but typical procedures involve 10- to 20-fold dilution with water, and calibration vs. standard solutions containing KCl and NaCl.[704,1068,1077,1082,1083,1084] Lower dilution factors, e.g., 1 + 4, are tolerable but call for matrix-matched standards (on pooled normal sera).[1070,1071] Urine samples usually contain 5 to 30 times more Li than serum,[1068] and, hence, require greater dilutions, say, 1:20 to 1:250, as well as calibration vs. spiked pooled normal urine or artificial urine or NaCl-containing standard solutions.[1067,1082]

Both concave (at lower concentrations, due to ionization) and convex (at higher concentrations) calibration curves can be obtained, and this may result in an over- or underestimation of Li, respectively. These adverse effects are easily dealt with by applying ionization buffers, appropriate dilution factors and flame conditions, as well as by checking for linearity of calibration curves.

In order to avoid certain difficulties when working with erythrocytes (special pipetting techniques, cell washing, corrections for trapped plasma, etc.), some authors practice indirect estimates of erythrocyte lithium[1070,1071] based on analysis of blood and plasma, as well as determining the hematocrit (Hct) and subsequent calculation according to the equation:

$$Li_{ER} = \frac{Li_{WB} - Li_P (1 - Hct)}{Hct}$$

If an ashing step is required, both wet[7] and dry[1061] ashing should be applicable.[91] Delves et al.[332] extracted Li from wet-digested blood with TTA-MIBK at pH 9.0 to 9.5, as part of a successive multielement analytical scheme.

IV. CONCLUSION

Both flame AES and AAS are simple, sensitive, routine techniques for monitoring lithium therapy by analyses of blood erythrocytes, saliva, serum/plasma, and urine. All that is needed is a simple dilution; standards contain the corresponding concentration of NaCl and KCl, or, alternatively, pooled normal serum or urine.

Flame emission (N_2O-C_2H_2 flame), ET-AAS, and IDMS are used at lower analyte levels.

Chapter 19

MANGANESE

I. DETERMINATION OF MANGANESE IN BIOLOGICAL SAMPLES

Mn concentrations in biological materials range broadly within four orders of magnitude, being about or less than 1 ng/mℓ in CSF, saliva, serum, and urine; around 1 μg/g in most samples of tissues, bone, hair, nails, and teeth; and well above 1 μg/g in food and feces (Table 1). Accordingly, various analytical techniques are used and different kinds of problems are encountered.

Few methods — ET-AAS, RNAA, ICP-ES, SSMS, as well as some catalytic procedures — are sensitive below the 1 ng/g level, and ET-AAS appears to be the best bet. At micrograms-per-gram levels, flame AAS is recommended, while the use of more expensive techniques — ET-AAS, ICP-ES, and XRF — is rarely justified.

NAA (often, RNAA[45]) has been widely applied to biological materials: blood,[685,686,1087,1102,1103] CSF,[687,1087] diet,[45,47,116] erythrocytes,[287,665,1087] hair,[9,11,48,101,290,635,670,1095] milk,[40,110] nails,[15,290,670] saliva,[671,672,687] serum/plasma,[45,186,287,665,687,1087,1095,1102] teeth,[23,24,51,103,676] tissues,[12,43,103,108,291,686] urine,[165,416,687,1102] etc.[45] INAA is very useful with specimens containing micrograms of Mn per gram of hair,[11,101,290,635,1095] nails,[15,290] teeth,[23,51,676] and tissues.[12] The inherent absence of a postirradiation contamination is a major asset of NAA.

DLs of the ICP-ES[29,35,289] are typically a few tenths of nanograms per milliliter (down to the impressive 0.01 ng/mℓ[29]); blood,[30,1104] bone,[392] milk,[35] serum,[30] tissue,[289,348,411,412,518,536,1104] and other biologicals[348] have been successfully analyzed.

The classical arc-source AES has been applied to blood,[36,1105] bone,[38] diet,[37,1105] feces,[37,1105] hair,[288] milk,[14,506] saliva,[41,380] serum,[36,413] sweat,[21] teeth,[26] tissues,[36,43,44,904] urine,[1105] etc.;[37,41] however, it lacks sensitivity in the low nanograms-per-gram range.

SSMS of blood,[3] hair,[54] nails,[16] plasma,[55] teeth,[25] and tissues[3,132,275,409] provides low DLs (e.g., down to 0.4 ng Mn per gram of soft tissue[3]).

PIXE is used by a few laboratories in studies on trace element concentration profiles in minute specimens.[128,407,852]

Catalytic[682,1098] and spectrophotometric[133,1105] assays, involving inexpensive apparatus, are incidentally performed.

II. ATOMIC ABSORPTION SPECTROMETRY OF MANGANESE

In flame atomization, the characteristic concentration and DLs are 0.02 to 0.06 μg/mℓ and 0.001 to 0.003 μg/mℓ, respectively.[56-59,294] The optimization of instrumental parameters involves minor compromises vs. the "maximum-signal" conditions, viz. somewhat broader slit width, a lower acetylene flow, and a slightly elevated observation height. The baseline stability and the signal-to-noise ratio are substantially improved by simultaneously measuring the three lines of the Mn triplet — 279.5, 279.8, and 280.1 nm — (with a bandpass of 1 to 2 nm), rather than by isolating the most sensitive line, 279.5 nm, from its close neighbors (with a bandpass of only 0.1 to 0.2 nm). A lean air-acetylene flame is recommended for better selectivity; in a more reducing flame and/or at a lower observation height, some minor matrix effects (acids, excess Fe, Si, PO_4^{3-}, SO_4^{2-}, $CaSO_4$, $MgSO_4$, etc.) are encountered. Dilute HCl and $HClO_4$ are appropriate media, but some other acids may also be used (HNO_3, H_2SO_4, etc.), provided their concentration is kept <10% v/v and is matched in both sample

Table 1
MANGANESE CONCENTRATIONS IN HUMAN BODY FLUIDS AND TISSUES

Sample	Unit	Selected reference values	Ref.
Blood	µg/ℓ	7.1 ± 2.6	244
		8.4 ± 2.6 (4.2—14.3)	1086
		(8.6—14.5)	1087
		9.0 ± 2.2 (3.8—15.1)	1088
		10 ± 4 (6—30)	1089
		11 ± 4.4 (3—21)	652
		12.2 ± 3.9	1090
		13 ± 2	687
Bone	µg/g	0.049 ± 0.007	268
		♂, (3.3—8.3);[a] ♀, (4.0—9.6)[a]	367
Brain	µg/g	(0.17—0.39)	1091
		0.2 ± 0.03	3
CSF	µg/ℓ	0.66 ± 0.10 (0.55—0.89)	1092
		<1	1093
		(0.83—1.50)	1087
Erythrocytes	µg/ℓ	14.7 ± 2.3 (9.0—20.8)[b]	665
		16 ± 1	287
Feces	mg/day	(0.86—3.40)	1094
Hair	µg/g	0.23 ± 0.11 (U.S.A.)	557
		x̄, 0.42 (0.075—50) (Japan)	11
		x̄, 0.15—0.41[c] (0.03—3) (U.S.A.)	9
		x̄, ♂, 0.425 (0.03—2.66) (Canada)	10
		x̄, ♀, 0.788 (0.156—3.72) (Canada)	10
		♂, 0.64;[d] ♀, 1.34;[d] (0.07—11) (New York)	288
		1.2 ± 0.6 (G.D.R.)	367
		1.4[d] (0.3—14) (G.D.R.)	1095
		2.33 ± 1.64 (0.17—8.17) (U.K.)	368
Kidney	µg/g	x̄, 0.073; 0.13 ± 0.16 (0.050—0.85)	105, 106
		0.62 ± 0.03	268
		4.5 ± 1.3[e]	367
Kidney cortex	µg/g	0.5 ± 0.06	3
		0.99	107
		3.21 ± 1.15[e]	373
Liver	µg/g	x̄, 0.41; 0.54 ± 0.17 (0.017—2.18)	105, 106
		0.5 ± 0.08	3
		0.53 ± 0.06	268
		1.41 (1.07—2.12)	108
		1.50 ± 0.79	275
		4.53 ± 1.25[e] (2.8—6.9)[e]	319
Lung	µg/g	x̄, 0.058; 0.04 ± 0.01 (0.016—0.11)	105,106
		0.080 ± 0.008	268
		(0.10—0.14)	109
		x̄, 0.19; 0.238 ± 0.079 (0.08—0.59)	103
		0.2 ± 0.03	3
Milk	µg/ℓ	15 (12—20.2)	1096
		20[f]	374
		21.4[b,e,g]	110
		8—29;[h] (0—62)	14
Muscle	µg/g	0.04 ± 0.007	3
Nails	µg/g	0.66 ± 0.51	670
		(0.2—2.2)[e]	16
Nails (toe)	µg/g	x̄, ♂, 2.5;[e] 4.3 ± 6.2[e]	15
		x̄, ♀, 3.4;[e] 5.2 ± 4.8[e]	15
		(0.5—27.2)[e]	15

Table 1 (continued)
MANGANESE CONCENTRATIONS IN HUMAN BODY FLUIDS AND TISSUES

Sample	Unit	Selected reference values	Ref.
Plasma or serum	μg/ℓ	0.58 (0.36—0.96)	1097
		0.64 ± 0.18 (0.44—1.06)	665
		(0.36—0.90)	1098
		1.02 (0.74—1.25)	1092
Saliva	μg/ℓ	2.4 (1.4—4.0)[i]	671
		2.8 ± 0.6 (0.5—7.9)[j]	672
		3 ± 1	687
		6.9 (4.9—8.6)[k]	671
Sweat	μg/ℓ	♀, 17 ± 6; ♂, 23 ± 14	381
Tooth (dentine)	μg/g	0.19 ± 0.06[e]	676
		0.63 ± 0.05	23
		1.11 ± 0.64;[l] 1.90 ± 1.57[m]	24
Tooth (enamel)	μg/g	0.28 ± 0.03 (0.08—0.64)	25
		0.59 ± 0.04	23
		x̃, <0.6; 2.2 (<0.6—63)	26
		x̃, 0.73;[e] 0.83 ± 0.36 (0.3—2.0)[e]	103
		0.99;[n] 1.87[o]	51
		1.42 ± 0.17;[l] 3.22 ± 0.92[m]	24
Tooth (whole)	μg/g	(0.45—0.87)	1099
Urine	μg/ℓ	0.54;[d] (0.08—2.67)	1100
		0.65 ± 0.53	1090
		0.7 (0.1—1.5)	1097
		(1—10)	1101
		(0.4—12.7)	896
	μg/day	0.8 ± 0.3 (0.09—2.9)[j]	416, 687

Note: Wet weight unless otherwise indicated. Mean ± SD (range in parentheses). x̃, median.

[a] Rib.
[b] ng/g (ppb).
[c] Range of medians.
[d] Geometric mean.
[e] Dry weight.
[f] Pool (New Zealand).
[g] Transitional.
[h] Range of mean values.
[i] Rest saliva.
[j] Nondialyzable.
[k] Parotid, stimulated.
[l] Permanent.
[m] Primary.
[n] European.
[o] African.

and standard solutions. Only bulk-matrix effects, transport interferences, and minor background are to be expected.

The ET-AAS of Mn is extremely sensitive and fairly selective. The characteristic concentration ranges between 0.004 and 0.05 ng/mℓ (0.4 to 5 pg), and DLs are down to 0.005 to 0.01 ng/mℓ (0.2 to 1 pg).[57,59-61,294] Pyro-coated graphite tubes are preferable to uncoated tubes because of better sensitivity (2 to 2.5 ×) and lower atomization temperatures (1800 to 2600 vs. 2600 to 2900°C).[60]

Manganese is readily determined, provided versatile equipment/programming is avail-

Table 2
AAS TECHNIQUES FOR Mn DETERMINATION IN BIOLOGICAL MATERIALS

Techniques	Biological materials
Flame AAS after dilution or digestion	Diet,[591,729-731] feces,[729] fish,[442,730] hair,[288,370,466,733] meat,[517,729,730] milk,[374,538,729] milk (evaporated),[712] tissues[43,268,319,442,462,467,517,538,541,739,789,856,1109]
Flame AAS after TAAH solubilization	Brain, muscle, plasma, tissues, etc.[741]
Microflame AAS	Tissue digests[760,761]
Flame AAS after chelate extraction	Blood,[332,473,1110] feces,[39] fish,[494] food,[39,659] milk,[745,1095] tissues,[473,1110,1111] urine[27,39,974,1094,1100,1101,1112-1114]
Direct ET-AAS (solid samples)	Fish (ash),[505] hair,[78-80,778] liver,[505,781,1108] teeth,[1099] tissues[655]
Direct ET-AAS with or without dilution	Blood,[330,652,1086,1089] CSF,[763,1092,1093] serum,[27,511,605-607,652,654,763,1086,1092,1093,1097,1115,1116] urine[27,330,763,1093,1097,1100]
Zeeman ET-AAS	Blood, serum[1088]
ET-AAS after acid extraction	Meat,[517] milk,[775] plasma,[1107] tissues[517,1107]
ET-AAS after digestion	Blood,[686,1117,1118] bone,[450] fish meal,[774] hair,[79,557,733,876] tissues,[67,516,518,686,774,793,1091,1119] tooth enamel,[84] urine[27]
ET-AAS after TAAH solubilization	Blood,[521] brain,[776] hair,[368] muscle biopsies,[92] plasma,[610] tissues[92,610]
ET-AAS after chelate extraction	Blood,[7,1090] serum,[7] tissues,[7,1120] urine[7,1090,1113]
Speciation studies	Rat liver fractions[793]
Miscellaneous	Enzymes,[1121] serum,[1121] urine[1101,1121]

able.[553,1106] Of major concern are ashing temperatures (up to 1000 to 1300°C, preferably in a "ramp" mode), heating rate during atomization, and calibration mode. ETA of Mn has been studied extensively.[311,553,706,707,1106,1107] The dilute HCl (1 to 2 *M*)[774,1107] and dilute H_2SO_4 ($\leq$4% v/v)[67,774,793] are claimed to be the most suitable acidic media, HNO_3 is tolerable (yet a slight depression is observed),[294,706,1107] while $HClO_4$ (>0.1 *M*) is a very strong depressant and should be avoided.[66,1107] Acid-matched calibration is typical, but with older graphite furnaces as well as with certain difficult matrices, more complicated calibration modes may be needed (matrix matched, standard additions, peak area).

Some salts of low/intermediate volatility still persist after ashing, giving rise to a minor background as well as to matrix effects; e.g., both enhancement [$MgSO_4$ (+ 25%), $Mg(NO_3)_2$ (+26%), $Ca(NO_3)_2$ (+24%)] and depression [$MgCl_2$ (−46%)] of Mn absorbance, due to as low as 500 μg/mℓ concentrations of the indicated concomitants, have been observed. With older (slow) graphite furnaces, peak-area measurements have proved not effective enough in eliminating these interferences.[706,707] The use of high heating rates,[60,509,553,1106,1107] gas-stop conditions,[1106] and L'vov platform atomization[553,1106] ought to be advantageous (STPF concept[553]).

Background correction likewise deserves attention: as the deuterium arc is not sufficiently intensive in this spectral region (279.5 nm), it may be needed to decrease the HCL current so as to match the intensities of the two light sources. Zeeman correction has proved to be highly efficient.[1088,1106]

In some assays (e.g., direct ET-AAS of solid microsamples[781,1108]), the less sensitive (10 to 20 ×) line at 403.1 nm is used. At least two important notes should be made here: (1) the bandpass should be as narrow as 0.2 nm so as to resolve this line from its close neighbor at 403.3 nm, and (2) the background should be checked by using a tungsten halide lamp or the Mo 404.1-nm line.[781]

III. AAS METHODS FOR ANALYSIS OF BIOLOGICAL SAMPLES

AAS is an established and currently the best method in routine practice. Analytical techniques are classified in Table 2.

Contamination control during all steps of specimen collection, storage, preparation, and measurement is of paramount importance. While the endogenous levels of the analyte are nanograms or even subnanograms per gram in certain biological liquids, laboratory ware, airborne particles, and stainless-steel needles or knives contain as much as five to eight orders of magnitude more Mn. Hence, it is very difficult to obtain and handle specimens of serum, plasma, urine, saliva, and CSF without extraneous Mn additions. In the past many researchers, including the author, have learned by experience that the "normal" levels of Mn in serum and urine are actually as low as the laboratory manganese background! Lievens et al.[108] found that the dust fallout in a normal and dust-free lab was equivalent to 0.063 and 0.002 ng Mn/cm^2/day, respectively; in many ordinary labs the situation is probably even worse. Halls and Fell[1097] used Venflon plastic cannulae (Viggo AB, Sweden) to collect blood samples; the first 10 mℓ of blood was still contaminated (0.9 ng Mn/mℓ of serum) and was discarded or used for other biochemical tests; serum Mn was significantly lower — presumably authentic — in the next blood portions (0.6 ng Mn/mℓ of serum), while the stainless-steel needles unduly raised the levels up to 1.3 ng/mℓ![1097] Heydorn et al.[186] applied the method of least duplicates so as to statistically account for the inevitable random contamination in serum Mn assays. Saliva may be contaminated by food debris[672] and dental metals, and teeth by debris and tartar,[23] as well as by carborundum separating discs[23] and other dentist instruments. Nail specimens require rigorous cleaning and washing,[15,623,670] and hair may also be externally contaminated.[733] Borosilicate glass,[1098] polyethylene,[163] and other labware materials may add Mn to samples, unless properly precleaned[1097] or replaced by more appropriate materials (quartz, PTFE, dense polyethylene/polypropylene, etc.). Plain grinding procedures are ruled out; van Grieken et al.[624] measured the contamination rate of animal tissues as 1.6 μg Mn/min of grinding (i.e., >500%/min) and 0.24 μg Mn/min (>120%/min) in corundum and agate, respectively. The brittle fracture technique[532] (in Teflon®, at liquid nitrogen temperature) has proved clean. At levels of nanograms of Mn, even reagents of special purity may introduce intolerable blanks. Manganese contents of Merck Suprapur® acids — HCl, $HClO_4$, H_2SO_4, and HNO_3 — were found to be 0.5, 0.5, 0.6, and 1 ng/mℓ, respectively,[311] i.e., below the levels claimed by the manufacturer, but still intolerable if, say, urine or serum had to be wet ashed. Much higher Mn levels have been encountered in Mg $(NO_3)_2$ ashing aid,[442] TCA,[1107] KOH,[1107] and other lab reagents.

Sampling may not be representative with certain specimens. While Mn levels in blood,[1086] serum,[186,1086] and urine[165] are quite stable, and the Mn distribution within liver,[108,407] brain,[776,1091] and hair[161,650] is relatively uniform, some sources of sampling errors have been documented: concentration differences between black and grey hairs[101] as well as along single hairs,[101] higher Mn in carious vs. sound teeth,[1122] post-mortem changes in autopsy tissues (up to 20%),[789] nonhomogeneous Mn distribution in diet and feces, and inadequacy of the arm-bag sweat collection technique vs. the whole-body washdown.[381]

Generally, all decomposition techniques can be applied to sample preparation, but the choice is governed by considerations of the AAS characteristics of the analyte as well as by the pronounced blank/contamination problems, as already discussed. Mn is readily put into solution. As early as 1962, Cotzias and Papavasiliou[1087] showed that simple acidification of plasma samples with HCl to pH 4.3 to 4.5 yielded protein-free supernatants containing 85 to 100% of total plasma manganese. It is reasonable, therefore, to use simplified preparatory steps, such as precipitation of milk proteins with HCl (1 *M* final concentration);[775] treatment of plasma or tissue homogenates with a dilute HCl (optimum 1 *M*, but 0.1 to 5 *M* HCl applicable) at 60°C for 1 hr;[1107] extraction of hair[52] or tissue[467] manganese with a 1% v/v HNO_3; and Mn extraction from feces, food, meat, milk, and tissues with 5% v/v HNO_3.[729] Single-acid (incomplete) digestions are also recommended; boiling with acids: HNO_3 (bone,[90] hair,[79,557,876] meat,[517] tissues[517,1119]), HCl (tooth enamel[84]), HCl + HNO_3

(liver[319] and feces, food, meat, milk, and tissues[729]), as well as pressurized decompositions with HNO_3 (bone,[90] tissues,[518,686] etc.) have all been practiced with few adverse comments.

Procedures for complete wet oxidation of organic matter have been described. The results of several comparison studies[43,442,462,729,1123] show that these digestions are straightforward, the choice being a matter of convenience. In ET-AAS, wet ashing with HNO_3 + H_2O_2,[442,505,789] H_2SO_4 + H_2O_2,[67,774,793,1091] and HNO_3 + H_2SO_4 + H_2O_2[170,462] is appropriate, but $HClO_4$-containing mixtures are to be avoided, as it is difficult to exactly control the final acidity, and $HClO_4$ is a severe depressant.[66,776] In flame AAS assays, however, HNO_3 + $HClO_4$[442,541,760,761,1107,1109] and HNO_3 + $HClO_4$ + H_2SO_4[43] are quite acceptable. A word of caution: procedures involving $HClO_4$, H_2O_2, and/or pressure are potentially hazardous!

LTA,[169,626] followed by a dissolution of the ashes in a dilute HCl[774,1118] or HNO_3,[516] is a clean ashing technique. Dry ashing may not be as satisfactory; while Mn volatization is not probable (at least up to 550 to 600°C[91]), minor retention losses on crucible walls as well as on silica or pyrophosphate residues have been observed. An evaporation of the ashes with HNO_3[174,374,731] or HCl,[442,729] or even a treatment with aqua regia[370,538] or H_2SO_4 and H_2O_2,[462] may improve the analyte recovery. Of course, such treatments, as well as ashing aids,[442,538,1117] give rise to higher blanks.

To conclude, complete ashing of the organic matter is rarely required and justified; hence simpler preparatory steps should be adopted.

There have been efforts to homogenize tissues with water[776] or aqueous Triton X-100,®[1107] which, however, resulted in poor precision, due to the incomplete release of Mn from the particulate matter. More efficient TAAH solubilization procedures have been described for hair and tissues;[92,368,521,610,776,1104] milk powder has also been successfully homogenized with aqueous detergents[712] (see Table 4).

Preconcentration (typically, chelate extraction) is included in some AAS procedures (see Tables 2 to 4). Dithiocarbamate extractions are, however, not particularly reliable: (1) pH should be $\geq$4, preferably 6 to 7; (2) other extractable concomitants may depress Mn extraction; (3) Mn dithiocarbamates are *very unstable* in the extracts and ought to be measured immediately (e.g., within 30 to 60 min);[76,657,794,1124,1125] and (4) oxidants depress extraction and should be eliminated by proper completion of digestion. The time stability of Mn extracts can be improved by completely eliminating dispersed aqueous droplets from extracts (i.e., by centrifugation or filtration)[657,1126] or by adding acetone to the extracts[1124] or by getting rid of the organic phase (e.g., back extraction, etc.).[76,659,1110] Extraction systems based on other reagents (cupferron, 8-hydroxyquinoline) pose their own problems, e.g., multiple extractions, a need for complete mineralization of the organic matter, etc.[1112,1114,1120]

A tempting approach to Mn biological assays is the direct ET-AAS of intact or simply diluted liquids (blood,[1088,1089] CSF,[763,1092] serum,[511,605,1088,1092,1097] and urine[763,1097]) and microsamples of solid tissue[505,655,781,1099,1108] and hair.[79,80,778] While the contamination risk is greatly reduced with such procedures, most of them call for a complicated calibration (e.g., vs. preanalyzed specimens or matrix-matched standards, or else by standard additions). Sample homogeneity may also prove to be a problem; in particular, blood samples should be completely hemolyzed (+ Triton X-100®[1089]) and tissue thoroughly powdered (approximately 1-mg samples!). Special, large volume graphite tubes,[505,1099] or cup versions of graphite furnaces and/or Zeeman background correction[1088] are used in some direct ET-AAS procedures. Most of them suffer from delivery errors, matrix residue accumulation, sensitivity drift, and prolonged heating cycles. Use of the L'vov platform, ashing in air,[605] or oxygen as an alternate gas, as well as efficient matrix modifiers (perhaps $Mg(NO_3)_2$ — up to 1400°C[553]), are yet to be further explored.

Table 3
EXTRACTION SYSTEMS FOR Mn FROM BIOLOGICAL LIQUIDS/DIGESTS

Extraction system	Matrix and conditions	Ref.
APDC-MIBK	Ashed milk; pH 4; 2×;[a] preextraction of interferents at pH 2 needed	745
K benzylxanthate-MIBK	Ashed tissues; pH 7.0—9.0	1111
Chinoform-MIBK	Ashed urine; pH 7.5	1112
Cupferron-MIBK	Ashed urine; pH 7; 5×[b]	1114
	Ashed blood/urine or pH-adjusted urine; pH 7.1 ± 0.1	1090
	Ashed tissues; pH 7.5; preextraction of Fe at pH 1 needed	1120
DPC-Py[c]-toluene	Ashed blood/tissues; pH 5.5—7.8	473
DPM[d]-n-butylamine-MIBK	Ashed blood; pH 10.5	1113
HHDC-$CHCl_3$	Ashed food; pH 6; 3×[b]	659
HHDC + 8-hydroxyquinoline (4 + 1) − $CHCl_3$ + i-butanol (3 + 1)	Biological digests (feces, food, urine); pH 6.5—7; 4—5×[b]	39
8-Hydroxyquinoline-MIBK	Ashed blood; pH 8.5—9	332
8-Hydroxyquinoline-$CHCl_3$ + MIBK (1 + 1)	Ashed urine; pH 8.3 (citrate); 2×[a]	27, 1101
SDDC-MIBK	Urine or ashed urine; pH 6 (acetate)	974, 1100
	Urine; pH 6—9	7
	Fish digests; pH 8.0	494
	Digests (blood, serum, tissues); pH 6.5—6.7	7, 1110
	Digested urine or milk; pH 5.2 (citrate)	1094, 1096

[a] Double extraction.
[b] Multiple/exhaustive extraction.
[c] Diphenylcarbazone-pyridine.
[d] Dipivaloylmethane (2,2,6,6-tetramethylheptane-3,5-dione).

Table 4
SELECTED AAS PROCEDURES FOR Mn DETERMINATION IN BIOLOGICAL SAMPLES

Sample	Brief procedure outline	Ref.
Blood	2 μℓ of heparinized, completely hemolyzed (+ Triton® X-100) blood injected into a Varian Techtron® CRA-63 graphite cup, and ramp dried to 90°C for 40 sec, ramp ashed to 530°C for 25 sec, ashed at 1200°C for 40 sec, and atomized at 2200°C; standards on blood; alternatively, 1 + 1 dilution with aqueous 0.4% v/v Triton® X-100	1089
Blood, serum	Zeeman ET-AAS of (1 + 1) diluted serum or (1 + 3) diluted blood: ramp-dried to 135°C, ramp-ashed at 350—450°C for 60 sec, and atomized at 2400°C; standards on blood or serum	1088
CSF, serum	ET-AAS: 125 (ramp, 120 sec), 800, and 1300°C; background correction; standard addition calibration	723, 1092
Diet, fish, hair, tissues	Dry ashing at 450°C for 24 hr; ashes evaporated with HNO_3 or HCl, and dissolved in, e.g., 10% v/v of HCl or HNO_3; flame AAS; background correction; acid-matched standards;	442, 591, 731, 733, 739
Feces, food, meat, milk	1- to 2-g sample boiled for 30 min with 15 mℓ of a (27:3:20, by volume) HCl-HNO_3-H_2O mixture, and diluted to 50 mℓ; flame AAS;[a] standard addition calibration	729
Food	HNO_3-H_2SO_4 digestion; flame AAS;[a] standards in 5% v/v H_2SO_4	730

Table 4 (continued)
SELECTED AAS PROCEDURES FOR Mn DETERMINATION IN BIOLOGICAL SAMPLES

Sample	Brief procedure outline	Ref.
Hair	HNO_3 digestion; ET-AAS; acid-matched standards[a,b]	79, 557, 876
	0.5- to 1-cm segments (ca. 0.1 mg) directly analyzed; ET-AAS: ramp to 1100°C and atomized at 2600°C; background correction; standards on spun silk or animal hair	79, 80, 778
	50 mg of hair + 0.1 mℓ of water + 2 mℓ of TAAH (Soluene® 350) heated for 1 hr at 55°C, cooled, and diluted to 10 mℓ with toluene; ET-AAS: 135, 1000, and 2600°C; background correction; standard addition calibration (24-hr solubilization at room temperature preferred)	368
Liver (small samples)	2-mg sample (dry weight) decomposed with 40 μℓ of a (5 + 1) HNO_3 + $HClO_4$ mixture in a Teflon®-lined bomb;[c] stepwise (!) heated for 3 hr at 90 and for 3 hr at 120°C, cooled, and diluted to 1 mℓ; 75-μℓ aliquots pulse nebulized in an air-acetylene flame; background correction; standard addition calibration	760, 761, 1109
Meat, tissues	1-g sample heated with 10 mℓ of 1 + 1 HNO_3 at 80°C, evaporated to ca. 3 mℓ, and diluted to 25 mℓ; flame AAS vs. acid-matched standards or ET-AAS: 150, 600, and 2500°C with standard addition calibration; background correction	517
Milk	Deproteinization with HCl (1 *M* final concentration); ET-AAS[a,b]	775
Muscle (biopsies)	10-mg sample (wet weight) solubilized overnight with 0.2 mℓ of TAAH (Lumatom®) at 60°C in a closed vessel;[c] ET-AAS: 150, 1000 (ramp), and 2400°C;[a] standard addition calibration; likewise References 521, 610, and 777 (tissues, etc.)	92
Plasma, tissue	Deproteinization with an eq vol of 4 *M* HCl at 60°C for 1 hr, and centrifugation; ET-AAS with Varian Techtron CRA-63 or CRA-90, and 105°C for 60 sec, 1300°C for 10 sec, and 2500°C; background correction; standards in 2 *M* HCl; 10% w/v tissue homogenates in 0.2% Triton® X-100 similarly deproteinized and analyzed (100, 900, and 2500°C)	1107
Serum	1 + 9 dilution with water; ET-AAS: 110, 1100, and 2850°C; background correction; standard addition calibration	57
Serum, urine	1 + 1 dilution with water; ET-AAS: ramp to 110°C, ramp ashing to 1100°C, and atomization at 2700°C (miniflow); background correction; standards in 0.1 *M* HNO_3[a]	1097
Tissues	HNO_3 digestion; ET-AAS: 120, 800, and 2400°C; background correction; standard addition calibration	1119
	TAAH solubilization (1 mℓ of Soluene® 100 containing 2% w/v APDC per 0.1 g of tissue) for 24 hr, and dilution with toluene; flame AAS[741,a,b] or ET-AAS[610,a,b]	610, 741
	LTA, and ashes dissolved in 10% v/v HCl or wet ashing with H_2SO_4 and H_2O_2;[c] ET-AAS; background correction; acid-matched standard solutions[b]	774, 1091
Tooth enamel (biopsies)	0.2-mg sample dissolved in 0.4 mℓ of 0.5 *M* HCl; ET-AAS; background correction; standard addition calibration	84
Urine	Cupferron-MIBK extraction at pH 7.1 ± 0.1 (imidazole-HCl buffer preferred) for 10 min, and centrifuged; ET-AAS: 85, 300 (ramp), 600, and 2600°C (miniflow); standard additions extracted similarly	1090
	ET-AAS: 50, 101, 1100 (ramp), and 2600°C;[b] background correction	763

[a] Background check advisabe.
[b] Standard addition check advisable.
[c] Safety precautions mandatory.

IV. CONCLUSION

AAS is the best method for trace Mn, with RNAA a highly reliable alternative.

Simple flame AAS procedures are used for the determination of micrograms-per-gram Mn levels in food, feces, hair, and tissues.

At nanograms-per-milliliter levels encountered in biological liquids, the direct ET-AAS is obviously the preferred technique. Contamination control is of paramount importance with specimens such as serum, urine, CSF, saliva, etc. Inasmuch as sample handling is preferably restricted to a minimum (e.g., addition of a composite diluent comprising a matrix/analyte modifier and a surfactant), versatile equipment/programming as well as somewhat complicated calibration are justified. Interferences are duly dealt with by means of a L'vov platform or STPF.

Chapter 20

MERCURY

I. DETERMINATION OF MERCURY IN BIOLOGICAL SAMPLES

Mercury levels in biomaterials range broadly within several orders of magnitude: from subnanograms to above micrograms per gram (Table 1). The cold-vapor technique of AAS (CVT-AAS) is the best routine method: it is easily accessible, sufficiently sensitive, and relatively simple.

NAA (often INAA) has been applied to Hg determinations in blood,[45,103,1128,1131] bone,[97,102] diet,[45,96,116,239,1128,1133] feces,[96] fish,[45,853,1147] hair,[9,11,45,48,98,99,103,104,118,184,290,1134,1148] milk,[110] nails,[103,290,560] serum/plasma,[111,656,1147] skin,[112] teeth,[103] tissues,[12,45,103,108,372,1128,1133] urine,[45,96,111,165,1128] and other biologicals.[45,121,1147]

Spectrophotometry (often after dithizone extraction) is an established technique for Hg,[7,133,414,591,1149-1152] which, however, is inferior to CVT-AAS as regards sensitivity, selectivity, and sample throughput.

Successful applications of MIP-AES[1141,1153] have been described, following CVT separation. ICP-ES[35,326] and arc-source AES (with chamber-type electrode)[130,1154] require preconcentration (e.g., by ion exchange,[326] coprecipitation,[1154] extraction,[130] reduction-volatization,[145] etc.). The direct XRF, too, lacks sensitivity[128,691,1155] and calls for an effective preconcentration step, e.g., by coprecipitation.[540]

Several other analytical methods have incidentally been applied: AFS,[1156] DPASV,[124,125,1157] catalytic,[577] kinetic-coulometric,[1158] enzymic-spectrophotometric,[1159] IPAA,[104] SSMS,[3,25,55,325] PIXE,[1155] substoichiometric isotope dilution,[1160] etc. A few laboratories are able to perform highly accurate and precise IDMS assays.[1161]

GC is an established method for the determination of both total Hg and CH_3HgCl in various biomaterials: blood,[1162,1163] feces,[1162] food,[1162,1164] fish,[853,1157,1165-1167] hair,[1155,1162,1168] milk,[1162] tissues,[1137,1162] urine,[1162] etc.[188,1164,1169] GC involves tedious and time-consuming preparatory steps (extractions, distillation) and an electron-capture detector (ECD) (a MIP-AES detector is even better, as it is independent of Hg chemical form[1167]).

Speciation schemes which are based on GC-MS, dithizone extraction-TLC,[1152,1165,1170] GC-ET-AAS,[283,284] extraction ET-AAS,[1163] TLC-CVT-AAS,[1170,1171] steam distillation CVT-AAS,[1172,1173] as well as many CVT-AAS procedures for selective reduction-volatization of organic, inorganic, and total Hg (see below) have been described. (For reviews see References 45, 1174, and 1175.)

II. ATOMIC ABSORPTION SPECTROMETRY OF MERCURY

The most sensitive spectral line, Hg 184.9 nm, is unfortunately beyond the working range of most commercially available AAS monochromators. A special monochromator as well as a nitrogen-purged optical path are therefore required. Recently, Hoffman et al.[1176] constructed an AA photometer, comprising a low-pressure Hg discharge lamp, a 26-cm absorption cell, and a photocell with a CsI photocathode (''blind'' above 200 nm).

The practical wavelength, although 50 times less sensitive, is 253.7 nm. EDL is recommended vs. HCL, as it provides higher intensity (50 ×) and lower DLs (1.4 to 3.6 ×);[138] additional drawbacks of HCL are a limited lifespan and pronounced dependence of sensitivity on lamp current.

In air-acetylene flame, characteristic concentration and DLs are 1.5 to 7.5 μg/mℓ and 0.1 to 1 μg/mℓ, respectively.[56-59,294] There is strong spectral interference from Co (due to the

Table 1
MERCURY CONCENTRATIONS IN HUMAN BODY FLUIDS AND TISSUES

Sample	Unit	Selected reference values	Ref.
Blood	μg/ℓ	3 ± 2	1127
		<5	96, 1128
		5	27
		6[a]	3
		(3—11)	1129
		9 ± 6	1130
		x̄, 11	1131
		14 (Japan)	94
		(12—24) (U.K.)	103
		17 ± 15 (Italy, rural)	1132
		20 ± 13 (4—62) (Italy)	1132
Bone	μg/g	x̄, 0.040 (0.04—0.096)[b]	97
		x̄, 0.30;[c] 0.45 ± 0.36 (0.03—1.04)[c]	103
Brain	μg/g	0.27[d] (0.06—10.6)	1133
		x̄, 1.07;[c] 2.94 ± 4.1 (0.12—15.2)[c]	103
Erythrocytes	μg/ℓ	<5	1128
Feces	μg/day	10[e]	8
		38.3 (2—109) (Italy)	96
Hair	μg/g	x̄, 0.45—0.66;[f] (0.1—2.3) (Austria)	99
		0.51 ± 0.91 (0.1—3.6) (Rome)	1132
		0.56 ± 0.88 (Italy, rural)	1132
		♂, 0.58;[d] ♀, 0.99;[d] (0.05—14) (New York)	288
		1.2 ± 0.6 (U.S.S.R.)	1134
		1.4 ± 2.1 (<0.1—12.8) (Iraq)	118
		x̄, 1.08 and 1.55; (0.2—40) (U.S.A.)	9
		x̄, 1.5; 2.4 ± 4.5 (0.3—32) (Italy)	98
		(0.01—2.5)	1135
		♂, 1—1.5;[g] (0—6) (U.S.A.)	814
		♀, 4—5;[g] (0—>18) (U.S.A.)	814
		(0.2—8)[c] (Canada)	104
		♀, 1.99;[d] ♂, 2.77;[d] (0.24—23.44) (Japan)	1136
		x̄, 3.9 (0.99—13.2) (Japan)	11
		4.1 ± 2.6 (Japan)	1137
		(<0.3—27) (U.S.S.R.)	102
		x̄, 4.2; 5.52 ± 5.21 (0.03—24.4) (U.K.)	103
		18.5 and 25 (U.S.A.)	1138
Kidney	μg/g	x̄, 0.091; 0.12 ± 0.04 (0.054—0.27)	105, 106
		x̄, 0.30; 0.762 (0.075—3.36)	1139
		x̄, 0.34; 2.0 ± 4.6 (0.018—18)	103
		0.71;[d] (0.02—27.3)	1133
		0.83	1140
		1.1 (Japan)	1137
Kidney cortex	μg/g	0.99 ± 0.84 (0.14—23)[c,h]	372
		9.4 ± 14 (0.35—53)[c,i]	372
Liver	μg/g	x̄, 0.037; 0.10 ± 0.02 (0.019—0.14)	105, 106
		0.077 (0.055—0.108)	108
		x̄, 0.12; 0.172 (0.026—0.545)	1139
		0.16 ± 0.01 (0.02—0.37)	187
		0.20;[d] (0.02—1.7)	1133
		x̄, 0.39; 1.0 ± 1.4 (0.042—5.56)	103
		0.47 (Japan)	1137

Table 1 (continued)
MERCURY CONCENTRATIONS IN HUMAN BODY FLUIDS AND TISSUES

Sample	Unit	Selected reference values	Ref.
Lung	μg/g	x̃, 0.008; 0.035 ± 0.015 (0.001—0.17)	105, 106
		x̃, 0.03; 0.03 (0.007—1.5)	1139
		0.24;[d] (0.03—6.2)	1133
		(0.046—0.094)	109
		x̃, 0.32; 0.52 ± 0.54 (0.066—2.3)	103
Milk	μg/ℓ	(1—6)	13
		8[a,c,j]	110
Nails	μg/g	♂, 0.39 ± 0.22 (0.04—0.9)[c,k]	560
		♀, 1.9 ± 1.8 (0.3—4.3)[c,k]	560
		1.01 ± 1.22 (0.25—8.09)	1132
		1.11 ± 1.18	1132
		x̃, 4.76;[c] 7.27 ± 8.39 (0.8—33.8)[c] (U.K.)	103
Placenta	μg/g	0.129 ± 0.161[c]	377
Plasma or serum	μg/ℓ	1.8 ± 0.6	111
		3	113
		<5	1128
Saliva	μg/ℓ	<5[l]	1141
Tooth (enamel)	μg/g	<0.1	25
Tooth (whole)	μg/g	(<0.5, often <0.1)[m]	1142
Urine	μg/ℓ	(1.13—9.8)	1143
		3.5 ± 2.3	1130
		4.3 ± 0.8 (1.9—9.0)	1128
		4.31 ± 2.33 (2.0—9.5)	1144
		(5—10)	27
		10	113
		<15	1145
		(<2—22)	96
	μg/day	1.0 ± 0.69 (0.16—2.4)	111
		1.4 (0—20)	1146
		3.7 ± 0.6 (1.3—6.1)	1128

Note: Wet weight unless otherwise indicated. Mean ± SD (range in parentheses). x̃, median.

[a] ng/g (ppb).
[b] Femur.
[c] Dry weight.
[d] Geometric mean.
[e] "Reference Man".
[f] Range of medians.
[g] Mode.
[h] Infants.
[i] Adults.
[j] Transitional.
[k] Thumbnails.
[l] Parotid.
[m] Roots.

Co 253.65-nm line), a pronounced effect of reducing agents, and background absorption. Absorbance strongly depends on the Hg oxidation state: $Hg^{o} \gg Hg(I) \gg Hg(II)$.

Both the Sampling Boat® and the Delves cup are of limited usefulness, owing to serious matrix effects (Co, PO_4^{3-}, etc.), background absorption, and inadequate sensitivity [DLs of

Table 2
THERMAL STABILIZERS OF MERCURY IN ET-AAS

Thermal stabilizer	Maximum dry/ ash temperature (°C)	Atomization temperature (°C)	Ref.
APDC-MIBK extracts	75	2000	7
Spheron Thiol® resin	100	>500 (opt. 1000)	1186
0.05 *M* KBr-0.5% v/v Br_2-1% v/v HNO_3	100	1100	1184
0.005 *M* $Na_2S_2O_3$	150	1200	1183
Ag foil lining	150	900	1182
Au foil lining	150	950	1182
4% HCl-4% H_2O_2	155	>1100 (opt. 2000)	1181
4% HCl-4% H_2O_2-4% HNO_3	155	>1100 (opt. 2000)	1181
$(NH_4)_2S$	<200	—	223, 1179, 1185
0.5% $AgNO_3$-0.1% $KMnO_4$-1% HNO_3	200	2000	1180
5% $(NH_4)_2S$-1% HNO_3	225	2000	1180
8% HCl-2% H_2O_2	250	2000	1180
0.05% $K_2Cr_2O_7$-1% HNO_3	250	2000	1180
0.1% $K_2Cr_2O_7$ − 5% HNO_3	275	750	1178
5% $(NH_4)_2S$ − 1% HNO_3	300	—	223
0.1% $KMnO_4$	<370	>2000	1179

0.02 μg/mℓ (10 ng)[415] and 0.1 μg/mℓ (0.4[1177] to 10 ng[415])]. Multiple peaks have been observed with Delves cup, presumably due to Hg-Ni interaction.[1177]

ET-AAS is not a particularly suitable technique for Hg, either. The characteristic concentration and DLs are down to a few nanograms per milliliter (0.1 to 0.2 ng),[61,223,1178-1180] the state-of-the-art figures being 0.05 ng/mℓ[59,1178] and 0.005 ng.[59]

Hg is readily lost during drying and ashing, so that thermal stabilization is essential. Several matrix/analyte modifiers have been tested and compared,[223,1178-1182] being effective between 75 and 370°C (Table 2); these reagents are believed to bind Hg^{2+} or CH_3Hg^+[1183] in relatively stable and nonvolatile species (K_2HgBr_4,[1184] $HgCl_2.H_2O_2$,[1181] chelates,[1183,1185] amalgams,[1179,1182] etc.). Drying/ashing temperatures are critical in view of accuracy and precision of assays. Atomization temperatures are relatively low, thus facilitating time resolution of the Hg peak from the ensuing background signal. Temperature-controlled heating is important; the "gas-stop"[7,1180,1183] and "peak-area"[1178] facilities are also very useful. A clean-out stage is required. The background absorption is severe.[1178,1184] Pronounced matrix effects should be expected and checked. The quality and "history" of the graphite tube appears to be important; better performance (sensitivity, precision, etc.) might be obtained with uncoated graphite tubes.

While all discussed techniques of AAS appear somewhat iffy in assays of nanogram-Hg levels, the CVT-AAS method is — quite the reverse — most popular, sensitive, inexpensive, simple, and selective. The DLs are usually around 0.1 ng/mℓ or 1 ng and can be cut down to an impressive 0.01 ng/mℓ or 0.1 ng[1175,1187,1188] or even to 0.003 to 0.005 ng/mℓ.[1189]

The principle of CVT-AAS is simple: mercury is reduced to its elemental state (Hg^o) either by heating/incinerating the sample at 500 to 1000°C (i.e., *thermal volatization/atomization*) or by adding a reductant to the sample solution/homogenate and sweeping the Hg^o out of the solution via a purge gas (i.e., *chemical reduction/volatization*). In both variants, the purge gas carrying Hg^o, while it may or may not pass through filters, drying agents, amalgamation attachments, condensers, etc. (so as to trap the concomitant water vapor, salt mist, etc. or to additionally concentrate the analyte), finally enters an absorption cell (gas cuvette) placed along the optical pass of a spectrophotometer, where the atomic absorption of Hg^o vapor is measured.

The CVT-AAS based on chemical reduction is more versatile and is commercially available as an attachment to AAS spectrophotometers. Portable, simple, and cheap photometers for Hg are also produced.

There are two types of CVT accessories. In the first type, the open-ended system, the Hg^o vapor is stripped out of the solution into the head space of the reaction vessel and then purged through the absorption cell to the scrubber; a peak signal is thus produced. In the second type, the closed system, the purge gas recirculates through the absorption cell, circulating pump, and the reaction vessel until a steady state is reached; absorbance thus reaches a plateau where height is measured, and then a valve is opened to sweep Hg vapor out of the system. Both systems have their pluses and minuses: the former is simpler, faster, more sensitive, and amenable to automation/semiautomation, while the latter appears to be less prone to interferences, and larger sample aliquots can be dealt with.

The dead volume of a CVT-AAS system should be kept low,[1190,1191] and the absorption cell should be rational in shape and dimensions. Inasmuch as convergent light beams are typical in most AA optical schemes, the optimum shape of the absorption cell may prove to be two truncated cones, as shown recently;[1129,1192] such shaping provides marked improvement in precision, sensitivity, and DLs.

Water vapor tends to condense on absorption cell windows and cause light losses. Drying tubes packed with $Mg(ClO_4)_2$,[1140,1145,1146,1193-1196] or, rarer, with $CaSO_4$ or silica gel, H_2SO_4 desiccators,[1197] or ice-cooled impingers[1198,1199] are often inserted between the reaction vessel and the absorption cell. These driers may need frequent change; moreover, it is now realized that a $Mg(ClO_4)_2$ drier, especially when getting damp, may be responsible for lower sensitivity, memory effects (due to trapping of some Hg), tailed peaks, prolonged measurements, and a gradual sensitivity drift.[1197,1200,1201] An H_2SO_4 desiccator needs occasional stirring.[1197] For these reasons, vortex mixing rather than aeration[1145] and heating the absorption cell[255,1161,1202-1204] (e.g., to 200°C) are becoming even more popular.

There is a choice between several reductants, depending on construction of CVT apparatus, preparatory step, expected interferents, and other factors. Comparison studies and discussion of these reductants have been made:[1141,1187,1202,1203]

1. Tin(II)chloride (or, better,[1205] sulfate[1191,1195,1205,1206]) in an acidic medium is the classical reductant of Hg(II), as proposed by Poluektov et al.[1188] and Hatch and Ott.[1195] Concentration of $SnCl_2$ or $SnSO_4$ is often between 5 and 20%, but somewhat lower concentrations — down to 1% — may be used in continuous-flow reduction systems[1202] or in systems with recirculation. Reagent solutions should be freshly (daily) prepared. Complete release of Hg from its chemical bonds with the matrix constituents (especially from Hg-C and Hg-S bonds) is essential. The reduction with acidic Sn(II) is more susceptible to interferences from I^-, Se, and Te than $NaBH_4$ reduction, but more tolerable to acids, Ni, Cu, As, and Bi.[1141,1203]
2. Various combinations of an acidic Sn(II) chloride or sulfate with other reagents — $CdCl_2$,[1145] NaCl,[1161,1195,1206] ascorbic acid,[1141] $NH_2OH.HCl$,[1194] etc. — are used, aimed at improving the time stability of the reagent and/or masking the -SH groups in solution.
3. Alkaline stannite by itself[1143,1173,1182,1207-1212] or in combination with Cd (II)[1198,1199,1208,1210,1215] or Cu(II)sulfate[1214] is used in speciation studies[1143,1208,1215] and in analyses of undigested biological samples/homogenates.[1163,1190,1208,1210-1212]
4. $NaBH_4$, a powerful reductant, is now increasingly used.[246,1127,1141,1187,1189,1203,1204,1216-1220] Its concentration is between 1 and 10%, often less than 5%, and 1 to 2% NaOH is added as stabilizer. Determinations are faster and more precise[1187] than with a Sn(II)

reductant (e.g., 1 to 3 vs. 3 to 5 min per assay),[1127,1141,1203] but are more liable to interferences from excess acids, hydride-forming elements, and foaming.[1141,1187,1203,1219,1220] It is hazardous to use $NaBH_4$ in a system which has not been thoroughly cleaned from the residue of Sn(II) reagent!

5. Ascorbic acid,[1182,1207,1221] in presence of excess halide ions, has been shown to reduce and volatize the mercury trapped in a 0.01 *M* I_2 absorbent.[1207]
6. Formaldehyde can reduce Hg(II) in presence of excess Se or Te.
7. Dihydroxymaleic acid at pH 9.2 to 14 is tolerable to Cl^-, Br^-, I^-, and IO_3^-.[1222]
8. Hydrazine hydrate (40% aqueous solution) has been incidentally applied as a reductant.[1182,1212]

Reductant solutions may be precleaned from traces of Hg by purging with an inert gas.

Certain parameters should be experimentally optimized with every piece of apparatus and type of samples: acidity and volume of the solution in the reaction vessel (the total volume being preferably low, e.g., 10 to 20 mℓ, and consistent); order of adding reagents; concentration and amount of reductant; reaction time; flow rate of purge gas; intensity of vortex mixing; etc.

The extent of interferences depends on many factors: design of apparatus, reductant, sample preparation procedure, acidity/pH, mode of calibration (peak height or peak area[1223]), as well as on whether an amalgamation attachment is used. Therefore, the published information on interferences[1141,1187,1188,1203,1205,1216,1219,1220,1224] should be used only as a guide:

1. Very strong suppressors are the ions of noble metals (Au, Ag, Pd, Pt, Rh, Ru),[1192,1212,1216,1224] which are readily reduced to their elemental state and retain Hg in the reaction vessel as an amalgam.
2. Among less pronounced interferers, let us mention some ions which are precipitated or reduced to hydrides: Se,[1141,1187,1188,1207,1216,1225] Te,[1188,1207] Cu,[1141,1187,1224] Sb,[1141,1187] As,[1141,1187] and Bi.[1116,1141] Their interference seems to be more pronounced in systems with $NaBH_4$ reductant,[1141,1187,1224] as well as in systems with an amalgamation attachment.[1187]
3. Some ions or ligands may strongly bind Hg in complexes or precipitates and thus depress Hg absorbance (especially in peak-height mode): S^{2-},[1175,1187,1207] I^-,[1141,1203,1212,1224,1226,1227] F^-,[1207] (excess) Br^-[1141,1188,1224] and Cl^-[1141,1175,1203] (usually in the order $I^- \gg Br^- > Cl^-$), $S_2O_3^{2-}$,[1175,1224] IO_3^-,[1227] cysteine,[1205,1218] excess $NH_2OH.HCl$,[1205,1228] residual Mn(IV) oxides,[1196] and not fully digested organic matter.[426,1218,1223,1229,1230]
4. Excessive foaming may occur, especially in presence of undigested or incompletely digested organic matter, excess NaCl or CH_3COOH,[1219] in particular with the $NaBH_4$ reductant in strongly acidic media. Additions of various antifoaming agents are therefore practiced, e.g., "Antifoam 60",[1231] Dow Corning® DB 110 A Emulsion,[246,1204] 1-octanol,[1198,1205,1210-1212,1223,1230] "Silicone Antifoam",[1199,1209,1232] tri-n-butylphosphate,[1127,1208] etc. Both foaming and antifoaming additions[1205,1231] cause signal depression (less pronounced in peak-area measurements).[1205]
5. The effect of acids was studied by Melcher,[1219,1220] Melcher and Welz,[1203] Rooney,[1216] and Poluekto et al.,[1188,1207] with most results being in good agreement. The $SnCl_2$ reduction system stands up to at least 2 *N* mineral acids,[1188,1207] with maximum concentrations as follows: 10% w/w HCl, 20% w/w HNO_3, up to 20 to 30% w/w H_2SO_4, and 20% w/w CH_3COOH.[1203] The $NaBH_4$ reductant is useful in 2 *N* HCl, H_2SO_4, $HClO_4$, and HNO_3,[1216] and up to 5% w/w HCl, 4% w/w HNO_3, 10% w/w H_2SO_4, and 2 to 3% w/w CH_3COOH are tolerable.[1203] In presence of HNO_3 or CH_3COOH, it is advisable to add also 3% w/w HCl so as to improve precision and make reduction

more tolerable to these two acids.[1203] It is worth noting that in presence of excess acids (e.g., 20% w/w HNO_3), the $NaBH_4$ reduction may become violent.[1219]

6. Background absorption is not likely to create problems if a heated absorption cell (>130 to 200°C) or a drying tube are employed. However, background check is advisable, as vapors of nitrogen oxides,[1193,1233] organic solvents (acetone, benzene) may give rise to some background. Continuum source (D_2 or H_2 lamp) or else adjacent spectral lines of other sources (Al, Co, Cu, Mg, Pt, Si)[302,629,1175] may be used for correction/check.

Calibration is typically performed with standard solutions containing the same amount of acids and antifoaming agents as the sample solutions. Peak-area measurements may improve accuracy in some cases.[426,1187,1205,1223,1234] Standard addition checks are easy to perform and are advisable with every newly adopted procedure, new type of samples, as well as in analyses of untreated or incompletely digested biomaterials.

III. AAS METHODS FOR ANALYSIS OF BIOLOGICAL SAMPLES

Literature on this subject is abundant (Table 3). Several important problems deserve discussion, as they govern accuracy, precision, and — more generally — usefulness of analytical data:

1. Representative sampling
2. Possible losses, contamination, and chemical transformations of the analyte during storage of specimens
3. Losses of Hg during preparation of samples for analysis — washing, drying, ashing, etc.
4. Efficiency of separation/preconcentration from the matrix
5. As regards AAS quantitation: (a) interferences (especially in procedures involving incomplete or no digestion), (b) volatility of Hg in ETA, (c) calibration, etc.

Mercury is rather nonhomogeneously distributed in some specimens, e.g., feces,[96,1280] food,[96] hair,[9,118,814,1148,1280] kidney,[372] milk,[1234,1280] teeth,[1142] and (aged) urine.[165,1196,1215,1276,1281,1282]

Urine specimens need vigorous shaking before aliquoting, as the precipitates formed contain mercury (even after storage at low pH[1257,1276]). Desquamated tubular cells in urine precipitate — especially in urine from Hg-exposed subjects[1281] — were also found to contain Hg.

Large spatial variations of hair mercury are known to exist, both within (along)[118,814,1148] and between[161] single hairs; the first (proximal) 1-cm hair segments are preferably analyzed[1168,1280] as indicative of the recent (last month's) exposure.

A relevant sampling procedure for parotid saliva has been described.[1239,1240]

During storage, Hg in biomaterials is likely to incur both partial losses[421,1215,1276,1283] and chemical transformations.[1162,1169,1189,1280] Bacterial contamination of specimens (blood, plasma, tissue homogenates, and urine) was found to be responsible for substantial volatization losses (as Hg^{o}[1283]) during storage[1276,1280,1283] and digestion[1276,1282] of samples. Moreover, mercury is readily adsorbed on all container/labware surfaces: glass, polyethylene, polypropylene, polystyrene, polycarbonate (see Reference 421), and even on silicone-treated glass,[1150] PTFE,[1141,1153] and quartz.[1141] The extent of adsorption depends on material and precleaning of the vessel, the particular organic matrix, storage time and temperature, pH, added conservants, etc. Minimum adsorption losses were found on glassy carbon and quartz.[1141] Glass containers are preferred to plastics.

Table 3
AAS TECHNIQUES FOR Hg DETERMINATION IN BIOLOGICAL MATERIALS

Technique	Biological materials
Flame AAS after extraction	Blood,[472] tissues,[472] urine[339,472]
Sampling Boat® after preconcentration	Urine[1235-1237]
CVT-AAS (only dilution or pretreatment directly in the reduction vessel)	Blood,[1127,1190,1198] urine[246,255,1204,1212,1215,1218,1232]
CVT-AAS after pretreatment in an open vessel	Blood,[27,421,1199,1210,1211,1216,1228,1238-1242] bone,[1243] fish,[271,853,1141,1166,1172,1199,1210,1212,1214,1226-1229,1244-1251] food,[271,1194,1214,1216,1247] hair,[370,1138,1168,1190,1198,1223,1228,1242,1253,1254,1435] milk,[1141] nails,[1228] placenta,[1243] plasma,[1146] saliva,[1239,1240] teeth,[1243] tissues,[187,271,1139,1141,1161,1172,1191,1198,1199,1210,1211,1214,1227,1228,1230,1242,1245,1247,1248] urine,[27,1143,1145,1146,1198,1210,1215,1216,1224,1228,1231,1239,1240,1242,1245,1254,1256] and other biomaterials[1244,1261,1262]
CVT-AAS after pretreatment in a closed vessel	Blood,[130,1132,1140,1254] bone,[1142] fish,[458,1160,1206,1244,1257] hair,[426,1132,1206,1243,1254] milk,[1243,1258] mussels,[1257] nails,[1132] teeth,[1142] tissues,[130,190,1140,1203,1211,1230,1254,1258,1259] urine[130,1204,1254,1260]
CVT-AAS after combustion	Fish,[1141,1244,1263,1264] hair,[288,1135,1193,1254,1263] milk,[1141,1265] nails,[1193] tissues,[255,1141,1254,1266] and other biomaterials[1141,1244,1254,1266,1267]
CVT-AAS with special apparatus	Blood,[1129,1268-1271] fish,[1222,1272] hair,[1136,1137,1284] meat,[1141,1222,1272] milk,[1141,1234,1272] plasma,[1269] tissues,[1137,1141,1269,1271] urine,[1144,1150,1196,1268,1269] and other biomaterials[1153]
ET-AAS after pretreatment	Fish,[1183] urine[1178,1184]
ET-AAS after chelate extraction	Blood, tissues, urine[7]
ET-AAS with special equipment	Fish,[1273] tissues,[1135,1273,1274] urine[1184]
Automated or semiautomated assays	Blood,[1275,1276] tissue digests,[1259,1276] urine[1178,1196,1259,1275,1276]
Speciation techniques	Blood,[1173,1199,1209-1211,1241,1275,1277,1278] egg,[1210] erythrocytes,[1277] feces,[1209] fish,[853,1167,1170,1172,1183,1189,1199,1210,1214,1279] food,[1164,1210,1214] hair,[1155,1168,1189,1253,1277] meat,[1214] mixtures of organomercurials,[1171] placenta,[1173] plasma,[1277] tissues,[1199,1208-1211,1225,1255,1278] urine,[1143,1189,1208-1210,1215,1277,1278] water[1189,1217]

Urine specimens are preserved by acidifying with HCl[1145,1196,1208] or HNO_3[1145,1204] (e.g., to pH 1 to 2 or <1% v/v of acid) or by alkalizing with NaOH (to pH ~ 13)[1215,1276] or else by adding toluene,[1276] thymol,[1280] sulfamic acid + Triton® X-100,[1215,1276] or cysteine + EDTA + NaCl.[1215] Storage times are extended by freezing[1204,1208,1215,1280] or at least refrigerating[1215] urine specimens.

Littlejohn et al.[1215] compared different storage conditions of a urine sample containing 40 μg Hg/ℓ for up to 24 days. Most suitable and recommended was the technique of Skare,[1276] in which 1 g of sulfamic acid and 0.5 mℓ of Triton® X-100 were added to 500 mℓ of urine, thus resulting in a time stability of at least 1 month at room temperature. (Littlejohn et al.[1215] recommended storage at 4°C). Suitable proved to be also the storage of alkalized urine (with NaOH to pH 12.7) at 4 and −10°C. In urines stabilized with L-cysteine + EDTA + NaCl, the losses at 4 and −10°C were confined to 10%, but untreated urine, even in a deep freeze, lost up to 20% of the Hg during the first few days of storage.[1215]

Méranger et al.[421] studied the effect of storage time (up to 672 hr), temperature (−70, −10, 4, and 22°C), and container type (Pyrex®, soda glass, polyethylene, polypropylene, polystyrene, and polycarbonate) on Hg levels of blood ($HgCl_2$-fortified, heparinized porcine blood). Greatest loss occurred at highest temperatures and from plastic containers; the only remedy was storage at the lowest temperatures in either glass or high density aliphatic plastic vessels; largest losses occurred after ⩾200 hr.[421] Other workers preferred to store blood[1140] and tissues[955,1140] at <20°C. *NG:* storage of tissues in Formalin®.[955]

Mercury vapors (from lab environment) readily penetrate polyethylene storage containers and may cause contamination of Hg-poor specimens (e.g., milk, bone, and even hair).

Various hair pretreatment procedures are documented: (1) simple "no-wash" approach;[104,625,1134,1136] (2) "mild" washing with water;[290,1148] (3) "natural" washing with hair shampoos[1168] or lab detergents;[1148,1168,1189] and (4) using organic solvents (acetone,[48,100,814] alcohol,[1148] benzene, ether,[9,100] etc.), either alone[1148,1168] or in combinations.[9,100,814,1253] Endogenously incorporated mercury is firmly bound and is not measurably extracted during washing[104,815,1148,1168] and experimental cosmetic treatment (shampooing, coloring, bleaching, and UV irradiation);[1168] even 1-hr "washing" with 1 *M* HCl did not extract any Hg.[1148] In vitro, 88 to 93% of external contamination ($^{203}HgCl_2$ and $CH_3{}^{203}HgCl$) was removed by a 15-min wash with either shampoo or a detergent solution, and 93% of the absorbed $^{203}Hg^{\circ}$ vapor was removed by a 30-min wash with a nonionic detergent.[1168] Recently, Suzuki and Yamamoto[1253] reanalyzed hair samples stored for 11 years and found no measurable changes in either organic or inorganic Hg.

Solid biological specimens (feces, food, hair, tissues) are safely homogenized by grinding/shattering at liquid-nitrogen temperature[426,532,533,944] or in an alkali[1162,1190,1199,1211,1214] (e.g., in 20 to 45% NaOH or 1.5 *M* KOH,[1228] either alone[1199,1228] or with addition of detergent[1276] or L-cysteine[1162,1169,1198,1211,1223,1253] or NaCl[1169,1190]). Alkaline homogenates are often used in direct and/or speciation procedures, so that their limited time stability should be reckoned with.[1168,1169,1189] Homogenations with water,[1162,1214,1255] or 1% saline[1198,1199] or 1% sucrose[1208] have likewise been reported, but they are not as effective as alkaline homogenation (involving partial digestion, as well).

Freeze drying is generally safe,[166,536,1233,1257] but — under improper operational pressure and/or temperature — some small, matrix-dependent losses cannot be ruled out. No significant losses of metabolized ^{203}Hg have been found with specimens of blood,[166,1233] brain,[166,1233] feces,[1233] fish,[1257] heart,[1233] kidney,[166,1233] muscle,[166,1233] mussels,[1257] serum,[166] spleen,[1233] and urine.[166] LaFleur[1233] found ~10% loss of Hg in feces of rats and guinea pigs fed with $C_6H_5{}^{203}HgCH_3COO$.

Ordinary oven drying is liable to Hg losses, depending on the matrix, heating rates, temperature, and duration. Iyengar et al.[166] established significant losses at 80°C from liver (up to 5%) and urine (~3%); at 105°C (tissue dependent, up to 10%, and up to 15% from urine); and at 120°C (tissue dependent, >10%, and up to 25% from urine). In certain specimens, however, viz. blood, kidney, and nail, mercury was fully retained after drying at 120°C.[166]

The complete decomposition of organic matter and the quantitative recovery of mercury as an inorganic Hg(II) is no easy achievement, perhaps more difficult than with any other element. On the one hand, mercury and its compounds are highly volatile and thus decompositions must be performed under strong oxidizing conditions but at lower temperatures. On the other hand, incompletely digested organic residues may interfere with the next quantitation step,[426,1138,1145,1146,1229,1230,1234,1258] by causing, e.g., complexation of Hg, foaming, inhibition of amalgamation, background absorption, etc. Therefore, there are certain typical digestion techniques, assigned to Hg. *NG:* ordinary dry ashing[91,266] and LTA,[169,275,276,626] as well as most common wet-ashing procedures (see below).

Ashing in (excited) oxygen in a dynamic system and with a liquid-N_2 cooled condenser ("cold finger")[1141,1153] is an efficient, fast, and very reliable technique, characterized by low blanks and complete recovery of Hg (and other volatiles); commercially available (e.g., Trace-O-Mat®, Kürner, Rosenheim, G.F.R.), it can be attached to apparatus for determining Hg (e.g., MIP-AES or CVT-AAS).[1153]

Large organic and biological samples (say, 5 to 20 g of food, meat, milk, tissue, etc.) have been successfully burnt in O_2 under pressure in the Bioklav® (Siemens AG, Karlsruhe, G.F.R.); mercury has been trapped in dilute HNO_3 and determined by CVT-AAS or other instrumental techniques.[273,274,540]

Another reliable technique is oxygen flask combustion,[91,110,266,288,1135,1137,1193,1265,1266] in particular with fatty samples (milk and milk products,[110,1265] fish, food, liver, etc.). It is important to use sample supports made of Ta, nichrome, or else stainless steel instead of Pt,[266,1175,1193] as platinum both amalgamates with Hg during ashing and interferes with the CVT-AAS assay. Various absorbing solutions have been used: dilute HCl,[110,266,1284] HCl + H_2O_2,[266] chlorine water,[1175] dilute HNO_3,[1193,1265] aqueous $(NH_4)_2Ce(NO_3)_6$,[1265] acidic $KMnO_4$,[266,1137,1265] and 0.025 *M* $K_2S_2O_8$ in 5% v/v H_2SO_4. Recently, Narasaki[1265] compared the last four absorbing solutions (in ashing milk and milk products) and recommended acidic peroxodisulfate.

Incineration of biological samples in air[1234,1244,1263,1264,1267,1272] or O_2[1129,1136,1150,1156,1254,1267-1269,1273,1274,1285] in special AAS apparatus of various design is also practiced. With efficient (Zeeman[1136,1273,1274]) background correction, the combustion gasses and Hg° vapor can directly enter an absorption cell; small solid samples (2 to 10 mg) are thus directly analyzed at an analytical rate of 1 assay per minute.[1273,1274] In another direct technique,[1150] the combustion gasses pass through appropriate condensers and filters to remove water vapor and salt mist, and then enter an absorption cell.

Both cited pieces of apparatus are rather complicated; an alternative approach is more popular, in which Hg° is separated and concentrated from combustion products by: (1) amalgamation with Au or another noble metal and subsequent thermal release;[1129,1160,1244,1268,1269] (2) trapping into a suitable absorbing solution which is then analyzed by CVT or another technique;[1156,1234,1254,1263,1264,1272,1285] and (3) condensing at liquid-N_2 temperature.[1267]

Amalgamation tubes are packed with Au,[1141,1156,1182,1187,1234,1244,1269,1270] Ag,[106,1141,1182,1230,1264,1285] Pt,[1182,1268] Cu powder (at 0°C),[1129] Pd,[1182] or else with alloys of these metals, in the form of wire, wool, folio, gauze, powder,[1129] "porous gold",[1270] as well as gold-plated inert substrates (sand,[1234] asbestos,[1182] quartz microspheres or wool,[1141,1160] etc.; these are preferable for their thermal and mechanical stability[1141] and less-pronounced retention/memory effects).

Factors to be considered in working with amalgamation attachments are the optimum temperature of thermal release (often, approximately 400 to 500°C); the maximum amount of Hg that can be trapped; the optimum gas flow rate; the possible inhibitory action of some concomitants (water vapor, S^{2-}, Sb, Se, etc.[1187,1269]); and others.

Mercury vapor can also be trapped in an absorbing solution of acidic $KMnO_4$ (e.g., 1 to 5% $KMnO_4$ in 10% v/v H_2SO_4),[1156,1254,1263,1264,1267,1285] from which it can then be reduced/volatized by the Hatch and Ott procedure[1195] or else extracted as dithiozonate.[1137]

The Hatch and Ott technique involves reduction of the excess $KMnO_4$ with $NH_2OH.HCl$, followed by adding Sn(II) to reduce Hg(II) to Hg°.[1195] What matters is (1) to use freshly (daily) prepared acidic $KMnO_4$[1194] and $SnCl_2$, (2) to reduce all residual $KMnO_4$ and Mn(IV) hydroxide before adding Sn(II), and (3) to avoid excess $NH_2OH.HCl$ which depresses Hg absorbance (both peak height and peak area[1205]). Some recent authors prefer oxalic[1205] or ascorbic acid[1174,1175] or else H_2O_2[1247] to NH_2OH.

Hg° may be alternatively trapped in 0.01 *N* I_2 in 1% KI,[1207,1272] but this requires ascorbic acid as a reductant instead of Sn(II).

Very serious problems are intrinsic to wet-digestion procedures preceding Hg assays: volatization losses, adsorption on digestion vessels, reagent blanks, incomplete oxidation of organic matter, and hazards (pressurized vessels, $HClO_4$, $HClO_3$, H_2O_2, fuming HNO_3, etc.). Analytical performance depends on numerous factors such as the properties of a particular organic matrix, amount of sample, volume and concentration of reagents, storage conditions of biomaterials (which may have undergone chemical transformations of the analyte to readily volatile species, say Hg°[1282,1283]) and other procedural details. Strict observance of original procedure is therefore recommended. Do not extrapolate the literature data to other specimens, conditions, etc. without thorough testing. Several important points are noteworthy:

1. Proper start of digestion to avoid losses at the very beginning [e.g., cooling in an ice bath while adding reagents (especially when H_2SO_4 and $KMnO_4$ are involved),[1145,1146,1160,1247,1282] moistening dry samples,[1247] etc.]
2. Keeping moderate digestion temperatures, often as low as 40 to 100°C,[27,271,1132,1196,1230,1238,1247,1248,1261,1276,1282,1286] (incidentally no heating is applied; some H_2SO_4-$KMnO_4$ digestions may even call for chilling in an ice bath[1140,1146,1282]); gradual, temperature-controlled heating is essential
3. Using effective reflux conditions, e.g., long-neck flasks,[1227] tall test tubes, reflux columns,[187,1254,1282] packed reflux columns,[1161] condensing systems with dry ice as refrigerant,[1261] or, incidentally, condensers for sample/analyte vapors, the condensate then being combined with the digest[91,266]
4. Digestion in closed or pressurized vessels (Tables 3 and 4)
5. Adding proper catalyst (often V_2O_5)[132,1246,1248,1250,1258,1261,1262,1283,1286]

In fact, the most effective procedures involve rational combinations of the indicated conditions (Tables 4 and 5). Obviously, some of these wet digestions do leave nonoxidized organics; while this has proved tolerable in urinalysis,[246,255,1231,1404] it tends to create difficulties with most other biological specimens (blood,[1127,1218,1231] hair,[426,1138,1155] tissues,[1229,1230] etc.). Thus, one compromises between a simpler preparatory step and a somewhat-complicated CVT-AAS quantitation.

Only a few most reliable wet-ashing techniques will be discussed below, while some more may be derived from Tables 4 and 6 and from References 1161, 1194, 1226, 1250, 1254, 1261, and 1276.

Decompositions in pressurized vessels (bombs, autoclaves, tightly capped test tubes or bottles, etc.) generally avoid volatization losses of Hg. They are very popular (Table 3) but little attention is paid to their *drawbacks:* (1) strict precautions are required so as to avoid violent reactions and excessive pressure build-up; (2) oxidation of organic matter is not complete,[426,1157,1258,1288,1289] especially when HNO_3 alone is used or ashing temperatures are low (<130 to 140°C) or else short digestion times are applied (e.g., <2 to 3 hr); and (3) Hg may penetrate through and adsorb on PTFE.[1141,1153] When higher temperatures (e.g., up to 170 to 180°C)[1142,1153,1290] or more effective combination of acids (e.g., HNO_3 + $HClO_4$ or HNO_3 + HCl)[1157] are used, the PTFE vessels are attacked[544,1153,1157] and pressure build-up becomes dangerous. Let us mention three important recent studies.

Kotz et al.[544] proposed glassy-carbon lining instead of PTFE so as to ensure higher digestion temperatures (up to 220 vs. 170°C), faster and more complete digestion, and less Hg retention. Applications to bone, hair, milk, serum, and tissues were described.[544]

Ahmed et al.[1157] (in a DPASV procedure for Hg in fish) used HNO_3 and $HClO_4$ (caution!)

Table 4
TECHNIQUES FOR (ACID) PRETREATMENT OF BIOLOGICAL SAMPLES PRIOR TO DETERMINATION OF THE TOTAL MERCURY[a]

Oxidation mixture and conditions[b]	Matrix
HNO_3 (autoclave/closed vessel)[a,c]	Blood,[1211,1254,1290] bone,[544] fish,[1160,1244,1257] food,[1290] hair,[544,1247,1257] meat,[1290] milk[544,1247] mussel,[1257] serum,[544] tissues,[190,544,545,1211,1254,1259,1290] urine[1254,1259,1260]
HNO_3 (closed vessel);[c] then completed with H_2SO_4 and $KMnO_4$	Fish, hair[1206]
HNO_3 + H_2SO_4 (autoclave/closed vessel)[a,c]	Fish,[458] hair,[426] tissue[1230]
HNO_3 + V_2O_5 (autoclave)[c]	Blood, hair, nails[1132]
HNO_3 + V_2O_5 (autoclave);[c] then completed with $K_2Cr_2O_7$	Kidney, liver, milk, etc.[1258]
HNO_3 + HF (autoclave);[c] H_3BO_3 added to final digest	Bone, teeth[1142]
HNO_3 + HCl (autoclave)[a,c]	Fish[1157]
HNO_3 + $HClO_4$ (autoclave);[b,c] subsequent photolysis by UV irradiation in a closed quartz flask at 70°C	Fish[1157]
HNO_3 + H_2SO_4 + $KMnO_4$ (autoclave)[c]	Serum, urine[1158]
HNO_3 + H_2SO_4 (closed vessel); then completed with $KMnO_4$	Blood, tissue[1140]
HNO_3 + $HClO_3$ (semiautomated,TCH)[c,d]	Fish, meat, milk, and other biological and organic samples[272,1141]
HNO_3 + $HClO_3$ + $HClO_4$ (TCH, tall tubes)[c,d]	Fish[1249]
H_2SO_4 + $K_2S_2O_8$ + $KMnO_4$ (in-line digestion in closed continuous-flow apparatus at 100°C)	Urine[1196]
H_2SO_4 + $K_2Cr_2O_7$ (10 min)	Blood, urine[27]
HNO_3 + $HClO_4$ + $K_2Cr_2O_7$ (TCH, air condenser)[c,d]	Fish[1167]
H_2SO_4 + $K_2Cr_2O_7$ (60°C, overnight)	Fatty food[271]
H_2CrO_4 or H_2CrO_4 + fuming HNO_3 (room temperature, 5—40 min)	Fish, meat, urine[1245]
HNO_3 + V_2O_5 at room temperature; then at 160°C;[c,d] cooled; completed with H_2SO_4 at 160°C[c,d]	Fish, environmental samples[1246]
HNO_3 + H_2SO_4 + V_2O_5 (70—80°C,[c] with efficient reflux condenser)	Biological samples,[1261] fish,[1248,1250,1283] tissue[1248,1283]
HNO_3 + H_2SO_4 + V_2O_5 (10 min at room temperature; TCH to 100°C;[c,d] finally completed with H_2O_2)	Fish[1286]
H_2SO_4 + $KMnO_4$ (ice cooling during adding reagents;[1145,1282] overnight at room temperature[1242] or ice bath[1146])[a]	Hair,[370] plasma,[1146] tissues,[1139] urine[1143,1145,1146,1149,1224,1242,1256]
H_2SO_4 + $KMnO_4$ (50—60°C)[a]	Blood,[1238] urine[1282]
KOH homogenation; $KMnO_4$ added, finally completed with H_2SO_4	Blood, fish, hair, liver, nail, urine[1228]
HNO_3 + $HClO_4$ + H_2SO_4 (TCH, packed reflux column)[c,d]	NBS SRMs(''Bovine Liver'', ''Coal'', ''Orchard Leaves'', etc.)[1161]
HNO_3 + $HClO_4$ + H_2SO_4 (TCH, long-neck flask)	Fish (liver, muscle), NBS SRM ''Bovine Liver''[1227]
HNO_3 + $HClO_4$ [(2 + 10), 70—75°C overnight in closed tall tubes][a-d]	Blood (small samples)[1276]
As above, but finally completed with H_2O_2[b-d]	Blood (larger samples)[1276]
HNO_3 + H_2SO_4 (reflux; 120—130°C)[a,c-e]	Fish,[187,1158,1229,1244] food[1194]

Table 4 (continued)
TECHNIQUES FOR (ACID) PRETREATMENT OF BIOLOGICAL SAMPLES PRIOR TO DETERMINATION OF THE TOTAL MERCURY[a]

Oxidation mixture and conditions[b]	Matrix
As above, but in a microwave oven (fast — 3 min)[a,d]	Fish (large samples, e.g., 5—10 g)[1251]
HNO_3 + H_2SO_4; finally HNO_3 + H_2O_2 (TCH, long-neck flask)[c,d]	Shark muscle powder and NBS "Bovine Liver" and "Albacore Tuna"[853]
HNO_3, H_2SO_4, and H_2O_2 (TCH, long-neck flask)	Large losses experienced with organic samples, fish, fatty food, milk, etc.[170,1214]
HNO_3, H_2SO_4,H_2O_2, and $KMnO_4$	Fish[1166]
H_2SO_4 (65°C for 60—90 min); then 50% H_2O_2 added dropwise[b,d]	Fish[271]
HNO_3 + HBr (reflux, 10 min)[d]	Fish[1226]
HNO_3 (room temperature, 3 min)[a]	Urine[1231]
HNO_3 (55°C, 2—4 hr)[a,c]	Bone, teeth[1243]
HCl + $NaNO_3$ (overnight)	Blood, tissue[1271]
HNO_3 + H_2O_2 (TCH, long-neck flask)[b-d]	Fish (liver, muscle)[1227]
HNO_3 + H_2SO_4 + $KMnO_4$ (room temperature overnight or longer; preferably start with HNO_3 and cool before adding H_2SO_4 and (dropwise) $KMnO_4$; gentle heating to 40—60°C before adding $KMnO_4$ may prove useful[27,1247])	Blood,[27,421] fish,[1247] food,[1247] hair,[1138] meat,[1247] tissues,[1230,1247] urine[27,246]
Deproteinization with TCA	Blood,[7,472] plasma,[7] tissue homogenates,[7,472,1255] urine[272,1235,1236]
HNO_3 (4 hr at room temperature); then $KMnO_4$ added (16 hr at 15—20°C)	Blood, hair, kidney, liver, muscle[1242]

[a] Certain techniques leave incompletely digested organic matter which may interfere with analysis.
[b] Strict safety precautions mandatory with most preparatory procedures, since strong oxidants and/or pressure build-up are involved.
[c] Gradual, temperature-controlled heating (TCH) required.
[d] Effective reflux conditions essential.
[e] Not applicable to blood- and/or $(CH_3)_2$ Hg-containing specimens.

Table 5
AAS PROCEDURES FOR MERCURY SPECIATION IN BIOLOGICAL MATERIALS

Sample	Species determined	Brief procedure outline	Ref.
Blood, fish, tissue homogenates (in 1% saline or 40% NaOH)	Inorganic Hg	Successively added: L-cysteine, HCl, sample solution/homogenate, 1% saline, standard addition, 16 *N* H_2SO_4, silicone antifoam, and finally, reduction is triggered by aqueous NaOH	1199
Blood	Inorganic Hg, organic Hg	*Inorganic Hg:* successively added: aqueous NaCl, aqueous L-cysteine, blood aliquot, aqueous NaOH, and aqueous $SnCl_2$ *Organic Hg:* aqueous $SnCl_2$-$CdCl_2$ reductant added to the solution remaining after previous step (described for inorganic Hg)	1210
	Inorganic Hg, total Hg	Automated Magos procedure	1275

Table 5 (continued)
AAS PROCEDURES FOR MERCURY SPECIATION IN BIOLOGICAL MATERIALS

Sample	Species determined	Brief procedure outline	Ref.
Blood, hair, tissue, urine	Inorganic Hg, total Hg	Modified Magos procedure (new reaction vessel); amenable to automation	1198
Blood, placenta	CH_3HgCl, inorganic Hg	CH_3HgCl selectively distilled from samples in presence of HCl and NaCl The separated MeHgCl and inorganic Hg reduced by Sn(II) in alkaline and acidic medium, respectively	1173
Blood, feces, kidney, urine	$HgCl_2$, methoxyethyl HgCl, C_2H_5HgCl, C_6H_5HgCl	Different procedures examined; to sample solution/homogenate added: (1) for the determination of $HgCl_2$ in feces, kidney, and urine — a drop of silicone antifoam and alkaline Sn(II) chloride solution; peak-area calibration; (2) for the determination of methoxyethyl HgCl in feces, kidney, and urine — acidic cysteine added; then proceeded as in (1); peak-area calibration; and (3) for the determination of $C_2H_5Hg^+$ in feces, kidney, and urine, and C_6H_5HgCl in blood — acidic cysteine added, heated for 1 hr on a boiling-water bath (loosely stoppered), and then proceeded as in (1); peak-area calibration	1209
Environmental waters	Inorganic Hg, organic Hg	*Total Hg* determined at pH 2—3 with $NaBH_4$ reductant in presence of 20 μg Fe(III)/mℓ; *inorganic Hg* selectively reduced at pH 10—12 with $NaBH_4$, and *organic Hg* calculated by difference	1217
Fish	Organic Hg	Sample extracted with toluene from 4 *N* H_2SO_4 in presence of $CuSO_4$ and KBr; the extract diluted 1 + 1 with 0.1% dithizone to thermally stabilize Hg in the following ET-AAS quantitation (ashing at 175—200°C, atomization at 950°C)	1183
	Alkylmercury	Digestion with aqueous KOH on a water bath for 30 min in presence of Cu(II); steam distillation of an aliquot of the alkaline digest in presence of $CuSO_4$ and HCl, and condensate trapped in a solution containing NH_4VO_3, H_2SO_4, and $K_2S_2O_8$; the resulting solution warmed on a water bath for 35—40 min, diluted, and analyzed by CVT-AAS	1172
	CH_3HgCl	CH_3HgCl distilled from homogenized samples by boiling with HCl-Cu(II) solution; double extraction of distillate with dithizone-$CHCl_3$, and third extraction with $CHCl_3$; combined extracts further concentrated and TLC separated; CH_3Hg dithizonate spot scrapped, ignited, and Hg° determined by AAS	1170
Hair	Inorganic Hg	Alkaline digestion; inorganic Hg determined after Magos[1199,1241]	1155, 1168
Hair, fish,urine, water	Inorganic Hg, organic Hg	KOH digestion at 90°C for 15—30 min in capped vials,[1168] and diluted with 1% NaCl; CVT-AAS: (1) *inorganic Hg* — $SnCl_2$ reduction, then $K_2Cr_2O_7$ and HNO_3 added (*caution:* consult the original paper for important details!), and (2) *organic Hg* reduced from the resulting solution with $NaBH_4$	1189

Table 5 (continued)
AAS PROCEDURES FOR MERCURY SPECIATION IN BIOLOGICAL MATERIALS

Sample	Species determined	Brief procedure outline	Ref.
Mixtures of organomercurials	E.g., $HgCl_2$, CH_3HgCl, C_6H_5HgCl, $CH_3C_6H_5HgCl$, etc.	TLC separations, combustion of the excised spots, and AAS determination	1171
Tissue homogenate (in 1% sucrose), urine	Inorganic Hg	Successively added: acidic L-cysteine, sample solution, standard addition, antifoam (tri-n-butylphosphate), $SnCl_2$, and finally aqueous NaOH	1208
Tissue homogenate	Inorganic Hg	Modified Magos technique (in presence of excess Se in test animals): tissue homogenates treated at 40°C for 30 min with an eq vol of 45% w/v NaOH containing 1% L-cysteine, and n-octylalcohol antifoam added; peak-area calibration	1211
Urine	Inorganic Hg	Successively added: acidic L-cysteine, sample aliquot, standard addition, acidic $SnCl_2$, purged with air, and finally aqueous NaOH	1215
	Inorganic Hg	Successively added: aqueous L-cysteine, aqueous NaCl, aqueous NaOH, octanol antifoam, acidic $SnCl_2$, urine aliquot; standard addition	1210
	Inorganic Hg, phenyl Hg	*Inorganic Hg:* successively added urine aliquot, L(+) cysteine solution, aqueous NaOH, and aqueous acidic $SnCl_2$ (i.e., reduction in 15% NaOH medium) *Inorganic Hg + phenyl Hg:* successively added urine aliquot, H_2SO_4, H_2O, and aqueous $SnCl_2$ (i.e., reduction in 32% H_2SO_4 medium)	1143

Note: See the original procedures for numerous important details.

Table 6
SELECTED AAS PROCEDURES FOR Hg DETERMINATION IN BIOLOGICAL SAMPLES

Sample	Brief procedure outline	Ref.
Biological NBS SRMs ("Bovine Liver", "Orchard Leaves", etc.)	Autoclave decomposition with HNO_3; automated CVT-AAS with amalgamation on Ag wool; 120 assays per day	1259
Biological NBS SRMs ("Bovine Liver", "Coal","Orchard Leaves", etc.)	HNO_3-H_2SO_4-$HClO_4$ digestion[a] using a TCH[b] and a packed reflux column; CVT-AAS with $SnCl_2$-H_2SO_4-NaCl reductant	1161
Blood, fish, hair, liver, nails, plant, urine, etc.	Homogenation with 1.5 *M* KOH under gentle warming; oxidation with $KMnO_4$ in an alkaline medium, completed with H_2SO_4, and finally oxalic acid added to reduce excess of oxidant; CVT; standard addition calibration	1228
Blood, fish, hair, tissue, urine, etc.	L-cysteine; HCl added to biological liquids or homogenates, then $SnCl_2$-$CdCl_2$ reductant added, and finally reduction triggered by aqueous NaOH; standard addition calibration	1168, 1190, 1199, 1215, 1241
Blood, hair, nails	HNO_3-H_2SO_4-V_2O_5 digestion in screw-capped tubes[a,b] at 70°C; CVT-AAS ($NH_2OH.HCl$ added, $SnCl_2$ reductant)	1132
Blood, hair, tissues, urine	Samples decomposed by: (1) HNO_3-H_2SO_4 under reflux (hair, tissues, urine), (2) autoclave digestion with HNO_3 at 140°C for 15 hr[a] (blood, hair, tissues, urine), and (3) combustion	1254

Table 6 (continued)
SELECTED AAS PROCEDURES FOR Hg DETERMINATION IN BIOLOGICAL SAMPLES

Sample	Brief procedure outline	Ref.
Blood, hair, tissues, urine	in Wickbold burner and trapping in $KMnO_4$-H_2SO_4 solution (hair, tissues); CVT-AAS ($SnCl_2$ reductant)	
Blood, tissues, urine	TCA deproteinization (2 ×) of blood or tissue homogenates; APDC-MIBK extraction at pH 3—4; ET-AAS: 75° for 40 sec and 2000°C; background correction; standards extracted similarly	7
Bone, hair, milk, serum, tissues	HNO_3 digestion at 200°C for 1—2.5 hr in a glassy-carbon lined autoclave;[a,b] CVT-AAS	544
Bone, teeth	Pressure digestion with HNO_3 and HF at 180°C for 90 min in a PTFE-lined autoclave;[a,b] cooled, opened, and saturated H_3BO_3 solution added; CVT-AAS ($SnCl_2$ reductant)	1142
Fish, meat, milk	Semiautomated $HClO_3$-HNO_3 digestion (long-neck flasks essential);[a,b] CVT-AAS with $SnCl_2$-ascorbic acid reductant	1141
Hair[c]	Pressure decomposition with HNO_3-H_2SO_4 or HNO_3-$HClO_4$[a,b] at 110°C; CVT-AAS with $SnCl_2$ reductant; acid-matched standards	426
Hair, nails	Oxygen-flask combustion;[a] taken up in (1 + 1) HNO_3, and CVT-AAS with $SnCl_2$ reductant; acid-matched standards	1193
Milk, tissues	Pressure decomposition with HNO_3 and V_2O_5 at 150°C;[a,b] cooled, opened, and $K_2Cr_2O_7$ added to oxidize NO_x and stabilize digests; CVT-AAS with $SnCl_2$-H_2SO_4 reductant; standard addition calibration	1258
Plasma, urine	H_2SO_4 digestion in an ice bath, aqueous $KMnO_4$ added, left at 0°C overnight, mixed, and centrifuged; CVT-AAS with $SnCl_2$-HCl reductant	1146
Tissues, urine	Wet digestion with H_2CrO_4 or H_2CrO_4 + HNO_3 at room temperature for 5—40 min; CVT-AAS	1245
Urine	1 mℓ of urine treated with 200 μℓ of 5% $KMnO_4$ directly in the reaction vessel of a Perkin-Elmer® MHS-1 or MHS-20 system, antifoaming agent added, and 10 mℓ of mixed acid (1.5% w/v HNO_3-1.5% w/v H_2SO_4) added; CVT with $NaBH_4$ or $SnCl_2$ reductant; acid-matched standard solutions	246, 1204, 1232
	H_2SO_4-$KMnO_4$ digestion at room temperature overnight loosely covered; 50% w/v $NH_2OH.HCl$ added; CVT-AAS with $SnCl_2$ and H_2SO_4 reductant	1224, 1242, 1256
	H_2SO_4 added under chilling in an ice bath, $KMnO_4$ added, and left overnight at room temperature; CVT-AAS with $SnCl_2$-$CdCl_2$-HCl reductant	1145
	HNO_3 and $K_2Cr_2O_7$ added to their final concentrations of 0.8 *M* and 0.1% w/v, respectively; ET-AAS; manual injection or (preferably) automated Instrumentation Laboratory FASTAC® aerosol deposition; temperature program: 150°C for 5 sec, 225°C for 10 sec, 275°C for 10 sec, atomization at 750°C for 5 sec, and clean-out at 2000°C for 10 sec; background correction; uncoated graphite tubes; peak-area calibration	1178
	1-mℓ sample treated with 10 mℓ of HNO_3, 50 μℓ of 1 *M* KBr, and 5 μℓ of Br_2, mixed and put aside for a few minutes; Zeeman ET-AAS: 100 and 1100°C; standard solutions containing 1% v/v HNO_3, 0.5% v/v of Br_2, and 0.05 *M* KBr	1184

[a] Safety precautions mandatory.
[b] Gradual, temperature-controlled heating essential.
[c] See also Reference 1435.

in a Teflon® bomb for 3 hr, and subsequently completed oxidation of dissolved organic residues by UV irradiation at 70°C in a closed quartz vessel.

Korunová and Dědina[1258] bomb decomposed samples of kidney, liver, milk, etc. with HNO_3 + V_2O_5 at 140°C for 2 hr; nondigested fat still persisted from some samples. Finally, oxidation was completed in an open vessel with $K_2Cr_2O_7$, which oxidized NO_x and stabilized digests for up to 20 days.

Oxidation mixtures containing chloric acid ($HClO_3$) were successfully applied in digestions of fish, meat, milk, and other biomaterials.[272,1141,1249] A semi-automated wet-digestion device with temperature-controlled heating, sample throughput of 10 samples per hour and total digestion time of 2.5 hr is available (VAO, Kürner, Rosenheim, G.F.R.); no Hg losses have been found up to 120 or 200°C [depending on the form of the (long-necked) digestion flasks], but fatty samples still need preextraction of fats.[1141]

Preconcentration of nanogram amounts of Hg is effected by several well-established techniques; the first three of the list are most often used, as already discussed:

1. Reductive volatization of Hg° from either acidic or alkaline media. This is essentially the basis of the CVT-AAS; obvious advantages of this technique are its simple and rapid performance, in-line combination with AAS quantitation, usefulness in speciation studies, and amenability to automation/semiautomation.
2. Amalgamation with noble metals (Au) or Cu.
3. Trapping Hg° in absorbents (e.g., $KMnO_4$, I_2-KI, etc.) or condensers (at liquid-N_2 temperature).
4. Adsorption of Hg° onto charcoal (which may then be analyzed by, e.g., the Sampling Boat® technique[1237]) or else onto Se-impregnated filter paper.[1147]
5. Chelate extraction of Hg(II) (in a broad range of pH, but more selectively at low pH) as dithiocarbamate[7,130,472,1235,1236] or dithizonate;[1137,1149,1185,1238,1239] e.g., from TCA-deproteinized blood, erythrocytes, plasma, tissue homogenates, and urine with APDC-MIBK at pH 3 to 4;[7,472] from (H_2SO_4-$KMnO_4$)-digested blood, saliva, and urine with dithizone-$CHCl_3$ at pH 0 to 1;[1137,1149,1238-1240] etc.[130,1185] Dithizone extractions are also used in speciation schemes.[152,1165,1170] Organic extracts are readily analyzed by flame AAS,[472] ET-AAS,[7,1185] Sampling Boat®,[1235,1236] or else by destructive distillation CVT-AAS.[1137,1238,1239]
6. Coprecipitation of Hg(II) with CdS or PbS (from [$KMnO_4$-HNO_3]-treated urine[1154]), or else with dithiocarbamate (from O_2-ashed food, tissues, etc.[540]).
7. Electrolytic deposition of Hg on Cu[1144,1291] or Ir[752] wire or else on Pt spiral filament[397] (from acidic solutions, e.g., 0.1 *M* HNO_3,[1291] or from urine acidified with HCl to pH 1 to 2[1144]). The filament is subsequently heated, releasing the plated Hg into a long-path absorption tube.[397,752,1144]
8. Ion exchange of Hg(II), e.g., on poly(dithiocarbamate) resin from pH-adjusted urine (pH 2, HCl),[326] or on Amberlite® IRA-400 resin from wet-treated blood/tissue.[1271]

A very attractive approach to Hg AAS assays is to analyze undigested or simply pretreated biological liquids/homogenates. Admittedly, ET-AAS has a limited potential, but the two recent applications to urine[1178,1184] (see Table 6), both of which involve modern methodology and apparatus, would seem promising. CVT-AAS has found numerous applications (Tables 3, 5 and 6) to untreated or homogenized (in an alkali plus suitable additives) or else to *in situ* pretreated (in the reaction vessel of CVT apparatus[1204]) samples. This approach, however, involves complicated calibration (e.g., by standard additions or peak-area measurements), foaming problems, longer measurement times, sensitivity/precision impairment, etc. Needless to say, it is invaluable in speciation studies.

Most speciation techniques as well as techniques for analysis of undigested biological liquids/homogenates are based on the Magos method:[1199,1208,1241] L-cysteine hydrochloride is added (so as to bind both inorganic and organic Hg into stable complexes and thus to keep all forms of Hg in solution until the reduction/volatization process is started, as well as to prevent chemical transformations of methylmercury to inorganic Hg), then a reductant is added which is either $SnCl_2$ (for inorganic Hg) or $SnCl_2$ plus $CdCl_2$ (for total Hg), and finally the reduction/volatization reaction is triggered by adding NaOH solution. Various applications and modifications of this technique have been proposed; some of them are briefly outlined in Tables 5 and 6. A few points are noteworthy: fresh samples or homogenates should be preferably analyzed, as gradual demethylation may occur in alkaline homogenates;[1162,1168,1169] calibration is done by standard additions (freshly prepared), often in presence of antifoaming agents, and incidentally by peak-area mode;[1209,1211] and many important details in original procedures must be strictly followed (e.g., amount and concentration of reagents, the maximum sample aliquot that can be handled, the order of additions made, etc.).

Finally, a few practical hints. Dilute Hg standard solutions should be prepared daily, since Hg(II) is readily lost due to reduction, disproportionation of Hg(I), and volatization of Hg°, as well as to hydrolysis, adsorption, and coprecipitation.[1141,1153,1292] If conservants are used (dilute HNO_3, aqueous $K_2Cr_2O_7$, etc.), one should be sure that they do not interfere with the next analytical steps, nor facilitate demethylation and redistribution of Hg species.[1189] Vessels of quartz, PTFE, or glass should be thoroughly cleaned by steaming with water and HNO_3,[1111,1153] or by boiling with (1 + 1) HNO_3;[1193,1292] BrCl has also proved efficient in the cleaning of glassware. Some reagents and samples (e.g., $KMnO_4$, $K_2S_2O_8$, $KI\text{-}I_2$, milk powder, hair, etc.) tend to trap Hg vapor from the atmosphere under improper storage conditions.

There are NBS SRMs certified for trace levels of Hg (liver, oyster, urine, etc.) and IAEA RMs and CRMs (fish, milk, muscle, etc.), whose use in quality control is encouraged.

The results of interlaboratory studies have been reported.[418,1123]

IV. CONCLUSION

CVT-AAS is the best technique for routine determination of nanogram amounts of Hg in biomaterials. NAA is the best alternative, as it is a nondestructive multielement method. A few more techniques based on ID, MIP-AES, DPASV, etc. would seem useful in comparative studies.

Representative sampling and thorough homogenation (with an alkali or at liquid-N_2 temperature) are important preanalysis considerations.

Contamination from the atmosphere and reagents is likely to occur at nanograms per gram levels. Volatization and adsorption losses during storage and mineralization of samples are very probable. Reliable preparatory procedures are ashing in O_2 (in a closed vessel or cold-finger apparatus); bomb decomposition (preferably with a glassy-carbon lining or with V_2O_5 catalyst, and/or with subsequent completion of oxidation in an open vessel); some wet-digestion procedures with temperature-controlled heating and efficient reflux (e.g., HNO_3 + $HClO_3$, HNO_3 + $HClO_4$ + H_2SO_4, etc.); as well as some incomplete digestions with $KMnO_4$= or $K_2Cr_2O_7$-containing mixtures.

CVT-AAS may suffer from interferences (by incompletely digested organic matter, masking ions, oxidants, foaming, etc.). Attention should be paid to procedural details and calibration mode; incidental standard additions or peak-area checks, though time consuming, are advisable.

ET-AAS is still of limited use, due to the extreme volatility of the analyte and background problems. Chelate extracts (dithizonate or dithiocarbamate) may be analyzed. Temperature programming and matrix/analyte modification are of critical importance.

Undigested or homogenized samples can be analyzed by the Magos method or its modifications, which permit to differentiate between total, organic, and inorganic mercury; standard additions calibration is typically employed. An alternative speciation method is GC (with MS, MIP-AES, ECD, or else with AAS detectors).

CVT-AAS is amenable to automation.

Chapter 21

MOLYBDENUM

I. DETERMINATION OF MOLYBDENUM IN BIOLOGICAL SAMPLES

Most of the published "normal" Mo concentrations in biological specimens seem positively biased, presumably due to inadequate sampling and handling of samples, as well as to analytical problems with this element. While Mo levels in liver, nails, hair, and feces are relatively high (up to 1 μg/g), its normal concentrations in body fluids appear to be only around 1 ng/mℓ or less (Table 1).

There are few reliable analytical techniques for nanogram amounts of Mo in biological samples. ET-AAS is the most sensitive method but, until recently, its use was limited owing to severe matrix effects and carbide formation.

At present, RNAA appears to be the most reliable and widely used method for the determination of nanograms-per-gram concentrations of Mo; applications to blood,[45,178,1293] diet,[45,116] hair,[48,1294] platelets,[119] serum,[45,111,1295] skin,[112] tissues,[45,108,1296] urine,[111] etc.[45,121,1293,1296] have been described.

SSMS analyses of blood,[3] hair,[54] nails,[16] teeth,[25] tissues,[3,132,275,409] etc.,[132] with DLs down to, e.g., 3 ng Mo/g in soft tissues,[3] have likewise been described.

Many early applications of the arc-source AES have been documented (food,[39] milk,[40] sweat,[21] teeth,[26] tissues,[44,904] urine,[39,320] etc.[36,39,41]), but some of these analyses seem to have been made on samples containing above-normal concentrations of Mo.[36,413] The DLs of ICP-ES are quite favorable, between a few and 0.2 ng/mℓ,[29,35,123,348,412] and this technique may become more important in the near future.

XRF is not sufficiently sensitive (DL ~1 μg/g)[3,123,691] and is applicable only to tissues of exposed test animals. PIXE[128] has been applied to liver biopsies.[407]

Methods involving simpler and inexpensive equipment include catalytic,[577,1297] spectrophotometric,[40,133,1298,1299,1306] and cathode-ray polarographic[1300] techniques which, however, are inadequately sensitive and require somewhat large samples (e.g., 10 mℓ of blood or plasma[1297,1300]).

II. ATOMIC ABSORPTION SPECTROMETRY OF MOLYBDENUM

In a reducing N_2O-C_2H_2 flame, the characteristic concentration is between 0.18 and 0.6 μg/mℓ, and DLs are down to 0.01 to 0.03 μg/mℓ.[56-59,294] Minor interferences from many concomitants have been observed, e.g., depression by (excess) Fe, Ca, K, Na, and H_2SO_4 and enhancement by Al, Cr, Mn, Cu, etc., depending on flame stoichiometry, observation height, and acid(s) present. It is hard to predict the nature and extent of these interferences. Dilute HCl or $HClO_4$ appear most suitable media, and HNO_3 is tolerable, while H_2SO_4 is inappropriate. Useful spectrochemical buffers are the salts of Al^{3+} (e.g., 0.1% Al^{3+} as chloride), NH_4^+ (NH_4Cl, NH_4F), HBF_4, etc. The standard solutions should typically be matrix matched.

In a reducing air-acetylene flame, sensitivity and DLs are impaired by factors of 2 and 5, respectively,[294] and numerous severe matrix effects are encountered.

In ETA, the characteristic concentration ranges between 0.06 and 8 ng/mℓ (6 to 84 pg), and the DLs are down to 0.015 to 0.07 ng/mℓ (2 to 7 pg).[59-61,294,1301]

Formation of stable, nonvolatile carbides of Mo with a graphite surface as well as with carbon residues of organic matrices is a serious impediment. Slow, tailed atomization peaks

Table 1
MOLYBDENUM CONCENTRATIONS IN HUMAN BODY FLUIDS AND TISSUES

Sample	Unit	Selected reference values	Ref.
Blood	μg/ℓ	1 (0.5—1.8)	1293
		1[a]	3
		3.3 ± 1.1[b]	178
Feces	μg/day	120[c]	8
Hair	μg/g	0.073 (0.02—0.13)	1294
Kidney	μg/g	x̃, 0.22; 0.19 ± 0.04 (0.072—0.67)	105, 106
		1.5 ± 0.59[b]	904
Kidney cortex	μg/g	0.23	107
Liver	μg/g	0.37 (0.16—0.72)	108
		x̃, 0.54; 0.62 ± 0.08 (0.34—1.24)	105, 106
		0.71	107
		1.30 ± 0.32	275
Lung	μg/g	x̃, 0.03; 0.12 ± 0.05 (0.011—0.27)	105, 106
		0.12 ± 0.01	3
Muscle	μg/g	~0.01	3
Nails	μg/g	<0.15[d]	133
Plasma or serum	μg/ℓ	0.58 ± 0.21 (0.28—1.17)	1295
Skin	μg/g	x̃, 1.4; 1.5 ± 0.58[e]	112
Sweat	μg/day	20[c]	8
		61	21
Tooth (enamel)	μg/g	x̃, <0.6; 0.7 (<0.6—22)	26
Urine	μg/day	49 (40—59)	320
		61[f]	13

Note: Wet weight unless otherwise indicated. Mean ± SD (range in parentheses). x̃, median.

[a] Pool.
[b] ng/g.
[c] "Reference Man".
[d] N = 1.
[e] Dry weight.
[f] Weighed mean (N= 46).

are typical, and a high-temperature clean-out stage is usually incorporated in the temperature cycle of the furnace. Pyrolytically coated tubes and "maximum power" are recommended for better sensitivity (e.g., by factors of 4 to 6 and 2, respectively[60]). With some furnaces, peak-area measurements may improve sensitivity by an order of magnitude.[551] *In situ* pyrolytic coating may[1302] or may not[72] improve sensitivity, depending on furnace construction and heating rates.

High ashing temperatures between 700 and 1900°C are tolerable, as they remove most volatile matrix constituents, thus eliminating or reducing matrix effects and background absorption.

Any residual background absorption should be checked. As the intensity of the deuterium lamp is low in this spectral region, a tungsten halide lamp or Pt 315.6- or Pd 314.0-nm lines are recommended.[629] "Furnace blank" may still persist at high atomization temperatures involved.[1303] Most acids strongly interfere;[294,1301,1304] dilute HCl and HNO_3 (<1 to 2 *M*) are appropriate, but $HClO_4$ and H_2SO_4 are intolerable, due to their depressive effect and attack on the graphite. Acid- and matrix-matched calibration is usually required.

Carbide-coated tubes and peak-area calibration would seem promising and should be further assessed.

III. AAS METHODS FOR ANALYSIS OF BIOLOGICAL SAMPLES

The literature on AAS of Mo in biological samples is scant; there are a few early flame AAS procedures which are not adequately sensitive,[350,973] as well as several complicated procedures involving ashing, chelate extraction, and flame atomization.[190,332,659,746] All these procedures seem to lack sensitivity, but some of them can be adapted to more sensitive ETA.

Few applications of ET-AAS have been described so far: a direct analysis of tissue ash;[1303,1307] digested erythrocytes, serum, tissues, and urine;[1305] and wet ashing and then coprecipitated with APDC trace elements in animal tissue.[446,533]

Several major problems have to be dealt with in ultratrace Mo assays: (1) to avoid contamination during specimen collection and ashing (e.g., severe contamination from stainless-steel needles and surgical knives); (2) to completely put Mo into solution and effectively preconcentrate it into a suitable final solution; and (3) to properly account for matrix effects in ETA. Few analytical laboratories seem to have solved these problems at present.

LTA is a clean and effective technique for oxidation of organic matter of large samples: blood,[169] erythrocytes,[1303] serum,[1303] urine,[1303] NBS SRMs "Bovine Liver",[1303] "Orchard Leaves",[626,1303] "Coal",[626] etc.; ashes are treated with a few drops of HNO_3, briefly re-ashed, and finally dissolved in (1 + 1) HCl or (1 + 1) HNO_3. Dry ashing at temperatures between 450 and 600°C is generally successful,[91,169,190,1294,1297,1300,1306] although some difficulties in taking up ashes may be expected.[91] Hence, more vigorous treatment of dry-ashed residues is advisable, e.g., with $HNO_3 + H_2SO_4$[1294] or $HNO_3 + H_2O_2$,[1306] or with aqueous NH_3.[190]

Some common wet-digestion procedures have been applied to biological samples with no adverse comments: blood ($HNO_3 + H_2SO_4 + HClO_4$);[332] plasma ($HNO_3 + HClO_4$);[1305] teeth ($HClO_4$);[746] and tissues ($HNO_3 + HClO_4 + K_2Cr_2O_7$).[446] The resulting digests cannot be directly analyzed by ET-AAS, owing to both matrix effects and limited sensitivity.

Liquid-liquid extraction is not particularly effective for Mo: (1) organic matter has to be completely oxidized; (2) the analyte must be present in a definite oxidation state, i.e., Mo (V) in thiocyanate extractions or Mo(VI) in most other extractions; (3) the optimum pH or acidity range is narrow; and (4) multiple extractions, say, 2 to 5 ×, are required with certain extraction systems. A few extraction systems for trace Mo in biological digests are summarized in Table 2.

Finally, two recent ET-AAS procedures are worth mentioning.

Bentley et al.[1303] dried overnight at 70°C 0.3- to 1-mℓ aliquots of serum or urine together with a few drops of HNO_3; dried residues were then ashed in a LTA apparatus for 8 to 24 hr; ashes were moistened with HNO_3, dried, and re-ashed (LTA) for another 8 to 24 hr; and finally, residues were dissolved in (1 + 1) HCl for at least 30 min. Erythrocyte samples were treated similarly, but were finally dissolved in (1 + 1) HNO_3. Digested samples were analyzed in a graphite furnace (pyro-coated tubes) programmed at 120°C for 50 sec, 1800°C for 40 sec, and 2800°C for 8 sec, without background correction, by standard additions calibration.

Khan et al.[1305] ashed 10-g samples of blood plasma with HNO_3 and $HClO_4$, extracted Mo with Zn dithiol in MIBK, from approximately 3% v/v $HClO_4$, and analyzed 20-μℓ aliquots of the extracts in a graphite furnace, programmed at 150°C for 30 sec, 1750°C for 30 sec, and 2700°C for 15 sec. Background correction and similarly extracted standards were employed.

IV. CONCLUSION

RNAA and ET-AAS are suitable analytical methods for trace Mo. Ashing (e.g., LTA or any other clean digestion technique) and effective preconcentration are usually involved.

Table 2
EXTRACTION SYSTEMS FOR Mo FROM BIOLOGICAL DIGESTS

Extraction system	Matrix and conditions	Method	Ref.
APDC-MIBK	Wet-ashed teeth; pH 1—1.4	Flame AAS	746
	Dry-ashed milk/tissues; pH <2	Flame AAS	190
α-Benzoinoxime-$CHCl_3$	Dry-ashed hair; 2 × [a]	RNAA	1294
BPHA[b]-toluene	Wet-ashed liver, plant, etc.; 5 *M* HCl—1.8 *M* H_2SO_4	RNAA	1296
Cupferron-MIBK	Wet-ashed blood; 2 *M* HCl; 3 × [c]	Flame AAS	332
Zn dithiol-MIBK	Wet-ashed plasma; ca. 3% v/v $HClO_4$	ET-AAS	1305
6 *M* HCl-pentylacetate	Dry-ashed serum/urine; 2 × [a]	Catalytic or CRP[d]	1297, 1300
HHDC + 8-hydroxyquinoline (4 + 1) in $CHCl_3$ + i-butanol (3 + 1)	Biological digests (feces, food, urine); 3—4 *M* HCl; 4—5 × [e]	Arc-source AES	39
8-Hydroxyquinoline-$CHCl_3$	Wet-ashed serum; 3 ×	RNAA	1295
TMBHA[f]-i-pentanol	Biological digests (tissue, plant) pH 1.5—2.5	Spectrophotometry	1306

[a] Double extraction.
[b] N-benzoyl-N-phenyl-hydroxylamine.
[c] Triple extraction.
[d] Cathode-ray polarography.
[e] Exhaustive extraction.
[f] N-o-tolyl-o-metoxy-benzohydroxamic acid.

Pyro-coated tubes, high heating rates, and matrix-matched calibration are a must. Contamination during specimen collection (e.g., from stainless-steel needles, knives, etc.) should be avoided.

Chapter 22

NICKEL

I. DETERMINATION OF NICKEL IN BIOLOGICAL SAMPLES

Normal concentrations of Ni in biological liquids (except for sweat) are very low, generally less than 0.01 μg/mℓ. Most tissues contain <0.1 μg Ni/g; somewhat higher levels are encountered in hair, nails, bone, teeth, food, and feces (Table 1).

ET-AAS is the best current method, with DLs below 1 ng/mℓ. Flame AAS can be applied to certain specimens, typically after ashing and preconcentration.

Other sensitive analytical methods are dimethylglyoxime-sensitized voltametry (DL down to 1 ng Ni/ℓ in aqueous solutions[1318,1319]), ID-GC-MS (according to F. W. Sunderman, Jr.[1320]), GC of volatile chelates,[1321] and SSMS[3,16,54,55,132,409] (DL 2 ng Ni/g of soft tissue[3]). Some early SSMS procedures[16,54] seem insufficiently sensitive. The same is true of several other instrumental techniques: NAA,[45,96,98,118,856] XRF,[128,292,293,636,691,695,1322] PIXE[128,697,852,1312] arc-source AES,[21,26,36,41,42,44,288,320,342,413,904] ICP-ES,[29,35,326,348,392,412,636,1104] and spectrophotometry,[133] which often need an effective preconcentration step (extraction,[908] ion exchange,[326,1322] etc.). See also the recent review[1320] and conference proceedings.[1318]

II. ATOMIC ABSORPTION SPECTROMETRY OF NICKEL

In flame AAS, the characteristic concentration and DLs are 0.04 to 0.16 μg/mℓ and 0.002 to 0.01 μg/mℓ, respectively.[56-59,294] What matters is to use a high-intensity, single-element HCL, as well as optimum (somewhat higher) lamp current and a narrow bandpass (e.g., 0.1 to 0.2 nm), in order to separate the resonance line at 232.0 nm from adjacent lines (231.1, 231.6 nm, etc.); bent calibration curves are still observed due to unresolved very close lines (231.98, 232.1 nm, etc.).

In a fuel-lean air-C_2H_2 flame and at a not too low observation height, Ni determinations are selective. Only excess acids (H_2SO_4, H_3PO_4) and high salt concentrations are liable to cause minor interferences. Background correction is often needed. Acid-matched standards are typical; suitable acids are dilute HCl, HNO_3, and $HClO_4$ (<10% v/v).

In ETA, the characteristic concentration is from 0.02 to 2 ng/mℓ (2 to 160 pg), and DLs are from 0.015 to 2 ng/mℓ (down to a few picograms).[57,59-61] A carbide formation takes place and, therefore, pyro-coated tubes[60,1323] or *in situ* pyro-coating[295,1302] are both advantageous, as they enhance sensitivity by factors of 5 to 6[60] and 1.4 to 3.6,[295,1302] respectively.

Marked background absorption accompanies most ET-AAS determinations; it should be lowered by applying appropriate ashing/atomization temperatures and corrected by a carefully aligned and balanced deuterium lamp.

Matrix interferences (often depression) are typical,[66,294,1324] due to excess acids (e.g., >1 to 2 *M* $HClO_4$,HNO_3, H_2SO_4) as well as to some ions (Fe, Cl^-, Cr, etc.). Ashing temperatures are between 500 and 1400°C, preferably in a "ramp" mode. $Mg(NO_3)_2$ acts as thermal stabilizer up to 1400°C.[553] Fast heating rates are required in the atomization step, so as to: (1) decrease atomization temperatures (to 2400 to 2600 vs. 2800 to 2900°C), (2) increase sensitivity, and (3) reduce matrix effects. Calibration by matrix-matched standard solutions or by standard additions is often mandatory.[553] STPF and Zeeman ET-AAS are promising with difficult matrices.[553]

Table 1
NICKEL CONCENTRATIONS IN HUMAN BODY FLUIDS AND TISSUES

Sample	Unit	Selected reference values	Ref.
Blood	µg/ℓ	4.8 ± 1.3 (2.9—7.0)	1308
		5 ± 1.4	27
		6.0 ± 1 (4.5—7.0)	1309
Bone	µg/g	0.9 ± 0.1	268
Feces	µg/day	258 ± 126 (80—540)	1310
		258 (219—278)	632
		<300	96
Hair	µg/g	0.22 (0.13—0.51)	1311
		♂, 0.47;[a] ♀, 1.14;[a] (0.045—11.0)	288
Kidney	µg/g	0.125[b]	1312
		<0.3	106
		♂ 0.57 ± 0.34[b]	904
		1.0 ± 0.3	268
Liver	µg/g	0.13[b]	1312
		0.14 ± 0.02	268
		<0.3	106
		0.33 ± 0.13[b]	904
		0.68 ± 0.49[b] (<0.5—1.8)[b]	319
Lung	µg/g	0.02	268
		<0.5	106
		0.73 ± 0.33[b]	904
Milk	µg/ℓ	20[c]	374
		20—83	13
Nails (finger)	µg/g	(1.8—24)[b] ?	16
Plasma or serum	µg/ℓ	2.13 ± 0.58	1313
		2.4 (1.0—4.1)	1314
		2.6 ± 0.8 (1.1—4.6)	1308
		3.1 ± 1.6 (0.6—5.3)	1315
		3.5 ± 1.7	27
		4.7 ± 1 (2.5—6.5)	1308
Saliva (parotid)	µg/ℓ	1.9 (0.8—4.5)	1316
Sweat	µg/ℓ	♂, 57 ± 26; ♀, 57 ± 21	381
		♂, 52 ± 36 (7—180)	675
		♀, 131 ± 65 (39—270)	675
	µg/day	83	21
Tooth (enamel)	µg/g	x̃, <0.6 (<0.6—13)	25
Urine	µg/ℓ	(0—2.5)	27
		2.04	1317
		2.3 ± 1.4 (1.0—5.2)	1308
		2.7 ± 1.6 (0.4—5.1)	1315
		4.45 ± 1.9	1313
	µg/day	1.7	1317
		2.4 ± 1.1 (1.0—5.6)	1308

Note: Wet weight unless otherwise indicated. Mean ± SD (range in parentheses). x̃, median.

[a] Geometric mean.
[b] Dry weight.
[c] Pool (New Zealand).

III. AAS METHODS FOR ANALYSIS OF BIOLOGICAL SAMPLES

Literature data on applications of AAS abound (Table 2); ET-AAS (typically after ashing

Table 2
AAS TECHNIQUES FOR Ni DETERMINATION IN BIOLOGICAL MATERIALS

Technique	Biological materials
Flame AAS after digestion or solubilization	Feces,[1331] food,[1326,1331] hair,[370,461] liver,[319] marine fauna (fish, oyster, etc.),[1327] milk,[374] tissues,[442,465,1331] urine[1331]
Flame AAS after chelate extraction	Blood,[27,319,471,473,482,1308,1335] feces,[39,1310] fish,[474,475,477] food,[39,477,659] hair,[473,482,1311] milk,[477,1336] serum/plasma,[27,640,657,1308,1335] serum proteins,[1333] tissues,[473,477,482,643,1334] urine[27,39,339,659,1308,1328,1333]
ET-AAS after dilution or digestion	Blood,[27,457] bone,[268] food,[1326] hair,[78-80,649,650,778,1325] liver,[457] lung,[516] serum/plasma,[27,296,582,1314,1323] tissues,[268] urine[27,457,1325,1329]
ET-AAS after chelate extraction	Blood,[27,473,1309,1337] hair,[473] saliva,[1316] serum/plasma,[27,1309,1313,1315,1330,1337] tissues,[473] tissue biopsies,[1332] urine[27,1309,1313,1315,1317,1325,1329,1330]

and extraction steps) is the best method available right now. Several important methodological points are to be considered in these assays: (1) representative sampling without extraneous Ni additions; (2) contamination control during ashing and preconcentration; (3) avoiding losses (due to adsorption/retention); (4) efficiency of chelate extraction and time stability of extracts; and (5) temperature programming and adequate calibration in ET-AAS.

Samples of blood/serum,[49,342,620,1315,1318] CSF,[621] and tissues/tissue biopsies[106,622] can easily be contaminated by Ni from stainless-steel sampling devices (needles,[621,622] surgical blades,[622] etc.) as well as from anticoagulants and storage containers;[1104,1323] to a lesser extent, such contamination might be expected for hair, nails, teeth, and other specimens. Blood samples are therefore drawn by means of plastic syringes[1315] and needles made of Pt[1318] or Pt-Ru alloy,[342] and the first several milliliters of blood are discarded[1315] (or used for other assays). Small or ''spot'' samples of hair,[80,649,1311,1325] nails, feces,[1310] food,[1326] tissues (e.g., lung[856,1318]), sweat,[381] etc. may not be representative. In particular, Ni in hair increases with the distance from the root[1311,1325] and can also be affected by dye stuffs,[1311] permanent-wave solutions, and other cosmetic treatments. The arm-bag technique of sweat collection[381,675] has been found[381] to give falsely high (5 to 6 $\times$) and variable results as compared with the whole-body wash-down technique.[381] Some ordinary grinding procedures tend to introduce intolerable Ni additions, e.g., 0.4 μg Ni/min of grinding in corundum and $<$0.2 μg Ni/min in agate;[624] cleaner homogenation techniques are thus advisable.[106,532,533,944] Urine samples are stored in precleaned polyethylene bottles at pH $<$1,[531,1315] so as to avoid analyte adsorption losses on particulate matter and container walls.[531] For long-term storage, urine should better be kept deep frozen.

In most Ni assays (except for a few examples of direct ET-AAS of hair segments[78-80,649,778]), the analyte is put into solution. Typically, organic matter is completely oxidized by means of wet or dry ashing or, rarer, by LTA[133,516,626,1325,1327] or ashing in O_2.[91,133,266] Solubilization of tissue samples with TAAH,[465,1104] applied incidentally, calls for further evaluation. Serum/plasma[27,640,657,1308,1311,1314] or urine[1313] proteins can be precipitated with TCA or, better, with TCA plus another acid (HCl or H_2SO_4).[640,657,1313,1328] Nickel is thus released in a protein-free supernatant and can then be chelate extracted. The protein precipitation technique is simple and suitable for routine work; it also involves lower blanks and better precision than wet digestion. TCA, however, does not quantitatively release the analyte from jackbean urease, a nickel metalloprotein;[1320] that is why its applicability to pathological sera was recently questioned.[1320]

The digestion techniques have been extensively studied and compared.[442,1315,1318,1329] The IUPAC Subcommittee on Environmental and Occupational Toxicology of Ni recommended[1318,1320,1330] wet digestion with HNO_3, H_2SO_4 and $HClO_4$ (according to Mikac-

Dević et al.[1315]) for urine and serum. This digestion is performed in tall quartz tubes, at temperature-controlled heating, and under reflux conditions; therefore, a small amount of acids is used, and blank levels are kept low. The next extraction step is performed in the same tubes.

Digestion with HNO_3, H_2SO_4, and $HClO_4$ has been widely applied to samples of blood,[27,332,1308,1315] feces,[1310] serum or plasma,[1313,1315,1318,1320,1330] tissues,[1308,1318,1320,1331,1332] and urine.[1308,1315,1317,1318,1320,1330,1331] Several other wet-digestion techniques have also been applied to biological samples with apparent success: HNO_3 and $HClO_4$ (fish,[474] tissues,[442] urine[1322]); HNO_3 and H_2SO_4 (fish,[477] meat,[477] milk[477] urine[1333]); and HNO_3 and H_2SO_4, completed with H_2O_2 (blood,[1319] fish,[475] hair,[1319] meat,[1319] milk,[1319] tissues,[170,1334,1390] urine[1319]). Certain safety precautions such as predigestion with HNO_3 and a gradual temperature-controlled heating[170,1315] are required in these procedures, especially when closed vessels or H_2O_2 or $HClO_4$ are involved.

Complete oxidation of the organic matter is not required in some procedures; instead, simpler preparation techniques are practiced: treatment with HNO_3 (either in a PTFE-lined pressure vessel[545,1325] or simple boiling of samples with the acid[27,268,457,1312]), incomplete digestion with a HCl-HNO_3 mixture,[319] etc. The resulting solutions ($<10\%$ v/v in HNO_3) are directly analyzed by flame or ET-AAS. These procedures, while simple and involving low blanks, require more versatile equipment and programming.

Dry-ashing techniques are less satisfactory for trace Ni in organic matter.[91,1317,1320] Partial losses of the analyte (e.g., 10 to 15%[1331]) are observed, due to the formation of insoluble silicates, oxides, etc. with silica crucibles and matrix residues.[91,1317,1320,1331] Thus, the use of Pt crucibles (up to 525°C),[1320] as well as vigorous treatment of the ashes (e.g., evaporation with HNO_3,[39,374,1326] HCl,[442] or even with aqua regia[27,370]) are resorted to. Ashing aids [HNO_3,[39,442] H_2SO_4,[91,288] $Mg(NO_3)_2$[442]], while improving the analyte recovery, also give rise to blanks.

Large samples are satisfactorily ashed (LTA) in excited oxygen.[133,516,626,1325,1327]

There are several preconcentration methods for trace Ni: chelate extraction,[76,647,1320] ion exchange[412] [e.g., on poly (dithiocarbamate) resin from acidified, filtered urine, at pH 2[326] or on Chelex® 100 column from wet-digested urine, at pH 5.4[1322]], and coprecipitation [with dithiocarbamates,[540] $Al(OH)_3$, $Fe(OH)_3$, etc.].

Chelate extraction of Ni from biological digests or protein-free TCA supernatants is straightforward; extracts are analyzed by flame or ET-AAS (Table 3). The efficiency of dimethylglyoxime, furildioxime, and APDC was thoroughly studied by Ader and Stoeppler[1317] and Mikac-Dević et al.;[1315] the APDC-MIBK system secures recoveries of $>99\%$.[1317] In view of the time stability of extracts, dithiocarbamate extractions from near-neutral media[427,428,640,657,794,1320,1330] are preferable to extractions from acidic solutions at pH 2 to 3. Under unfavorable conditions — incomplete digestion, low pH, incomplete separation of the organic and aqueous (acidic) phase,[428,657] extraction in presence of excess Mn,[340,657] etc. — the extracts may be stable for only a few hours (down to 1 hr[1317]). Excess iron tends to interfere with some extractions. The determination of Ni in organic extracts does not pose any significant problems; ashing and atomization temperatures of 400 to 1000 and 2600 to 2700°C, respectively, are practical. Background correction may be required.

Finally, the direct approach to ET-AAS assays of Ni in diluted biological liquids deserves discussion. Attempts have been made to directly determine Ni in diluted serum:[27,296,582,1314,1323] samples were diluted (1 + 1 to 1 + 5) with water,[296,582] or an aqueous detergent (0.01% v/v Triton X-100®[1323]), or with an aqueous 1 to 2% v/v HNO_3,[27,1314] ramp dried and ashed up to 800 to 1400°C, and atomized with as high a heating rate as available. Similar procedures have been developed and applied to the HNO_3 digests of various biomaterials.[27,457,516,517] While such procedures are very attractive owing to their low blanks, simplified preparatory step, and low reagent consumption, they are beset by some iffy conditions; modern, versatile

Table 3
EXTRACTION SYSTEMS FOR Ni FROM BIOLOGICAL LIQUIDS/DIGESTS

Extraction system	Matrix and conditions	Ref.
APDC-i-butylacetate	TCA-deproteinized serum/plasma; pH 2.5	1335
APDC-2-heptanone	Acidified urine; pH 2.8	339
APDC-MIBK	Urine; pH 2.5	1328
	Ashed (LTA) urine; pH 2.5 (phthalate)	1325
	Wet-ashed feces; pH 2.5	1310
	Wet-ashed blood; pH 2.8—3.2	332
	Wet-ashed organic samples (milk, oil, kale, etc.)	1336
	Wet-ashed tissue; pH 9	1332
	Wet-ashed serum/urine; pH 7[a]	1330
	Wet-ashed fish/liver; 4% v/v H_2SO_4	475
	TCA-deproteinized serum or wet-ashed blood, tissue, urine; pH 2.5 (phthalate)	1308
	Deproteinized (with TCA and H_2SO_4 blood; pH 9)	1313
APDC-xylene	Acidified urine (10 μℓ of 1 *M* HNO_3/1.4 mℓ of urine)	1325
APDC + DDDC-MIBK	Wet-ashed food (fish, meat, milk, etc.); 5% v/v H_2SO_4	477
Bis(trifluoroacetylacetone)-ethylene-diimine-benzene	Dry-ashed tissues; GC procedure (volatile chelate!)	1321
Dimethylglyoxime-$CHCl_3$	Wet-digested urine; NH_4 citrate added	1333
Dimethylglyoxime-MIBK	Ashed blood, serum, urine; pH 9 (citrate) 2×[c]	1309, 1337
5 m*M* diphenylcarbazone + 2.5 *M* pyridine in toluene	Ashed biomaterials (blood, hair, tissues, etc.); pH 6.5—7 (acetate)	473, 482
Dithizone-$CHCl_3$	Wet-ashed fish; pH 8—9.5 (citrate and aqueous NH_3 added); 3×[d]	474
Furildioxime-MIBK	Wet-ashed blood/urine; pH 9.0	1315
K ethylxanthate-MIBK	Tissue digests; pH 5.5—9.8 (optimum 8, acetate)	643
HHDC-butylacetate	Deproteinized (with HCl and TCA) serum; pH 6 (acetate)	640, 657
HHDC-$CHCl_3$	Ashed excreta/food; pH6; 4—5×[e]	659
HHDC-xylene + diisopropylketone (3 + 7)	Wet-ashed tissues; pH 4 (preextraction of Fe and Cu with cupferron-$CHCl_3$ required)	1334
HHDC-$CHCl_3$ + i-pentanol (3 + 1)	Biological digests (feces, food, urine); pH 4.8—5.0; 4—5×[e]	39
SDDC-$CHCl_3$	Ashed blood/urine; pH 6.5	342
SDDC-MIBK	Hemolyzed blood (severe suppression by EDTA)	471

[a] Recommended procedure.
[b] Preextraction of Fe(III) from deproteinized blood required.
[c] Double extraction.
[d] Triple extraction.
[e] Exhaustive extraction.

equipment is required [in particular, effective background correction, pyro-coated tubes, very high heating rates (''maximum power'', temperature-controlled heating), ashing with oxygen[582] or air [296] as an alternate purge gas, very slow drying/ashing ramps,[1323] L'vov platform,[457] etc.], the whole heating cycle is as long as 5 to 6 min per assay, matrix-residue accumulation takes place and, finally, the calibration is complicated (standard additions or matrix matched). Hence, only a few successful applications are described (see Table 4), and the current techniques involve wet ashing and chelate extraction (see the IUPAC reference method[1330]).

IV. CONCLUSION

ET-AAS is the best current technique for nanograms-per-gram concentrations of Ni. Dimethylglyoxime-sensitized voltametry appears a promising alternative.

Table 4
SELECTED AAS PROCEDURES FOR Ni DETERMINATION IN BIOLOGICAL SAMPLES

Sample	Brief procedure outline	Ref.
Blood, liver, urine	HNO_3 digestion; $NH_4H_2PO_4$ added to digests (0.8% w/v); ET-AAS with L'vov platform (pyrocoated, tantalized): ramp to 900°C, and atomization at 2600°C (''maximum power''); background correction; standards containing HNO_3 and $NH_4H_2PO_4$	457
Feces	HNO_3-H_2SO_4-$HClO_4$ digestion;[a] flame AAS;[b,c] alternatively, APDC-MIBK extraction at pH 2.5 [1310d]	1310, 1331
Food	Dry ashing at 450°C for 18 hr, 1 mℓ of HNO_3 added and re-ashed for 1 hr, finally dissolved in 4 *M* HNO_3, and diluted; flame or ET-AAS[b,c]	1326
Hair	Solid sampling of 1-cm hair segments; ET-AAS: ramp to 1100°C, atomization at 2500°C; calibration against spun silk or animal hairs (preleached with dilute HNO_3 and soaked in standard solutions)	78—80, 649, 650, 778
	Pressure decomposition with HNO_3,[a] and dilution to ca. 1 *M* HNO_3; ET-AAS: 100, 420, and 2700°C (gas stop); background correction	1325
Meat, tissues	1-g sample digested with 10 mℓ of (1 + 1) HNO_3 at 80°C, evaporated to ca. 3 mℓ, and diluted to 25 mℓ; ET-AAS: 150, 600, and 2700°C; background correction; standard addition calibration	517
Serum, urine	HNO_3-H_2SO_4-$HClO_4$ digestion (tall tubes, temperature-controlled heating);[a] APDC-MIBK extraction at pH 7.0; ET-AAS[b,e]	1330
Serum	4- or 6-fold dilution with aqueous 0.01% Triton® X-100; ET-AAS: 100 (ramp 20 sec + hold 35 sec), 130 (93 + 10 sec), 1000 (25 + 45 sec), and 2650°C (0 + 3 sec); pyro-coated tubes; background correction; standard addition calibration	1323
	1 + 1 dilution with water; ET-AAS: 120 (40 + 30 sec), 600 (30 + 40 sec, O_2 introduced), 600 (1 + 25 sec), 1000 (15 + 35 sec), 2700 (0 + 6 sec, ''maximum power'', miniflow), and 20°C (1 + 4 sec); background correction; standard addition calibration	296, 582
Tissue (biopsies)	HNO_3-H_2SO_4-$HClO_4$ digestion;[a] APDC-MIBK extraction at pH 9; ET-AAS: 140, 420, 1060, and 2600°C; background correction	1332
Tissues (exposed test animals)	50- to 100-mg samples (dry weight) solubilized with 2 mℓ of 46% w/v TAAH in toluene at 60°C for 24 hr and diluted with toluene; flame AAS; standard addition calibration[b]	465
Urine (exposed humans)	HNO_3 added to pH ~ 2; ET-AAS: 69, 632, 1200 (ramp), and 2627°C (gas stop); background correction; standard addition calibration	1325

[a] Safety precautions mandatory.
[b] Background correction advisable.
[c] Standard addition check advisable.
[d] Limited time stability of extracts.
[e] IUPAC reference method.

Contamination and nonrepresentative sampling are likely to occur. The best ashing techniques are HNO_3-$HClO_4$-H_2SO_4 digestion in tall quartz tubes and (incomplete) pressurized digestion with HNO_3. APDC-MIBK extraction at pH 7 is effective (in absence of oxidants). Direct ET-AAS has yet to be perfected; versatile equipment/programming, pyro-coated tubes, efficient background correction, high heating rates, and matrix-matched calibration are all important in direct ETA of Ni. Simple pretreatment with HNO_3 and L'vov platform (STPF) atomization would seem very promising, as well.

Chapter 23
PALLADIUM

I. DETERMINATION OF PALLADIUM IN BIOLOGICAL SAMPLES

Normal concentrations of Pd in biological specimens are extremely low — about or below the nanograms-per-gram level (Table 1), and literature data on the analysis of biomaterials are very scant, revealing a limited interest in this element.[942,1338-1341]

ET-AAS, after preconcentration, appears to be the best technique for subnanograms-per-gram Pd assays.[942,1338]

Few applications of RNAA,[1342] arc-source AES,[36,380] and SSMS [25,55,132] have been found in the literature; some of these procedures appear to be insufficiently sensitive.

II. ATOMIC ABSORPTION SPECTROMETRY OF PALLADIUM

In a fuel-lean air-acetylene flame, the determination of Pd is very selective and fairly sensitive. The characteristic concentration is between 0.05 and 0.3 μg/mℓ, and the DLs are down to 0.01 to 0.02 μg/mℓ.[56-59,294]

The analytical line at 247.6 nm is preferred for better linearity; when the somewhat more sensitive line at 244.8 nm is used, the calibration curves are generally bent because of neighboring lines (e.g., Pd 244.6 nm).

An optimization of lamp current, bandpass, and burner height is essential. Under optimum conditions, only transport interferences and bulk-matrix effects, as well as a minor depression from Pt should be expected. The use of acid-matched standards may prove appropriate, and a background check is advisable.

In ETA, the characteristic concentration ranges between 0.1 and 2 ng/mℓ (10 to 73 pg), and the DLs are down to 0.05 to 1 ng/mℓ (several picograms).[59-61,1343-1345]

The use of pyrolytically coated tubes is recommended both for better sensitivity (by a factor of 4)[60] and lower atomization temperatures, e.g., 1900 to 2200 instead of 2700 to 2900°C;[60,1343,1345] applying higher heating rates during the atomization step is beneficial as well. The ashing temperatures can be between 800 and 1200°C, depending on the particular matrix, so that some matrix constituents of high or moderate volatility can be eliminated before the atomization starts. Simultaneous background correction is still needed, and standard addition check is advisable. Dilute HCl appears to be a tolerable medium,[1341] while $HClO_4$[1339] and excess HNO_3 should be avoided. Platform atomization and peak-area calibration, as well as aerosol deposition, are all promising in view of reducing interferences.[1344]

III. AAS METHODS FOR ANALYSIS OF BIOLOGICAL SAMPLES

Direct injection of biological liquids or digests into a graphite-tube furnace obviously lacks sensitivity in analyses of blood (DL 0.01 μg/mℓ[1339]), urine (DL 0.003 μg/mℓ[1339]), and biological tissues (DL 0.2 μg/g[1341]). A major preparatory step, including wet digestion, preconcentration by liquid-liquid extraction, and ET-AAS, is needed in order to determine normal Pd amounts in blood, feces, hair, and urine.[942]

Dry ashing should be avoided, as both retention losses and difficulties in dissolution of Pd metal are to be expected.[91]

The wet-ashing procedures are successful provided a final treatment with HCl is included and evaporation to dryness is avoided. Jones[1339] digested 5-g blood or 50-mℓ urine samples with HNO_3 and $HClO_4$ until fumes of $HClO_4$ appeared (to a final volume of approximately

Table 1
PALLADIUM CONCENTRATIONS IN HUMAN BODY FLUIDS AND TISSUES

Sample	Unit	Selected reference values	Ref.
Blood	μg/ℓ	<9	1338
		<10	1339
Feces	ng/g	<1	1338
Hair	μg/g	<0.02	1338
Saliva	μg/ℓ	(5—10)	380
Tooth (enamel)	μg/g	<0.05	25
Urine	μg/ℓ	<0.3	1338
		<3	1339

Note: Wet weight unless otherwise indicated. Mean ± SD (range in parentheses).

1 mℓ), and finally added 5 mℓ of 1.2 *M* HCl; 50-μℓ aliquots were injected into the graphite furnace programmed at 100°C for 20 sec, 500°C for 20 sec, 1900°C (to expel the interfering $HClO_4$ and salts), and 2700°C; background correction and similarly processed standards were used.[1339]

Miller and Doerger[1341] predigested the tissue samples with HNO_3, added NaCl (5 to 10 mg) to convert the analytes (Pd and Pt) into their chlorides, and further treated the samples with aqua regia and with HCl; the graphite furnace was programmed at 150°C for 45 sec, 1200°C for 45 sec, and 2600°C for 9 sec; background correction and acid-matched standards were used.[1341]

So far, the most sensitive ET-AAS procedure for Pd in biological samples appears to be that of Tillery and Johnson:[942] the samples were ashed with HNO_3 and $HClO_4$, evaporated to near dryness, and treated with HCl; the analyte was extracted as a chlorostannous Pd(II) complex with tri-n-octylamine in xylene, and analyzed by ET-AAS. The DLs in nanograms per gram were blood, 0.4; feces, 0.02; hair, 0.6; and urine, 0.007.[942]

Byrne,[1342] in an RNAA procedure, extracted Pd from wet-ashed (H_2SO_4 + HNO_3) biological samples with dimethylglyoxime-$CHCl_3$ in presence of EDTA. Some other extraction systems might also prove useful, e.g., as halide, dithiocarbamate, etc.[76]

IV. CONCLUSION

Ultratrace (subnanograms per gram) palladium determinations involve wet digestion (not to dryness) with a final HCl treatment, preconcentration (extraction), and ET-AAS or NAA.

Chapter 24

PLATINUM

I. DETERMINATION OF PLATINUM IN BIOLOGICAL SAMPLES

Normal levels of Pt in biomaterials (~nanograms per gram or less, Table 1) are very difficult to determine, being below the DLs of direct ET-AAS and SSMS.[3,25,55,325] An efficient preconcentration step is required in ET-AAS assays at subnanograms-per-gram levels.[942]

Cis-dichlorodiamine platinum(II) (DDP, *cis*-Pt) is now extensively studied in connection with chemotherapy of cancer. There is therefore a current need to determine elevated Pt levels in biological specimens from patients[188,926,1342-1354] and test animals.[1350,1355,1356] The therapeutic concentrations of Pt in plasma (e.g., 0.1 to 0.7 μg/mℓ[188,1347]) can be routinely monitored by ET-AAS. Other techniques, such as XRF following ion exchange[1347] or NAA,[1350,1355] do not seem to offer any advantages over ET-AAS.

II. ATOMIC ABSORPTION SPECTROMETRY OF PLATINUM

In a lean air-acetylene flame, the characteristic concentration and DLs are 0.9 to 2 μg/mℓ and 0.04 to 0.1 μg/mℓ, respectively.[56-59,294] In a N_2O-C_2H_2 flame (with a narrow red zone, caution!), determinations are more selective but less sensitive (~ 4×).

Dilute HCl or HNO_3 (e.g., ≤1 *M*) are appropriate acidic media. Higher concentrations of acids, noble metals, masking agents, salts, etc. markedly interfere. Acid- and matrix-matched standards as well as background correction are called for; standard addition checks are advisable.

In ETA, characteristic concentration is between 0.45 and 40 ng/mℓ (45 to 460 pg), and DLs are down to 0.1 ng/mℓ (~50 pg).[57,59-61,294,819] Higher atomization heating rates as well as pyro-coated tubes are recommended for greater sensitivity (e.g., by a factor of 2 to 4).[60] Many potential interferents can be eliminated by higher ashing temperatures (1200 to 1900°C). Atomization temperatures are often around 2600 to 2700°C, but may be somewhat lower (say, 2350°C[1356,1357]) with pyro-coated and/or rapidly heated furnaces.[60,1345,1348,1356,1357] Most acids depress absorbance and attack the graphite surface; dilute HNO_3, HCl, and $HClO_4$ (<0.5 *M*)[1358] are tolerable. Acid-matched standards and background correction are always required; at times, more complicated calibration is called for.

III. AAS METHODS FOR ANALYSIS OF BIOLOGICAL SAMPLES

Most ET-AAS procedures are aimed at determining therapeutic Pt levels and/or studying *cis*-Pt pharmacokinetics. Analyses of bile,[1350] blood,[1338,1345,1355,1358] CSF,[1350] erythrocytes,[1350] plasma,[926,1350,1357,1358] serum,[188,1346,1351,1352,1358] serum fractions,[1352,1353] tissues,[1341,1350,1355-1357] urine,[1339,1350,1357,1358] and urine fractions[1353] have been described. The binding of *cis*-Pt to serum proteins was studied by means of ultrafiltration[1348,1352] and ion exchange;[1348] this assay appears to be important as the protein-bound Pt species in plasma may be inactive.[1348]

Most recent applications are briefly outlined in Table 2. These assays raise few problems. Specimens are quite stable on storage;[1339,1351,1356] standards are easily prepared by spiking pooled serum with *cis*-Pt or in saline or else in the same dilute acid as sample solutions; ramp ashing to high temperatures tends to eliminate many potential interferents. Adverse effects like matrix residue accumulation and sensitivity drift may occur in analyses of undiluted[1346] or (1 + 1) diluted serum;[1345,1346] heating cycles in such direct procedures are somewhat prolonged.[1346,1350]

Table 1
PLATINUM CONCENTRATIONS IN HUMAN BODY FLUIDS AND TISSUES

Sample	Unit	Selected reference values	Ref.
Blood	μg/ℓ	0.4 and 1.8[a]	1338
		<30	1339
Plasma	μg/ℓ	<45	55
Tooth (enamel)	μg/g	<0.09	25
Urine	μg/ℓ	<3	1339

Note: Wet weight unless otherwise indicated. Mean ± SD (range in parentheses).

[a] Two California populations.

Table 2
SELECTED AAS PROCEDURES FOR Pt DETERMINATION IN BIOLOGICAL SAMPLES[a]

Sample	Brief procedure outline	Ref.
Blood	HNO_3-$HClO_4$ digestion;[b] evaporated to almost dryness, and finally taken up in 0.35 *M* $HClO_4$; ET-AAS: 100, 1400, and 2700°C (gas stop); background correction; standards in 0.35 *M* $HClO_4$	1358
Blood, feces, hair, urine	HNO_3-$HClO_4$ digestion;[b] evaporated to near dryness, and dissolved in HCl; extraction of chlorostannous Pt(II) complex with tri-n-octylamine in xylene; ET-AAS	942
Blood, tissues	1-g sample + 1 mℓ of HNO_3 heated under reflux for 24 hr, the fatty phase separated and extracted with 0.25—0.5 mℓ of HNO_3 under reflux for 15 min (this step is essential), both aqueous phases combined, and diluted to 10 mℓ; ET-AAS: 125, 1500 (ramp), and 2700°C (gas stop); background correction; acid-matched standards	1355
CSF, plasma	Dilution; ET-AAS: 250 (ramp), 650 (ramp), and 2700°C (ramp); background correction; matrix-matched standards	1350
Plasma/serum, urine	ET-AAS: 100 (ramp), 600 (ramp), 1400, and 2700°C (gas stop); background correction; matrix-matched standards	1358
Plasma, urine (binding of *cis*-Pt drug)	Ultrafiltration, cation exchange, and ET-AAS	1348, 1353
Serum	1 + 4 dilution with a (0.01 *N* HCl—0.1% Brij®) solution; ET-AAS: 90, 1350, and 2700°C; standards on serum	1351
	1 + 1 dilution with water; ramp drying and ashing up to 1500 — 1730°C; serum-based standards	1345, 1346
Tissues	1 + 1 homogenation with aqueous 0.25% Triton® X-100; CRA-90 program: 95, 1300, and 2350°C; standards in saline	1357
	Dry ashing in porcelain; gradual temperature increase up to 700°C, ashes soaked with aqua regia for 3 hr, and diluted; CRA-90 program: 90, 1000, and 2350°C; standards in dilute HCl	1356
Tissues, urine	Intact urine; tissue digested with HNO_3 and (finally) $HClO_4$;[b] evaporated with HC1, and taken up in 0.1 *M* HCl; ET-AAS: 250 (ramp), 1300 (ramp), and 2700°C (ramp); background correction; acid-matched standards (standard addition check advisable)	1350

[a] Most procedures applicable to higher-than-normal levels.
[b] Gradual, temperature-controlled heating. Safety precautions mandatory.

Modern methodology tends towards elimination of sample digestion. However, determinations of lower Pt levels call for major preparatory steps, including ashing and preconcentration.[942] Dry ashing is prone to retention losses and calls for gradual temperature increase and severe treatment of ashes with aqua regia or repeated evaporation with HCl in order to convert Pt° to soluble chloride species.[1356,1359] Wet ashing is therefore more successful, simpler, and faster, provided evaporation to dryness and baking are avoided. Digestions with single HNO_3,[1352,1355] HNO_3-$HClO_4$,[942,1339,1350,1354,1358] as well as more complicated treatments (HNO_3; NaCl; aqua regia; HCl)[1341] have been described; many of these have involved final treatments with HCl.[942,1339,1341,1350]

Ultratrace Pt(II) may be preconcentrated by liquid-liquid extraction (as chlorostannous Pt(II) complex with tri-n-octylamine in xylene,[942] APDC-MIBK,[1352] etc.[76,1354]) or ion exchange.[926,1347,1348]

IV. CONCLUSION

Normal concentrations of Pt (~ nanograms per gram) are determined by ET-AAS after a major preparatory step, involving, e.g., wet ashing and extraction.

Direct ET-AAS is the best routine technique for monitoring therapeutic levels of *cis*-Pt in biological media. Attention should be paid to temperature programming and calibration mode.

Chapter 25

RUBIDIUM

I. DETERMINATION OF RUBIDIUM IN BIOLOGICAL SAMPLES

Rb concentrations in most biological specimens are relatively high, generally between 0.1 and 10 μg/g (Table 1), and their determination does not pose any special difficulties.

Flame AAS, flame emission photometry,[1363] NAA (both INAA and RNAA),[45,96,108,111,112,116,165,840] and XRF[3,128,636,691,851] are well-established techniques for the quantitation of Rb in biological samples.

Both flame AES and AAS are sufficiently sensitive (DLs down to a few nanograms per milliliter) and easily accessible techniques.

Several other analytical methods offer the advantages of multielement, instrumental performance but require rather expensive equipment: XRF,[128] PIXE,[128,407,696-698] SSMS,[3,16,25,52,54,55,132,275,409] and INAA.[45]

Certain NAA assays (INAA) can be performed without radiochemical separations, e.g., analyses of CSF,[621] erythrocytes,[631,840] hair,[45,48,101,184] placenta,[377] plasma or serum,[631,840] platelets,[119] skin,[112] tissues,[12,45,631] etc.[45]

II. ATOMIC ABSORPTION SPECTROMETRY OF RUBIDIUM

In air-acetylene flame, the characteristic concentration is between 0.03 and 0.1 μg/mℓ, and DLs are down to 0.002 μg/mℓ.[56-59,294] EDL is a preferred light source vs. HCL,[138] due to its higher intensity, stability, and longer lifetime. The photomultiplier used should be sufficiently sensitive in the respective spectral region (Rb 780.0-nm line). Acid-matched standard solutions and ionization buffer additions (e.g., 0.1% of Cs^+ or Na^+, as chlorides) are essential. Concave calibration curves at lower concentrations, if observed, are indicative of nonsuppressed ionization. The ionization degree can be lowered by any one of the following approaches or their combinations: (1) applying lesser dilution factors; (2) using a three-slot burner; (3) optimizing the burner height; (4) employing cooler flames (e.g., air propane); and (5) using a fairly fuel-rich flame.

Every AA spectrophotometer can also be used for atomic-emission determinations of Rb. All that is necessary is to switch to an AES mode (i.e., to modulate the flame emission signal) as well as to employ a thinner flame [i.e., to rotate the standard air-acetylene burner to 90° or use an emission (''Meker'') burner]. In AES assays, the standard solutions should preferably be matrix matched, and require an appropriate blank solution for zero adjustment.

In ETA, the characteristic concentration and DLs are down to a few tenths of nanograms per millileter (~10 pg absolute),[57,59,138] depending on volume of tube furnace, volume of solution injected, intensity of light source, and emission noise due to the incandescent graphite tube.

Ashing temperatures are up to 800°C. The matrix residues may give rise to background absorption and interferences; hence it is advisable to check for these effects by standard additions as well as by background correction. A tungsten halide flame or a Ba 778.0-nm line[302] can be used for background correction. The purge gas should be argon, as N_2 may involve excessive background and noise due to CN formation.

Table 1
RUBIDIUM CONCENTRATIONS IN HUMAN BODY FLUIDS AND TISSUES

Sample	Unit	Selected reference values	Ref.
Blood	mg/ℓ	1.97	1360
		2.49[a]	13
		2.7 ± 0.4	3
		(0.77—8.8)	96
Brain	μg/g	4.0 ± 1.1	3
CSF	μg/g	0.06	52
		0.11 ± 0.04 (0.05—0.20)	621
Erythrocytes	mg/ℓ	4.18	1361
		4.18 ± 0.62 (3.16 — 5.51)	1360
		4.28 ± 0.98 (2.66 — 7.24)[b]	840
		(2.5 — 5.5)	1362
Feces	mg/day	0.3[c]	8
		3.6 (1.7 — 6.6)	96
Hair	μg/g	x̄, <1 (<1—2.5)	98
		0.22	54
		1.25 ± 0.83 (0.5—5.32)	101
Kidney	μg/g	x̄, 3.01 (2.44 — 8.40)	106
		3.2 ± 0.3	105
		x̄, 3.0; 3.3 ± 1.4 (1.0 — 6.0)	12
Kidney cortex	μg/g	3.5	107
		5.2 ± 0.5	3
Liver	μg/g	4.9 (2.9 — 6.3)	108
		x̄, 5.2; 4.5 ± 0.5 (3.1 — 9.1)	105, 106
		x̄, 5.6; 6.4 ± 2.8 (2.7 — 9.7)	12
		6.5 ± 2.6	275
		7.0 ± 1.0	3
Lung	μg/g	x̄, 2.99 (2.50 — 7.49)	106
		3.5 ± 0.4	3
		4.2 ± 1.1	105
Milk	μg/g	0.63 (0.60 — 0.66)	182
Muscle	μg/g	5.0 ± 0.5	3
Nails (finger)	μg/g	(1.4 — 2.9)[d]	16
Placenta	μg/g	25.0 ± 5.8[d]	377
Plasma or serum	mg/ℓ	0.12 (0.07— 0.18)	1362
		0.16	1361
		0.16 ± 0.03 (0.12 — 0.21)	1360
		0.16 ± 0.04[e] (0.09 — 0.20)[e]	1360
		0.17 ± 0.04 (0.09 — 0.27)	840
		0.20 ± 0.06[b]	119
		0.217	1363
		0.27 ± 0.19	111
		0.3	1364
Skin	μg/g	x̄, 7.8;[d] 9.1 ± 4.0[d]	112
		3.47 ± 1.43;[d,f] 6.62 ± 2.32[d,g]	565
Urine	mg/ℓ	1.5	1364
		1.52	1361
	mg/day	2.4 ± 0.6 (1.3 — 3.4)	111
		2.27[h]	13

Note: Wet weight unless otherwise indicated. Mean ± SD (range in parentheses). x̄, median.

[a] Weighed mean (N = 243).
[b] μg/g.

Table 1 (continued)
RUBIDIUM CONCENTRATIONS IN HUMAN BODY FLUIDS AND TISSUES

[c] "Reference Man".
[d] Dry weight.
[e] Variations in the same individual within a period of several weeks.
[f] Dermis.
[g] Epidermis.
[h] Weighed mean (N = 21).

III. AAS METHODS FOR ANALYSIS OF BIOLOGICAL SAMPLES

The flame AAS has been applied to determinations of Rb in blood,[1360,1364] erythrocytes,[1360-1362,1365] food,[731,1366] plasma,[1361,1362,1364,1365,1367] serum,[1360,1364] tissues,[462,1366,1367] and urine.[1360,1361,1364,1365,1367]

The sampling stage may involve certain errors due to nonrepresentative sampling (e.g., feces, food, etc.) or hemolysis in serum samples or else post-mortem changes in autopsy tissue specimens.[789] Rubidium concentrations are slightly but significantly higher in serum than in plasma (e.g., by an average of 12 ng/mℓ, i.e., by less than 10%).[656] Inasmuch as the erythrocyte Rb is about 26 times as high as the serum Rb,[840] very delicate serum preparation techniques[656,1364] should be performed, especially when determining therapeutic concentrations of Rb-containing drugs in erythrocytes, plasma, and serum.[1364]

Cornelis et al.[165] found no significant changes of Rb concentration in urine after 3-day storage at room temperature in precleaned polyethylene containers.

Iyengar[789] found a significant post-mortem decrease of Rb in rat livers (up to 30 to 40%, depending on autolysis time and storage conditions), and stressed the importance of standardizing the sampling time for autopsy cases.

Prior to flame AAS analysis, liquid samples are usually diluted with water or with an aqueous ionization buffer. Undiluted plasma[1362] or urine[1364] can be directly sprayed, but this might create difficulties with calibration and burner blockage; this approach is to be recommended only if pulse-nebulization and matrix-matched or standard-addition calibration are applied.

Organic matter may be destructed by dry ashing (up to 500 to 550°C)[462,731] or by LTA[275,462,626] or else by any common wet-digestion procedure.[91,462] Matrix-matched standards may be necessary with high-salt digests (e.g., >1% w/v). In the multielement scheme of Locke,[462] the sample solutions contained 1% w/v of tissue ash (LTA), 1% w/v of phosphate (Na_2HPO_4 + KH_2PO_4, 1 + 4), and 5% w/v of H_2SO_4; all three preparation procedures evaluated (LTA; dry ashing overnight at 500°C and then 2 hr at 550°C; and wet digestion with HNO_3 and H_2SO_4, finally completed with H_2O_2) proved to be equally applicable, though the use of LTA was recommended in view of its applicability to most elements studied.[462]

The calibration standards in blood and plasma/serum analyses should contain the corresponding amounts of NaCl *and* KCl[1364] or, alternatively, be based on pooled blood or serum.[1364]

IV. CONCLUSION

Flame AAS and AES would seem to be the best methods for a routine analysis of biological samples, at both normal and therapeutic levels. XRF and INAA are relevant, nondestructive, multielement, but also more expensive, techniques.

Rubidium is determined in air-acetylene flame, preferably with a three-slot burner. Ionization suppression is a must.

Chapter 26

SELENIUM

I. DETERMINATION OF SELENIUM IN BIOLOGICAL SAMPLES

The concentration of Se in most human biological materials is between 0.01 (e.g., in milk, urine) and 1 μg/g (hair, nails, kidney) (Table 1). The best analytical techniques for determining nano- and microgram amounts of Se in biological samples are flameless AAS (both HG and ETA), fluorimetry, and NAA.

Fluorimetric assays usually involve complete digestion of the organic matter, reduction of the analyte to Se(IV), formation of piazselenol and its solvent extraction in presence of masking agents, and, finally, fluorimetric assay. These procedures are well established but are somewhat tedious and time consuming, prone to significant blank corrections, and involve potentially cancerogenic organic reagents. Applications to various biomaterials: blood,[1299,1380,1381] food,[1373,1382-1385] hair,[1373,1386] milk,[40,1381,1382,1385-1387] nails,[1377] tissues,[187,1373,1382-1384] urine,[1377,1379,1386] etc.[1372,1380-1383,1386] have been described.

NAA is a very sensitive and popular technique for trace Se in blood,[45,178,573,1128,1370] bone,[97] CSF,[621] erythrocytes,[45,622,631,1370]food,[45,47,96,116,1376,1385,1389] hair,[9,11,45,48,98,101,102,118,184,1128] milk,[110,182] nails,[45,560,1377] placenta,[377] plasma or serum,[45,111,167,186,631,840,1370,1376] plasma/serum protein fractions,[120,1370] platelets,[119] skin,[112,565] teeth,[45,50,51] tissues,[45,51,106-108,631,1370,1389,1390] urine,[96,111,165,1128] etc. Many of these assays can be performed by INAA.[45,1391]

The sensitivity of XRF[128] is confined to tissue samples,[636,1392-1394] and a preconcentration step is usually required.

Some laboratories are able to perform PIXE assays[128] on small-size samples of blood,[696] blood fractions,[698,1395] hair,[128] and tissue biopsies (of only 10 to 100 μg dry weight).[407]

SSMS[3,16,25,54,55,132] provides DLs down to 2 ng/g.[3]

GC determinations of total Se and speciation of some oxidation states or methylated species[1371,1375,1396,1397] are performed after solvent extraction of the analyte as volatile piazselenol[1371,1375,1378] or after evolution of volatile hydrides.[1396]

Incidentally, some other techniques: ASV (after an ion-exchange separation[1398] differential pulse cathodic stripping voltametry (DL 30 ng/g),[1399] ID-GC-MS,[1400] ICP-ES (after preconcentration from biological liquids by ion exchange on poly(dithiocarbamate) resin[326] or by distillation as $SeBr_4$[192]), and flame AFS (pulse nebulization of diluted blood,[952] or hydride-generation AFS[144,1401]) have been applied.

Some of the indicated techniques (NAA, XRF, PIXE, SSMS) require expensive equipment and highly skilled personnel and are, therefore, not readily available in routine practice.

Likewise, see the recent reviews.[45,123,1388,1402]

II. ATOMIC ABSORPTION SPECTROMETRY OF SELENIUM

The flame AAS is not sensitive; the characteristic concentration and DLs are 0.26 to 1 μg/mℓ and 0.05 to 0.5 μg/mℓ, respectively.[56-59]

Due to the low wavelength used — 196.0 nm — significant light losses are typical of this element. An EDL source is highly recommended because of its higher intensity (e.g., by an order of magnitude vs. the HCL[138]), longer lifetime, and lower DLs attainable (e.g., 2 to 2.6 ×[138] or even 5 to 6 ×).

The absorption of flame gases as well as the background absorption of matrix constituents are marked; thus, simultaneous background correction is recommended.

The assays are usually made in a slightly fuel-rich air-acetylene flame. Alternatively, a

Table 1
SELENIUM CONCENTRATIONS IN HUMAN BODY FLUIDS AND TISSUES

Sample	Unit	Selected reference values	Ref.
Blood	mg/ℓ	0.077 ± 0.011 (0.034—0.166)[a]	1128
		(0.011—0.372) (New Zealand)	1368
		0.12 (0.08—0.18) (U.S.A.)	1369
		0.182 ± 0.036 (Canada)	1370
		0.171[b,c]	13
		0.21 (0.181—0.249) (Japan)	1371
Bone (femur)	μg/g	x̄, 0.057 (<0.06—0.12)	97
Brain	μg/g	0.09 ± 0.02	3
Erythrocytes	mg/ℓ	0.105 ± 0.017 (0.038—0.240)[a]	1128
		0.119 ± 0.015 (0.098—0.140)	1372
		0.16 ± 0.03 (0.09—0.21)	840
		0.236 ± 0.060	1370
		0.32 (0.255—0.404)	1371
Feces	μg/day	20[d]	8
		26 (8—77)	96
Hair	μg/g	x̄, <0.2; 1.4 ± 3.1 (<0.2—14.8) (Italy)	98
		0.303[e] (0.025—1.58) (New York)	288
		0.332 ± 0.061 (0.218—0.505)[a]	1128
		x̄, 0.44; 0.42 (0.24—87.5) (Canada)	10
		0.57 ± 0.038[f] (U.S.A.)	1373
		x̄, 0.69 (0.135—66) (Japan)	11
		x̄, 1.1 (U.S.A.)	9
		1.28 ± 0.84 (0.2—6.91) (India)	101
		(<0.1—98) (U.S.S.R.)	102
		1.4 ± 1.6 (<0.2—8.2) (Iraq)	118
			105
Kidney	μg/g	0.79 ± 0.07	1373
		1.09	
Kidney cortex	μg/g	x̄ 0.70 (0.37—1.03)	106
		x̄ 0.755; 0.755 ± 0.207 (0.37—1.03)	181
Liver	μg/g	x̄ 0.19; 0.245 ± 0.122 (0.14—0.46)	106, 181
		0.26 (0.22—0.33)	108
		0.30 ± 0.13 (0.16—0.57)	187
		0.30 ± 0.1	3
		(0.18—0.66)	1370
Lung	μg/g	0.1 ± 0.02	3
		x̄, 0.11; 0.095 ± 0.017 (0.044—0.22)	105, 106
		x̄, 0.120; 0.165 ± 0.182 (0.044—0.700)	181
		0.15[b,g]	13
Milk	μg/ℓ	15 (11—22)	182
		(13—21)	13
		18 (7 —60)	1374
		21.2 ± 8.7 (8—36)	1372
		22 (13—62)[d]	8
		(10—83)	1375
		(98—155)[f,h]	1376
Muscle	μg/g	(0.26—0.59)	1370
Nails	μg/g	♂, 0.58 ± 0.27 (0.3—0.9)[i]	560
		♀, 0.88 ± 0.59 (0.3—1.5)[i]	560
		1.005 ± 0.288 (0.656—1.545)	1372
		1.14 ± 0.06 (0.70—1.69)	1377
Placenta	μg/g	0.34 ± 0.01 (0.23—0.48)	1378
		1.70 ± 0.61[f]	377

Table 1 (continued)
SELENIUM CONCENTRATIONS IN HUMAN BODY FLUIDS AND TISSUES

Sample	Unit	Selected reference values	Ref.
Plasma/serum	mg/ℓ	0.046 ± 0.004 (0.038—0.068)[a]	1128
		0.103 ± 0.018	631
		0.12 (0.100—0.139)	1371
		0.122[b,l]	13
		0.13 ± 0.02 (0.09—0.18)	840
		0.144 ± 0.029	1370
		0.19 ± 0.06	111
Saliva	ng/mℓ	3.62 ± 1.10 (2.0—4.9)	1372
Skin	μg/g	0.345 ± 0.072;[f,j] 0.518 ± 0.163[f,k]	565
		(0.12—0.62)	1370
		x̃, 1.1; 1.1 ± 0.52[f]	112
Tooth (dentine)	μg/g	0.28	22
Tooth (enamel)	μg/g	0.012;[m] 0.08[n]	51
		0.27 ± 0.02 (0.12—0.50)	25
		♂, 0.44; ♀ 0.52	22
Urine	μg/ℓ	5 ± 2 (2—11)	1378
		7.4 ± 1.3 (2.5—17.6)[a]	1128
		12.3 ± 8.2 (2.6—47)	1379
		49 ± 6 (15—94)	1377
		51.7 ± 20.8 (28—92)	1372
	μg/day	6.2 ± 1.0 (2.6—13.6)[a]	1128
		6.2, 12, 14, and 22;[o] (2.3—75.2)	96

Note: Wet weight unless otherwise indicated. Mean ± SD (range in parentheses). x̃, median.

[a] Se-deficient area in Italy.
[b] Weighed mean.
[c] N = 617.
[d] "Reference Man".
[e] Geometric mean.
[f] Dry weight.
[g] N = 18.
[h] ng/g (ppb).
[i] Thumbnails.
[j] Dermis.
[k] Epidermis.
[l] N = 507.
[m] European.
[n] African.
[o] Mean values of four different locations in Italy.

N_2O-C_2H_2 flame can be used, providing higher selectivity but lower sensitivity (~ 4 ×).

In ETA, the characteristic concentration is between 0.1 and 3 (10 to 300 pg) and DLs are 0.05 to 2 ng/mℓ (5 to 50 pg).[57,59-61,228,1403]

Several *serious* problems are encountered in ET-AAS of Se: (1) severe background absorption; (2) premature volatization of the analyte; (3) matrix and acid effects; and (4) sensitivity drift.

Background absorption is particularly pronounced due to the low wavelength (196 nm) and high volatility of the analyte. Many salts and even nonvolatile acids,[228,1404] which during the ashing stage cannot be completely eliminated from the graphite tube, give rise to unspecific light losses during the atomization stage. This situation is somewhat improved by adding matrix/analyte modifiers, which thermally stabilize the selenium species and/or facilitate volatization of some concomitants during the ashing stage. Important factors are the choice of the modifier and the optimization of the furnace program (ashing temperature, atomization temperature, heating rate, gas-flow rate).

The use of a thoroughly aligned and balanced deuterium background correction system is of paramount importance in all Se assays by ET-AAS. Moreover, the correction capability of the system for dynamic signals and a structured background must always be checked; e.g., the background correction is not accurate in the presence of excess iron due to partial absorption of the D_2-lamp radiation by neighboring iron lines.[1405] This effect can be checked by measuring the background-corrected absorbance with several different slitwidth settings. It is radically eliminated by either Zeeman correction[927] or chemical separation.

Matrix effects are typical and numerous; many cations,[226,228,1404,1406,1408] anions,[226,1407] and acids[228] exhibit enhancement or depression of the selenium signal. There are a number of interference studies,[226,228,1180,1404,1406] but their results are not consistent. All acids in excess are depressants.[226,228,1409,1410] Many ions enhance Se absorbance due to thermal stabilization of the analyte during the ashing as well as the early atomization stage.

Ediger[223] was the first to propose the "matrix modification" technique: adding Ni salts stabilized the inorganic selenium up to 1200°C, probably as nonvolatile nickel selenide. This technique is now very popular,[226,1180,1390,1409-1416] the concentration of Ni being between 0.03 and 1%, typically 0.1 to 0.2%. The added Ni does not necessarily eliminate the interferences,[226] but it produces major improvements: (1) higher ashing temperatures; (2) leveling off of the enhancement effect of other ions; (3) sensitivity improvement (by a factor of 1.9 to 6); (4) better precision; and (5) less critical temperature programming effect.

Several other ions have also been used as analyte modifiers: Cu,[223,1180,1369,1408,1417] Ag,[1180,1406,1414] Rh,[1403] Mo,[1404] etc.[1180,1406] The efficiency of these matrix/analyte modifiers as well as the extent of matrix effects depend on numerous factors: the chemical form of Se (e.g., selenite, selenate, organically bound selenium[1406,1414]); the concentration of the modifier[1404] and its oxidation state[1180] [e.g., Cr(VI) and Mn(VII) thermally stabilize selenite, while Cr(III) and Mn(II) do not[1180]]; the acids and anions present and their concentration;[1404,1409,1410] the age,[1409,1410] (pyro) coating,[60,1418] and pretreatment[1404,1418] of the graphite tube; and the temperature program.

Szydlowski[1417] compared the efficiency of three modifiers, Ni, Mo, and Cu, and recommended copper (600 μg/mℓ) as the best solution for selenium-2,3-diaminonaphthalene complex in extracts. He rejected nickel for higher blanks, multiple peaks, and for higher atomization temperatures associated with this modifier; Mo was found to depress the Se absorbance.

To conclude, the ET-AAS determinations of Se are prone to numerous strong interferences from ions, salts, and acids; the extent of these interferences is difficult to predict, as it depends on many factors. Thus, separation of Se from the matrix and its thermal stabilization by a suitable matrix/analyte modifier are needed. Calibration should be by acid- and matrix-matched standards. The efficiency of the background correction system should be checked. Thoroughly optimized furnace programming, with as high as tolerable ashing temperatures (e.g., 800 to 1200°C in presence of a matrix/analyte modifier), and high heating rates during atomization are essential. Precoated tubes are preferred for greater sensitivity and lesser matrix effects, viz pyro-coated[60,1413,1418,1419] and/or metal (carbide)-coated tubes (Zr,[1418] Ta,[1418] Mo[1404]). Gas-stop atomization mode is beneficial — if not beset by excessive background.

The hydride-generation AAS is a rapid and sensitive technique for nanogram amounts of Se. The DLs are in the range of 0.3 to 60 ng,[231,232,1402] being typically about 1 to 5 ng Se. Depending on the maximum volume of sample solution which can be introduced into the reaction vessel, the relative DLs are between 0.02 and 2 ng/mℓ (often, several tenths of nanograms per milliliter). These figures are comparable with the DLs of some other instrumental techniques involving HG: MIP-AES (25 ng or 1.2 ng/mℓ),[131] ICP-ES (0.4 to 0.8 ng/mℓ),[142,145] AFS (0.06 ng),[144] and nondispersive AFS (0.3 ng or 0.015 ng/mℓ).[1401]

There are several important reviews[232,242,1402] and optimization[1420-1422] and interference[149,155,226,237,238,1401,1422,1423] studies on HG-AAS of Se.

The reductant of choice is aqueous $NaBH_4$ which provides higher reaction yields, faster reduction rates, lower blanks, and easier automation/semiautomation vs. other reducing systems: $NaBH_4$ pellets,[237,238,1424] Zn-HCl,[237,1401] Al-HCl, and Mg-$TiCl_3$-HCl.[811]

The analytical performance (accuracy, precision, day-to-day variability, sensitivity, DLs, etc.) depends on numerous chemical and instrumental parameters:

1. The oxidation state of Se is of vital importance; *only* Se(IV) is reduced to H_2Se, and therefore any selenate [Se(VI)] must be quantitatively reduced to selenite prior to HG. This prereduction is often done by heating the analyzed solution with HCl (3 to 6 *M*) for 5 to 30 min at 100°C.[1369,1371,1379,1380,1396,1408,1417,1426,1427] Prolonged heating may entail partial losses of the analyte as, e.g., volatile $SeO.Cl_2$, $SeO_2.2HCl$, etc.[1375] Sinemus et al.[139] avoided losses by heating their environmental water samples with 5 *M* HCl for 10 min at 80°C in a closed PTFE-lined autoclave. Some authors prefer reducing selenate to selenite with H_2O_2[1415] or with H_2O_2 and subsequently $NH_2OH.HCl$;[1427] this reduction is believed to be less prone to volatization losses, and has been successfully practiced prior to liquid-liquid extraction of Se(IV).[1415,1427]
2. If methylated selenium species are present, they will be reduced to alkylselenium hydrides and stripped from the analyzed solution at different rates (and therefore will give completely different response!) as compared with the H_2Se evolved from selenite (see, e.g., Reference 1396).
3. The nature and concentration of the acid used is very important in view of sensitivity,[142,238,812,1420,1424] interferences,[1422,1428] blank levels,[1401,1420] lifetime of the absorption cell, and background absorption.[139,1424] The acids may be ranged in the following order according to their usefulness: HCl > H_2SO_4 > H_3PO_4 > $HClO_4$.

   The HCl medium provides the best sensitivity and lower blanks; it is most convenient since the prereduction to selenite is usually done by HCl. The optimum HCl concentration depends on construction and operating parameters of the particular HG system. Most authors have recommended higher HCl concentrations (so as to ensure better selectivity), e.g., 8 *M*,[250] >5 *M*,[812] 5 *M*,[139,142,149] 4 to 4.5 *M*,[1424] 3 to 5 *M*,[142] 3 *M*[1399,1421] and 2 *M*,[1429] whereas others have been successful with much lower acidities — 0.2 *M*,[140] 0.3 *M*,[1422] 0.4 to 0.5 *M*,[1420] 0.5 *M*.[1401] Anyway, acidity exhibits a strong effect on peak heights, interferences,[1422,1428] etc. and should be optimized for every particular system and matrix.

   The drawbacks of HCl media are the etching of the inner surface of the quartz absorption cell[1424] and the background absorption of the excess HCl (e.g., >4 to 4.5 *M*).[1424]

   The H_2SO_4 is quite tolerable, but it gives a slightly lower sensitivity[142,226] and higher blanks.[1420] Even small amounts of HNO_3 (say, 1% v/v) strongly depress the signal.
4. The concentration of the reductant should be optimized for every particular HG system[139,1420] in order to provide greater sensitivity, precision, and accuracy of the assays.[1402,1420] It is in the range of 0.5 to 8% w/v $NaBH_4$[1402] (often, 2 to 5% w/v), alkalized with 1 to 2% w/v NaOH. Somewhat lower concentrations proved better with the open-ended atomization system (Perkin-Elmer® MHS-10) than the closed system (MHS-1) — 3.5 vs 5.0%.[1420] The higher the reductant concentrations, the greater the sensitivity, blank levels, and RSD.[1420]

   The reductant activity (and therefore the sensitivity, day-to-day reproducibility, etc.) depends on its purity[1368] and storage conditions. Verlinden et al.[1420] filtered a $NaBH_4$

solution, stabilized with 2% w/v NaOH, in a polyethylene bottle and stored under refrigeration for at least 8 weeks until use.

5. The higher the volume of the analyzed solution, the lower the absorbance; therefore, this volume should be kept approximately constant and, preferably, low (e.g., 10 mℓ) in all assays.[1420]
6. The optimum temperature of the quartz atomization cell should be about 850 to 900°C,[139,1420,1421] but somewhat lower (800°C)[1429] or higher (1000°C)[1399,1430] temperatures have been reported as well.
7. The hydride collection time[1402,1422] and the flow rate of the carrier gas[1420] are to be optimized for better sensitivity, accuracy, and precision.
8. The linearity of calibration curves may be restricted to, e.g., 70[150] or 100 ng Se;[1420,1421] it is better in open-ended systems (e.g., up to 300 ng in MHS-10[1420]) or in a peak-area calibration mode (e.g., up to 500 vs. 100 ng[1421]).
9. Finally, one of the foremost chemical parameters of the assay is the presence of potentially interfering concomitants in the sample solution. The results obtained in the numerous interference studies cannot be directly compared, because the nature and extent of these interferences depend on many factors. Among these factors are the make of the apparatus;[1420] the acidity of the analyzed solution;[1422] the $NaBH_4$ concentration,[1424] age and purity; the mode of measurements — peak height or peak area;[1401] the oxidation state of the interferent; the absolute amount of the interferent and — to a lesser extent — the concentration ratio interferent:selenium;[1422] the presence of masking agents and other cross-effects between concomitant species;[1422] etc.[1402] Both reductants, $NaBH_4$/HCl and Zn/HCl, are prone to depressive interferences albeit to a different extent for every particular interfering ion, as shown by Yamamoto and Kumamaru[237] and Nakahara et al.[1401]

Strong depressants are the noble metals (Ag, Au, Pd, Pt, Ru, Rh),[155,1401,1422] other hydride-forming elements (Bi, Sn, Sb, Ge, As),[1401,1402,1413,1422] Hg,[149,1401] oxidants (HNO_3, $HC1O_4$, MnO_4^-, $Cr_2O_7^{2-}$, $S_2O_8^{2-}$ VO_3^-, MoO_4^{2-}),[179,226] KI,[179,250] and some ions that might react with the analyte (Cu, Co, Ni, Fe, Pb).[149,155,238,1401,1422] Some of these interferences can be dealt with by dilution of the sample solution,[1422] optimizing (increasing[1422,1428]) the HCl concentration (e.g., Cu^{2+}, Ni^{2+}, Fe^{3+}, and Ag^+ interferences[1422,1428]) using peak-area measurements (e.g., the suppression from Hg[1401]), adding masking agents (not very effective), and finally, by chemical separation. Calibration by standard additions is often needed.

III. AAS METHODS FOR ANALYSIS OF BIOLOGICAL SAMPLES

The HG-AAS has been widely applied to analyses of biomaterials: blood,[1368,1399,1431] fish,[141,238,239,412,1399,1428] food,[141,249,250,1385,1389,1430] liver,[238,239,254,412,1389,1420,1429,1431,1432] milk,[250,1399,1425] nail,[1399] serum,[1431] tissues,[141,238,412,1389,1399,1431,1432] urine,[159,1407,1429] etc.[238,1252,1396,1399,1406,1424,1430] The results of several collaborative studies have been published.[249,250,1123]

ET-AAS has been applied to analyses of biological SRMs (NBS "Bovine Liver",[223,1415,1431-1434] "Oyster Tissue",[1373] etc.[1415,1431]), blood,[156,1369,1403,1411-1413,1416,1431] erythrocytes,[1427] fish,[1408-1410,1434] food,[1385,1408-1410,1417] meat,[1408-1410] milk,[1431] plasma or serum,[1414,1416,1427,1431] tissues,[1390,1408,1412,1416,1432] urine,[1406,1431] etc.[228,1407,1408] Most of these procedures involve digestion and separation of Se from the matrix, and a few attempts have been made to directly determine Se in biological liquids[1403,1406,1413,1414] and digests:[223,1390,1416] blood,[1403,1406,1413,1416] serum,[1414,1416] tissues,[223,1390,1416] and urine.[1406]

The sampling step involves no particular problems. Selenium is quite homogeneously distributed in most samples (e.g., in hair,[10,101,1373] liver,[108] serum,[840] etc.) and only some "spot" samples of food and feces,[96] and perhaps urine (unless normalized to creatinine excretion[1370]) may not be representative. Clean homogenation procedures, at liquid nitrogen temperature, have been described.[532,533,944] Incidentally, scalp hair may be severely contaminated by Se from some shampoos[118] or scalp medications;[9] both endo–[815] and exogenously deposited[162] selenium is difficult to wash away.

There are three main groups of problems which are of major concern in Se assays:

1. Partial losses of the analyte during drying and ashing of biological specimens should be expected.
2. Separation of Se from the biological digests is mandatory, and the efficiency of this step depends on chemical factors such as the selenium oxidation state, presence of residual oxidants, non-decomposed organic species, interfering ions, etc.
3. ET-AAS is prone to serious matrix effects, background absorption, volatization losses, etc.; thus, many parameters of this assay should be carefully optimized and versatile equipment be used.

It is well documented[91,166,423,535,1386,1390] that selenium may be partly lost during drying of biological specimens. These losses are obviously tissue[166,1386] and technique dependent.[166,423,1390]

Fourie and Peisach[423] found losses of -15 and -5% during drying at 120°C and freeze drying, respectively, of tracer-labeled oysters. Koh[535] dried samples of carp, liver, alfalfa, and reconstituted NBS SRMs ("Bovine Liver", "Orchard Leaves") in a microwave oven at 60°C; this procedure has been found to be cheaper and faster (15 min vs. 48 hr) than ordinary oven drying. Hoffman et al.[1386] found small but significant losses of Se from animal liver and muscle during freeze drying, but de Goeij et al.[536] did not. Raie and Smith[1390] established that up to 40% of Se could be lost on drying the liver tissue at room temperature as well as during freeze drying unless the samples were maintained at <-35°C and at 0.05 Torr. In an extensive recent study, Iyengar et al.[166] found no losses of the metabolized ^{75}Se tracer in rat tissues, blood, erythrocytes, nails, and serum during freeze drying up to 72 hr under pressure of 0.05 Torr and during oven drying at 80 or 105°C, but small and significant losses (up to 5%) were found for specimens of blood, brain, lung, and muscle at 120°C.

Selected techniques for the decomposition of organic matter prior to Se determinations (by AAS or by other techniques) are summarized in Table 2. Ordinary dry ashing is quite out of the question. Dry ashing at 450 to 500°C, in presence of ashing aids [$Mg(NO_3)_2$ or $Mg(NO_3)_2$ + MgO or $Mg(NO_3)_2$ + HNO_3], is a simple and useful technique; thorough mixing of the sample with an excess ashing aid and gradual heating are essential.

LTA in excited oxygen[169] may involve severe losses of Se[275,276,1373] (e.g., recovery (R) of only 18% for liver samples[275]), depending on the particular matrix and input power applied. At least two laboratories, however, have reported complete recoveries: Lutz et al.[626] found R = 99% and R = 105% for the NBS SRMs "Coal" and "Orchard Leaves", respectively, and Behne and Matamba,[167] by INAA, obtained identical results for endogenous selenium (104 ng/g) in three sample aliquots of human blood serum which were freeze dried, dried for 3 days at 90°C, and ashed in excited oxygen (LTA), respectively.

Combustion in oxygen, in a closed bomb, or in cold-finger apparatus has proved to be fast, clean, and reliable for Se.[91,168,1153,1380,1394,1399,1411,1432]

Wet-digestion procedures are most universally applied in Se assays. What matters is to have smooth, safe, and complete digestion and to avoid losses due to foaming, bumping, volatization, adsorption, or violent oxidation; blank levels and sample throughput are also a consideration.

Very important are the early stages of digestion. Ihnat[1382] moistened and dispersed his samples with water prior to HNO_3 addition in order to avoid charring during the nitric acid

Table 2
SELECTED ASHING PROCEDURES FOR BIOLOGICAL MATERIALS PRIOR TO DETERMINATION OF THE TOTAL SELENIUM

Oxidation mixture[a]	Matrix[b]
HNO_3, $HClO_4$, and H_2SO_4[a]	Blood,[131] "Bovine Liver" (NBS SRM),[141,239,1415] feed,[1425] fish,[141,238,239,1409,1428] food,[141,1409] hair,[1386] milk,[1386,1425] tissues,[141,238,1386] urine[159,1386]
HNO_3, $HClO_4$, H_2SO_4, and (finally) H_2O_2[a]	Blood,[1431] fish,[1382,1383] food,[1382,1383] milk,[1382,1383,1431] serum,[1431] tissues,[1382,1383,1431] urine[1431]
HNO_3 and $HClO_4$[a]	Blood,[156,1368,1369,1380,1381,1398] "Bovine Liver" (NBS SRM),[277,1398,1429] feed,[1380,1384] fish,[1381] food,[1401,1409,1410,1417] hair,[1398] milk,[1381] tissues,[535] urine[1379,1429]
HNO_3, $HClO_4$, and $HClO_3$[a]	Biological SRMs (fish, flour, etc).[1394,1430]
HNO_3 and H_2SO_4	Biological materials (egg, meat, rice, etc.),[1408] muscle[1392]
HNO_3, H_2SO_4, and (finally) H_2O_2[a]	Biological materials,[170] erythrocytes,[1427] plasma[1427]
HNO_3, H_3PO_4, and H_2O_2	Blood, egg, liver, meat, serum, urine, etc.[1252,1400]
HNO_3 (autoclave),[a] then completed by dry ashing in presence of $Mg(NO_3)_2$	Food, evaporated milk, tissues, etc.[250]
Dry ashing in presence of $Mg(NO_3)_2$	Blood,[1431] "Bovine liver" (NBS SRM),[1389,1431] food,[1389] serum[1431]
Dry ashing in presence of $Mg(NO_3)_2$ and MgO	Marine samples, plants, tissue, etc.[1393]
Dry ashing in presence of $Mg(NO_3)_2$ and HNO_3	Blood,[1378,1399] fish,[1399] hair,[1378] liver,[1378,1399] milk,[1399] nails,[1399] placenta,[1378] urine,[1378] etc.[1378]
Combustion in O_2 (closed vessel)[a]	Biological SRMs (fish, plants, etc.),[1394,1430] blood,[1380] plants[1380]
Combustion in O_2 under dynamic conditions in a cold-finger apparatus	Biological and organic samples[168,1153,1399]

[a] Some of these techniques are potentially hazardous (in particular those involving $HClO_4$ or H_2O_2 or pressure of fatty samples, etc.!). Gradual, temperature-controlled heating is essential.

[b] Procedures other than AAS are References 1379 to 1384 and 1986 (fluorimetry); 1392 and 1294 (XRF); 131 (MIP-AES); 1398 (ASV); 1401 (AFS); 1378 (GC); and 1400 (ID-GC-MS).

digestion stage; this contributed to digestion time but was mandatory for some specimens (e.g., meat and flour). Many authors[239,1380,1382,1409,1410,1417,1427,1431] prefer soaking their samples overnight with HNO_3 and applying heat subsequently. Excess HNO_3 is needed to maintain oxidizing conditions until all organic matter is completely decomposed. In order to keep lower blank levels, reflux conditions are recommended,[170,1380,1398] e.g., tall test tubes, condensers, long-neck flasks, etc. Temperature should be raised gradually (temperature-controlled heating);[170,239,1368,1384,1394,1401,1408] the prechosen temperature ramp may proceed automatically or semiautomatically.[170,1381,1384]

Raptis et al.[1394] recently found that up to 50% of the selenium may be adsorbed onto the glass surface during the digestion of biological materials with HNO_3-$HClO_3$-$HClO_4$ mixture, and strongly recommended the use of quartz vessels.

Finally, the digestion should be properly completed, i.e.,: (1) traces of organic matter should be completely oxidized by heating to fumes of $HClO_4$ or H_2SO_4; (2) nitric acid decomposition products should be expelled as they may interfere with the next extraction or hydride-generation step; and (3) the analyte should be quantitatively converted into a definite oxidation state — selenite, Se(IV).

It should not be overlooked that some decomposition techniques may not completely oxidize all organic matter and all selenium species (e.g., the decompositions with sole HNO_3

in pressurized vessels[254,544,1290,1399]). The digestion with HNO_3, either in an autoclave or in an open vessel, should be useful only as a predigestion stage,[250,1409,1431] with indispensible wet[1409,1431] or dry (in presence of $Mg(NO_3)_2$ ashing aid)[250] ashing as a finishing touch. Efforts have been made to use HNO_3 digestion in speciation studies on environmentally important oxidation states of Se,[1371,1375,1431] but the residual undigested organic matter has proved to inhibit the subsequent extraction of the analyte.[1431]

Eventually, a few words of caution: some wet-decomposition techniques, under certain conditions (e.g., excessive pressure build-up, insufficient acid-to-sample weight ratio, large samples, fatty samples, rapid heating, etc.), are potentially dangerous. Care should be taken in procedures involving $HClO_4$, or pressurized vessels,[250,254,544,1290,1390] (especially the bomb decomposition with HNO_3 and $HClO_4$!),[1390] as well as the ashing technique involving $HClO_4$ and H_2O_2![1416] The original procedure should always be consulted for important instructions.

There are several techniques for preconcentration of Se from biological digests:

1. Evolution of volatile hydride, H_2Se (or alkylhydrides in selenium speciation studies[1396])
2. Liquid-liquid extraction (as piazselenol[1369,1371,1375,1378,1380,1382,1383,1386,1415,1417,1425,1427,1431,1433] or dithiocarbamate[172,1408,1426] or ion-association complex[1436])
3. Distillation as volatile halide — $SeBr_4$[192] or $SeCl_4$
4. Precipitation or coprecipitation with ascorbic acid[1409,1410] or NaI[1394] or $SnCl_2$ + $NH_2OH.HCl$ (in presence of Te and Cu)[1392] or $La(OH)_3$[149] or As or $Fe(OH)_3$
5. Electrochemical preconcentration by controlled-potential electrolysis,[1437,1438] or internal electrolysis on copper wire;[1424] this technique is attractive in view of its combination with flame AAS
6. Ion exchange: sorption of Se on poly(dithiocarbamate) resin[326] or stripping the interfering ions from solution by cation-exchange resin[1398,1407,1411,1412]

These techniques are not equivalent: precipitation is better combined with XRF;[1392,1394] distillation, ion exchange, and electrolysis are more time consuming; therefore, HG and extraction are most popular. In these two techniques it is very important to have the analyte in a definite oxidation state — selenite.

Some extraction systems for Se(IV) are compiled in Table 3. Strong oxidants and some metal ions interfere with the extraction or next ET-AAS or both. Coextracted metal ions (e.g., Cu, Zn, Co, Ni, Fe[1408]) enhance the selenium absorbance in extracts in the ET-AAS assay; therefore, masking (during extraction) and adding of a matrix/analyte modifier (after injection and drying of the extract in a graphite furnace) are strongly recommended. Nickel[1411,1412,1415,1427,1431] and copper[1369,1408,1417] salts have proved to be the best thermal stabilizers of Se in organic extracts. The time stability of the 2,3-DAN-toluene extracts of Se(IV) has been at least 24 hr in ET-AAS assays vs. 1 hr in fluorimetric measurements.[1369]

The HG technique is prone to depressive interferences from oxidants, acids, and concomitant ions. One should also expect excessive foaming unless all surface-active organics are fully oxidized. An antifoam agent may have to be added. Acid-matched calibration is mandatory, and standard addition checks are highly recommended.

Direct analyses of undigested[1406,1413] or digested[1409,1410] biological samples by ET-AAS, though very attractive, cannot as yet be considered as reliable. Ihnat[1409] found severe interferences; Alexander et al.[1406] and Saeed et al.[1414] studied the thermal stabilization of an endogenously incorporated selenium tracer in blood,[1406] serum,[1414] and urine,[1406] as compared with aqueous standard solutions.[1406] Selenium was stabilized in aqueous standards by Sb, Cd, Mn, Ni, Mo, KI, KIO_3, Ag, Th, Tl, Zn, and Zr; in urine by Mo (up to 900°C), Ni (1100°C), and Ag (1300°C); and in serum by Ag (1250°C) and Ni (1050°C); but in blood only Mo was partly useful.[1406,1414] Tada et al.[1403] studied 35 elements as thermal stabilizers of blood selenium and found Rh (100 μg of $Rh(NO_3)_3$/mℓ, in 0.01 *N* HNO_3) to be the most

Table 3
EXTRACTION SYSTEMS FOR Se FROM BIOLOGICAL DIGESTS

Extraction system	Matrix[a] and conditions	Ref.
2,3-DAN[b]-cyclohexane or decalin	Feed, milk, etc.; pH 1.8	1425
2,3-DAN-cyclohexane	Blood/plant; pH 1	1380
	Urine; pH 1 (EDTA and $NH_2OH.HCl$)	1379
	Feed, hair, milk, tissues, urine, etc.; EDTA added	1386
	Food (meat, milk, etc); EDTA added	1382, 1383
2,3-DAN-decalin	Feed, food, liver, etc.; pH 1.8	1384, 1417, 1433
2,3-DAN-toluene	Blood; pH 1—2 (EDTA and NH_2OH)	1369
4-Cl-1,2-DAB[c]-toluene	Erythrocytes/plasma; pH 1—2 ($NH_2OH.HCl$ added)	1427
	Blood, milk, liver, serum, urine, etc.; pH 1—2 ($NH_2OH.HCl$ added)	1431
1,2-Diamino-4-nitrobenzene-toluene	Biological SRMs (liver, etc.); 4 o-diaminobenzene derivates compared: 4-CH_3; 4-Cl; 4,5-dichloro; and 4-NO_2, the latter being the best	1415
APDC-MIBK	Egg, fish, meat, rice, water, etc.; pH 4.5 (EDTA and Cu(II) added)	1408
Dithizone-CCl_4	Blood, tissues; cation-exchange separation of interfering ions; 3 *M* HCl ($NH_2OH.HCl$ added)	1411, 1412
3.2 *M* HCl — 50% v/v $HClO_4$-acetophenon-$CHCl_3$	Feed, liver, kidney; $(C_6H_5.CO.CH_2)_2SeCl_2$ complex extracted	1436

[a] Wet-ashed biomaterials.
[b] DAN diaminonaphthalene.
[c] DAB diaminobenzene.

efficient matrix/analyte modifier. Nevertheless, direct analyses of diluted biological liquids/digests may be impaired by a serious, structured background (due to iron[927,1405,1406,1414] and phosphates[1439]). The Zeeman AAS equipment, the L'vov platform technique, and the STPF concept may appear the most promising approaches to the direct ET-AAS of Se in biological samples.

IV. CONCLUSION

Fluorimetry, NAA, and flameless AAS are relevant techniques for the determination of Se in biological samples. Both HG- and ET-AAS require major preparatory work (digestion, reduction to Se(IV), separation from the matrix). Some important parameters of these assays have to be optimized and, more generally, the AAS of Se calls for further perfection.

Many common preparatory procedures (drying, dry ashing, some wet–ashing conditions) are liable to pronounced analyte losses. Reliable ashing techniques are combustion in oxygen (in closed-vessel or cold-finger apparatus), HNO_3-H_2SO_4-$HClO_4$ digestion (under carefully controlled conditions), dry ashing in presence of ashing aids, and bomb decomposition with HNO_3, subsequently completed in an open vessel. It is essential to expel oxidants from the final digest and reduce the analyte to selenite.

Prior to ET-AAS quantitation, Se(IV) is often extracted with o-diamines or dithiocarbamates. Direct ETA is prone to severe background and matrix effects; thermal stabilization (with Ni, Cu, Ag, etc.) and thorough optimization are essential. Zeeman STPF AAS would seem very promising.

HG-AAS may suffer from interferences (many ions, acids, oxidants, foaming agents, etc.); standard addition calibration is encouraged.

Table 4
SELECTED AAS PROCEDURES FOR Se DETERMINATION IN BIOLOGICAL SAMPLES

Sample	Brief procedure outline	Ref.
Blood	HNO_3-$HClO_4$ digestion;[a] finally HCl added [→Se(IV)]; HG-AAS	1368
	HNO_3-$HClO_4$ digestion;[a] reduction to Se(IV) with HCl; EDTA and NH_2OH added; 2,3-DAN-toluene extraction at pH 1 to 2; ET-AAS: extract injected, followed by 20 μℓ of aqueous 0.1% $Cu(NO_3)_2$; standards extracted similarly	1369
	1 + 4 dilution with aqueous 100 μg/mℓ Rh (as nitrate in 0.01 *M* HNO_3); ET-AAS: 270°C for 300 sec, 1150°C for 60 sec, and 3000°C for 5 sec; background correction; standard addition calibration	1390
	1 + 1 dilution with 0.1% v/v Triton® X-100; 20-μℓ injection, followed by 10 μℓ of standard solution and 20 μℓ of 1 mg Ni/mℓ solution; pyro-coated tubes: 100°C (30 + 15 sec), 110°C (20 + 10 sec), 120°C (10 + 5 sec), 300°C (20 + 10 sec), 500°C (10 + 20 sec), 1200°C (20 + 10 sec), and 2400°C (1 + 7 sec, gas stop); background correction; standard addition calibration	1413
Blood, serum, tissues	Wet digestion of a 0.1-g sample with $HClO_4$ and H_2O_2 (0.5 + 1 mℓ)[a] at 70°C for 6—7 hr; Ni(II) added (0.5%); ET-AAS: 125, 550, and 2500—2700°C; background correction	1416
Blood, erythrocytes, milk, serum, tissues, etc.	HNO_3-$HClO_4$-H_2SO_4 digestion;[a] and finally completed with H_2O_2; aqueous $NH_2OH.HCl$ added; Se(IV) extracted with 4-Cl-1,2-DAB (or with 4-NO_2-1,2-DAB,[25]) in toluene at pH 1—2; extracts injected into a Varian Techtron® CRA-63 furnace, follwed by aqueous 0.25% $Ni(NO_3)_2$: 240, 950, and 2400°C; background correction	1415, 1427, 1431
Blood, fish, liver, milk, nails, etc.	Ashing in O_2 under dynamic conditions in apparatus with a cold finger (Trace-O-Mat®), and ashes dissolved in (1 + 1) HCl under reflux; HG-AAS from 3 *M* HCl	1399
Blood, tissues	Oxygen flask combustion, and ashes taken up in 0.01 *M* HCl; solution passed through a cation-exchange column (Amberlite® IR-120); eluate treated with $NH_2OH.HCl$ and HCl [→ Se(IV)]; dithizone-CCl_4 extraction, extract injected into a graphite furnace, dried, and 0.1% Ni^{2+} (as nitrate) added; 100, 600, and 2500°C; background correction	1411, 1412
Fish, liver	HNO_3 digestion overnight, and completed with H_2SO_4 and $HClO_4$;[a] semiautomated HG-AAS	239
Food (fish, tissue, etc.)	HNO_3-$HClO_4$-H_2SO_4 digestion;[a] semiautomated HG-AAS from 5% H_2SO_4-30% HCl	141
Food (fish, meat, milk, etc.)	Bomb decomposition with HNO_3;[a] an aliquot dry ashed in presence of $Mg(NO_3)_2$ at 450°C, dissolved in 8 *M* HCl, and heated for 10 min on a steam bath [→Se(IV)]; HG-AAS	250
Food (egg, fish, meat, etc.)	HNO_3-H_2SO_4 digestion;[a] digest heated for 20 min with 4 *M* HCl [→Se(IV)]; APDC-MIBK extraction at pH 4.5 in presence of acetate, Cu^{2+}, and EDTA; ET-AAS: 200, 1050, and 2600°C	1408
Food, tissues	HNO_3 digestion, completed with $HClO_4$;[a] digest heated for 15 min with 1 *M* HCl; 2,3-DAN-decalin extraction; ET-AAS: 20 μℓ of extract, followed by 20 μℓ of aqueous 0.06% Cu(II), ramp dried to 125°C, ramp ashed to 800°, and atomized at 2700°C (gas stop)	1417; 1433
Liver, urine	HNO_3-$HClO_4$ digestion,[a] completed with H_2SO_4; digest treated with HCl [→ Se(IV)]; HG-AAS from ca. 2 *M* HCl	1429
Milk	HNO_3-$HClO_4$ digestion,[a] completed with HNO_3 + H_2SO_4, and finally HCl added; HG-AAS	1425
Serum	1 + 1 dilution with aqueous 0.5% Ni^{2+} (as nitrate); ET-AAS: 150 (ramp), 1050 (ramp), and 2200°C (''maximum power'', miniflow); background correction; standard addition calibration	1414

Table 4 (continued)
SELECTED AAS PROCEDURES FOR Se DETERMINATION IN BIOLOGICAL SAMPLES

Sample	Brief procedure outline	Ref.
Tissues	Dry ashing in presence of $Mg(NO_3)_2$, and ashes dissolved in 6 *M* HCl (for 48 hr!); HG-AAS	1389
Urine	HNO_3-$HClO_4$-H_2SO_4 digestion;[a] HG-AAS	159

[a] Temperature-controlled heating. Safety precautions mandatory.

Chapter 27

SILICON

I. DETERMINATION OF SILICON IN BIOLOGICAL SAMPLES

Silicon is present in all human biological materials, generally above the 1 μg/g level, being somewhat lower (a few tenths of micrograms per gram) only in CSF, milk, and plasma. Some specimens of tissues, food, feces, teeth, hair, and nails contain as much as several tens of micrograms Si/g (Table 1). Nevertheless, the accurate and precise determination of these relatively high concentrations is difficult, due to the lack of sensitive analytical methods for Si, as well as to certain methodological problems such as nonrepresentative sampling, inadequate handling of specimens, severe contamination, and difficulties in keeping silicon in solution. Nondestructive methods or procedures with as few as possible analytical steps are to be preferred. Unfortunately, the choice is very limited.

The sensitivity of XRF[3,128] and arc-source AES[14,26,38,40,904] is insufficient, the DLs in soft tissues being as high as 11 and 0.5 μg/g, respectively.[3,123] SSMS[3,16,25,52,54,55] and MS[53] are very sensitive (with DLs down to 0.2 ng/g[3]), but are expensive and are not easily accessible in routine work.

NAA is rarely used to determine Si in biological specimens.[45,1442] Electron-probe microanalysis has been applied to study the distribution of Si in subcellular regions.[1444]

Among the methods requiring liquid or digested samples, the most sensitive and promising analytical techniques are the ICP-ES[32,33] and the ET-AAS;[1440] their DLs are a few nanograms per milliliter[29] and a few tenths of nanograms per milliliter,[59,60] respectively. The classical spectrophotometric procedures for trace Si are tedious, time consuming, and prone to contamination and high blank corrections.[414,1441]

II. ATOMIC ABSORPTION SPECTROMETRY OF SILICON

The flame AAS of Si is fairly selective but not sensitive. The characteristic concentration and DLs are 0.8 to 2.2 μg/mℓ and 0.02 to 0.2 μg/mℓ, respectively.[56-59,294] Certain instrumental parameters should be carefully optimized: the bandpass should be only 0.2 nm in order to resolve the resonance line at 251.6 nm from its several neighboring lines (in spite of this the calibration curves are often bent); the acetylene-to-nitrous oxide ratio should be adjusted to provide a red zone of ~1 cm height, and the light beam should pass precisely through this red zone. An ionization buffer, 0.1% K^+ or Na^+ (as chlorides), should be added to both standard and sample solutions. Suitable spectrochemical buffers proved to be HBF_4 or HF or NH_4F or $LiBO_2$ in 3% v/v HNO_3 or $LiBO_2$ + 1% w/v La^{3+} in 0.3 *N* HCl. Background correction and standard addition check are advisable.

In ETA, the characteristic concentration is 0.1 to 7 ng/mℓ (10 to 146 pg), and the DLs are down to 0.1 to 0.2 ng/mℓ (20 to 50 pg).[58-60,294] At ashing temperatures between 1300 and 1900°C, the analyte may be partially lost (e.g., as volatile SiO); therefore, nitrogen purge gas probably acts as a thermal stabilizer of silicon (→ Si_3N_4).[1445] A serious drawback of ETA is the formation of stable silicon carbides. The extent of both carbidization and premature volatization (as SiO, etc.) can be lowered by:

1. Using pyro-coated graphite tubes for better sensitivity and lower atomization temperatures.[60]
2. Applying higher heating rates during the atomization stage, e.g., >1200°C/sec,[1445] "maximum power",[60] and temperature-controlled heating;[60,1445] as a result, the sen-

Table 1
SILICON CONCENTRATIONS IN HUMAN BODY FLUIDS AND TISSUES

Sample	Unit	Selected reference values	Ref.
Blood	mg/ℓ	1	52
		1.49 (0.88—3.04)	1440
CSF	mg/ℓ	0.21	52
Feces	mg/day	10[a]	8
Hair	μg/g	13	54
		90[b]	1441
Kidney	μg/g	23 ± 17[b]	904
		40.0 ± 11	3
Liver	μg/g	24 ± 8.5[b]	904
		33.6 ± 13.8	3
Lung	μg/g	57.4 ± 10.7	3
		390 ± 280[b]	904
Milk	mg/ℓ	0.342 ± 0.050	13
		0.34[a]	8
		0.28—0.65[c] (0.015 — 1.72)	14
Muscle	μg/g	41 ± 0.9	3
Nails	μg/g	56[b]	1441
Nails (finger)	μg/g	(74—200)[b]	16
Serum	mg/ℓ	0.77 (0.60—1.24)	1440
Sputum	mg/ℓ	(1.8—4.2)	1442
Tooth (enamel)	μg/g	106.4 (26—1155)[b]	26
Urine	mg/ℓ	4.7	1443
		10[a]	8
		12.8;[d] 5.92—20.0[d,e]	1440

Note: Wet weight unless otherwise indicated. Mean ± SD (range in parentheses).

[a] "Reference Man".
[b] Dry weight.
[c] Range of mean values.
[d] N = 1.
[e] Variations over 6 days.

sitivity has been improved by a factor of 2.6,[60] while keeping lower atomization temperature (2600 to 2700°C).

3. Adding other carbide-forming elements: 0.1% Ca^{2+} (as nitrate, three-fold gain in sensitivity[72]), 1 μg/mℓ Ca^{2+}, + 25 μg/mℓ La^{3+},[1446] impregnating graphite tubes with Ta,[1447] W,[1447] or else with Nb.[1448]

Frech and Cedergren[1445] stressed the importance of the ashing stage: the concomitants (O, S, and Cl) must be removed prior to atomization by 1-min ashing at 1327°C (1600 K). Lo and Christian[1440] also ramp ashed the diluted biological liquids for 70 sec to 1250°C, eliminating most of the interfering species; the maximum tolerable temperature was 1400°C.[1440,1446]

III. AAS METHODS FOR ANALYSIS OF BIOLOGICAL SAMPLES

So far, the work of Lo and Christian[1440] appears to be the only published ET-AAS procedure for determining Si in biological liquids.

Aliquots (10 μℓ) of (1:1) diluted blood or serum or milk, or (1:7) diluted urine were directly injected into a graphite-tube furnace, ramp dried at 100°C for 30 sec, ramp ashed to 1250°C, and atomized at 2250°C for 15 sec. It was essential to apply a slow dry/ash ramp so as to avoid foaming of blood samples. Standard addition calibration and background correction were used.

Alder et al.[78-80] directly introduced 1-cm hair segment (of approximately 100 μg) into a graphite furnace, obtaining DL of 1 μg Si/g of hair.

Contamination control is of paramount importance in trace Si assays. Let us recall that silicon is the most abundant element after oxygen, and that its concentration ratio in the crust of the earth and in biological samples is as high as 10^5 to 10^6! Accordingly, serious contamination by airborne dust, containers, and reagents is to be expected. Most of the common grinding procedures are quite inapplicable.[624] As few as possible sample treatments in precleaned plastic ware (not in glass!) and deionized water must be used. The analyte is nonhomogeneously distributed in some samples (e.g., in diet, feces, urine[33]). Clean homogenation procedures should be adopted.[33,532,944]

Concerning the ashing of biological tissues, dry ashing up to 600°C (in Pt dishes, slow temperature increase), followed by fluxing the ashes with Na_2CO_3[33] or $LiBO_2$, and wet ashing in PTFE-lined autoclaves, appear to be the most reasonable techniques.

IV. CONCLUSION

External contamination is severe in trace silicon assays; hence, *direct* ET-AAS procedures with versatile equipment/programming, pretreated graphite surfaces (pyrolytic/carbide coatings), simultaneous background correction, and matrix-matched calibration are advisable. Strict contamination/blank control and use of plastic ware are a sine qua non.

Chapter 28

SILVER

I. DETERMINATION OF SILVER IN BIOLOGICAL SAMPLES

The concentrations of Ag in biological samples are very low (nanograms per gram, see Table 1). Accordingly, very sensitive analytical techniques — ET-AAS, NAA, AES, SSMS — and some catalytic procedures are applicable.

The most powerful technique, NAA, has been widely used to determine Ag in various biological specimens: blood,[45] bone,[97] CSF,[621] diet,[45] hair,[11,45,48,98,118,184] nails,[560] platelets,[119] serum or plasma,[45,111,120,167] serum protein fractions,[120] teeth,[23,51] tissues,[45] urine,[45,111] etc.[45,121] Some of these procedures involve INAA.[11,23,45,51,98,118-121,167,560,621]

AES analyses, with an arc[26,36,44,288,380,413] or plasma[32,1450] excitation source, have been described for samples of blood,[32,1450] hair,[288] milk,[1450] saliva,[380] teeth,[26] tissues,[36] etc. The DLs of ICP-ES are down to 2 ng/mℓ.[29,35]

SSMS analyses of hair,[54] nails,[16] plasma,[55] teeth,[25] tissues,[132] etc. have been described.

Catalytic procedures for trace Ag (e.g., down to 10 ng/mℓ in saliva[1451]) are documented, as well.[577,1451]

II. ATOMIC ABSORPTION SPECTROMETRY OF SILVER

The flame AAS of Ag is straightforward. The characteristic concentration and DLs are 0.02 to 0.06 and 0.002 μg/mℓ, respectively.[56-59] A lean air-acetylene flame is used, and instrumental parameters (lamp current, bandpass, burner height) are not critical; a somewhat higher lamp current and broader slitwidth are advisable for better signal-to-noise ratio. Flame AAS determinations are selective; only bulk-matrix effects, transport interferences, and minor background absorption should be expected. Acid-matched calibration is typical.

The Delves cup[294,415] and the Sampling Boat®[294] accessories give DLs of 1 ng/mℓ (0.1 ng) and 0.2 ng/mℓ (0.2 ng), respectively, but need a more complicated calibration as well as a simultaneous background correction.

The Fuwa-Vallee long-path absorption tube for flame gases is sensitive down to 2 ng Ag/mℓ.[57]

In ETA, the characteristic concentration is very low, down to 0.0002 to 0.05 ng/mℓ (between a few and 0.1 pg absolute).[57,59,61,819] The lowest DL is claimed as 0.0006 ng/mℓ.[59]

Ashing temperatures are kept as low as 300 to 500°C in order to avoid losses of the analyte. While these temperatures are only 300°C for organic extracts of Ag,[7,427] they may be up to 850°C in a nonvolatile bone matrix.[1099] Rowston and Ottaway[819] found the appearance temperature of Ag in 0.01 *M* HCl to be about 700°C, i.e., the sublimation temperature of metallic Ag. A dilute (0.01 *M*) HNO_3[57] seems to be a suitable acidic medium. Higher concentrations of HNO_3 and H_2SO_4 depress Ag absorption. L'vov platform[229] and gas-stop[819] atomization should prove useful in reducing premature evaporation and matrix effects.

Background correction is always necessary. The intensity of the D_2 lamp is low in this spectral region (328.1 nm), and one may have to decrease the HCL current below its optimum so as to balance the intensities of the two light sources. A tungsten halide lamp is therefore preferable to a D_2 lamp. Several other spectral lines can be used to check for background contribution: Ne 332.4,[302,781] Mo 337.9,[629] Pt 337.3,[629] or Sn 326.2 nm.[302]

Table 1
SILVER CONCENTRATIONS IN HUMAN BODY FLUIDS AND TISSUES

Sample	Unit	Selected reference values	Ref.
Blood	μg/ℓ	<2.7	1449
		<10	96
Bone	μg/g	1.1 ± 0.2[a,b]	3
Brain	ng/g	4 ± 2	3
CSF	ng/g	(<1—9)	621
Feces	μg/day	30 (1.1—202)	96
		60[c]	8
Hair	μg/g	x̃, 0.07 (0.083—2.6)	11
		♀, 0.15; ♂, 0.18; (0.007—4.30)	288
		x̃, 0.2 (<0.1—9.9)	98
		0.60 ± 1.32 (<0.2—9.2)	118
		0.68 ± 0.11 (0.04—8.33)	101
		1.2 (0.4—2.0)	438
Kidney	ng/g	2 ± 0.2	3
		(<5—45)	106
		<30[d]	904
Kidney cortex	ng/g	1 ± 0.2	3
Liver	ng/g	6 ± 2	3
		(<5—32)	106
		150 ± 39[d]	904
Lung	ng/g	2 ± 0.1	3
		(<5—60)	106
		<30[d]	904
Milk	mg/ℓ	0.01	1450
Muscle	ng/g	2 ± 0.5	3
Nails (finger)	μg/g	(0.34—1.1); [d,e] (0.76—3.4)[d,f]	3
Nails (toe)	μg/g	♂, 0.34 ± 0.20 (0.1 — 0.5)[d]	560
		♀ 0.74 ± 0.74 (0.1—1.5)[d]	560
Plasma or serum	μg/ℓ	0.4	120, 167
		0.90 ± 0.40	111
Saliva	μg/ℓ	(5—100)	380
Skin	μg/g	0.72 ± 0.68[d,g] and 0.052 ± 0.055[d,h]	565
Tooth (dentine)	μg/g	2.18 ± 0.84	23
Tooth (enamel)	μg/g	x̃, <0.1 (<0.1—9)	26
		0.03—1.30	25
		0.11[i] and 0.23[j]	51
Tooth (whole)	μg/g	(0.009—0.086)	1099
Urine	μg/ℓ	(0.42—3.8)	673
	μg/day	0.55 ± 0.28 (0.32—1.1)	111
		<3	96

Note: Wet weight unless otherwise indicated. Mean ± SD (range in parentheses). x̃, median.

[a] Ash.
[b] Rib.
[c] "Reference Man".
[d] Dry weight.
[e] Adults.
[f] Children.

Table 1 (continued)
SILVER CONCENTRATIONS IN HUMAN BODY FLUIDS AND TISSUES

[g] Epidermis.
[h] Dermis.
[i] African.
[j] European.

III. AAS METHODS FOR ANALYSIS OF BIOLOGICAL SAMPLES

Both flame[319,474,590,1452-1454] and ET-AAS[7,80,648,778,781,1099] have been applied to various biomaterials: blood,[7,1452,1454] bone,[1099] feces,[1454] food,[590] hair,[80,649,778] milk,[590] teeth,[1099] tissues,[7,319,474,590,781,1452-1454] and urine.[7,1452,1454] Blood silver has been directly determined by Delves cup AAS.[1449] Applications of flame AAS are confined to samples from exposed/treated animals,[1454] unless a preconcentration step is involved.[474,590,1452,1453]

The following four problems in ultratrace Ag assays should be dealt with: (1) representative sampling; (2) losses during storage and preparation of samples; (3) efficiency of the preconcentration step; and (4) graphite furnace programming.

Silver may be expected to be inhomogeneously distributed in some specimens (feces, food, hair, nails, skin, tissues, urine); incidental contamination of hair, nails, skin, and saliva from jewelry and dental materials should also be expected. Bate[162] found no suitable washing procedure for removing exogenous silver, deposited on hair after its soaking in a solution containing Ag^+ and approximating the composition and pH of perspiration.

Drying and ashing of biological specimens may not be safe for ultratrace Ag. This element is easily reduced to its metallic form and thus is prone to volatization losses (above 350 to 400°C), as well as to retention by silica crucibles, dry-ashed residues, $CaSO_4$ precipitates, plastic and glass surfaces, filter paper, etc. Some contradictory statements can be found in the literature, depending on differences in experimental conditions. Iyengar et al.[1445] observed no losses of metabolized ^{110m}Ag during freeze drying up to 72 hr (under 0.05 Torr) as well as during oven drying at 80 and 105°C in rat blood, feces, fur, and various tissues; a somewhat lower recovery (96 ± 3.35%) was established in rat kidney samples oven dried at 120°C for 1 day. Behne and Matamba[167] managed to lyophilize, oven dry at 90°C for 3 days, and plasma ash (LTA) human serum containing 0.4 ng Ag/mℓ. Gleit and Holland[169] found silver recovery rates of only 72 and 65%, during LTA and oven ashing at 400°C for 24 hr, respectively, the Ag tracer being added to blood as an inorganic salt. Plain dry ashing should therefore be regarded on the whole as unsuitable.[91,169,1454]

The addition of a bulky ashing aid — $Mg(NO_3)_2$ — was found to be appropriate with samples of blood and tissues.[1452]

Wet-oxidation techniques seem far superior: $HNO_3 + HClO_4 + H_2SO_4$,[91] $HNO_3 + HClO_4$,[474,1450,1454] $HNO_3 + H_2SO_4$,[7,590] and $HNO_3 + HCl$[319] digestions have been described in the literature with few adverse comments. Evaporation to dryness should be avoided, and digests cannot be stored.

Chelate extraction with sulfur-containing ligands, usually dithiocarbamates[7,590,1452] or dithizone,[474,1453] is the preferred preconcentration technique for trace Ag. Silver pyrolydinedithiocarbamate was found to be more stable after extraction at pH 1 to 2 than at pH 3 to 6 (20 vs. 1 day, respectively).[427] Higher pH, e.g., 5 to 7, is needed in SDDC-MIBK extractions.[7] Analyses of organic extracts are simple and sensitive, provided that ashing temperatures are restricted to 300°C[7,427] and simultaneous background correction is applied.

A few selected procedures for Ag in biological materials are briefly outlined in Table 2. Note that some of the reported procedures are of limited sensitivity and thus are applicable only to treated test animals[1449,1453,1454] or to argyria patients.[1452]

Table 2
SELECTED AAS PROCEDURES FOR Ag DETERMINATION IN BIOLOGICAL SAMPLES

Sample	Brief procedure outline	Ref.
Blood	50-$\mu\ell$ aliquot injected into a Delves cup, dried at 150°C;[a] standards on blood; DL 2.7 ng/mℓ	1449
Blood, tissue	HNO_3-H_2SO_4 digestion; SDDC-MIBK extraction at pH 5 — 7; ET-AAS: 100, 300, and 2500°C;[a] standards extraction similarly	7
	Dry ashing in presence of $Mg(NO_3)_2$; SDDC-MIBK extraction at pH 2.8; flame AAS[b]	1452
Blood, feces, tissue, urine	HNO_3-$HClO_4$ digestion;[c] ashes dissolved in HNO_3, then tartaric acid and aqueous NH_3 added; flame AAS[b]	1454
Bone, teeth	Solid sampling (5—10 mg of powdered specimens diluted 1 + 1 with graphite), and ashed up to 850°C;[a] peak-area calibration vs. hydroxyapatite standards	1099
Food, milk, meat	HNO_3-H_2SO_4 digestion; DDDC-MIBK extraction at pH 4.5 ± 0.05; flame AAS[b]	590
Urine	SDDC-MIBK extraction pH 5—7, and centrifugation; ET-AAS: 100, 300, and 2500°C[a]	7
	SDDC-MIBK extraction at pH 2.8 (CH_3COOH), and centrifugation; flame AAS;[b] standard addition calibration	1452

[a] Background correction.
[b] Most flame procedures insufficiently sensitive.
[c] Safety precautions mandatory.

IV. CONCLUSION

Silver (nanograms-per gram levels) is best determined by ET-AAS and NAA. L'vov platform atomization, background correction, and gas-stop mode are useful. Liquid-liquid (dithiocarbamate) extraction may be involved at subnanograms-per-milliliter levels. Retention/adsorption losses during preparation/storage are highly probable.

Chapter 29

STRONTIUM

I. DETERMINATION OF STRONTIUM IN BIOLOGICAL SAMPLES

The normal concentrations of Sr range broadly within four orders of magnitude in different biomaterials, being generally below 0.1 μg/mℓ in biological liquids and soft tissues, and well above the 1 μg/g level in hair, feces, bone, and teeth (Table 1). In cases of Sr gluconate therapy, the serum and urine levels are significantly higher (e.g., ~1 μg/mℓ),[1459,1460] and the determination of these Sr levels by flame AAS is simple and reliable.[1460]

AAS, AES, NAA, and SSMS are the most sensitive instrumental methods.

NAA has been applied to analyses of blood, bone, diet, erythrocytes,[287] hair[48,290] nails,[290] saliva,[671] serum or plasma,[287] teeth,[23,50,51,676] tissues,[291] urine, etc. (see also Bowen).[45] Some Sr determinations (hair, nails, teeth, etc.) are effected by INAA.

Few laboratories have been able to perform very sensitive SSMS analyses of blood,[3] bone,[3] hair,[54] nails,[16] plasma,[55] teeth,[25,1457] and tissues;[3,132,409] the DL in soft tissues has been as low as 0.8 ng/g.[3]

The arc-source AES has been widely applied to analyses of biological specimens: blood,[36] milk,[14,40] serum/plasma,[41,42] teeth,[26] tissues,[36,44,904] etc.[36,41]

The DLs of ICP-ES are below 1 ng/mℓ[35] and down to 0.02 ng/mℓ;[29] this technique is very promising, especially in simultaneous multielement assays.[35,289] The pulse-nebulization (dipping technique) N_2O-C_2H_2 flame AES provides DLs of 3 ng Sr/mℓ in undiluted serum.[1060]

The XRF method is not sufficiently sensitive with respect to some specimens,[128] but samples of blood,[691] bone,[851] hair,[293] tissues, and urine[695] have been successfully analyzed. The more sensitive PIXE technique is a powerful analytical tool in distribution studies of Sr (and simultaneously of many other elements) in small sample regions (e.g., in hair,[128] tooth enamel,[852] tissues,[128] etc.). The DL of Sr in tooth enamel is down to 0.19 pg (22 μg/g relative).[852]

II. ATOMIC ABSORPTION SPECTROMETRY OF STRONTIUM

In flame atomization, the characteristic concentration is in the range of 0.04 to 0.12 μg/mℓ, and the DLs are between 0.001 and 0.006 μg/mℓ.[56-59,294] A fuel-rich air-acetylene flame or a stoichiometric N_2O-C_2H_2 flame can be used, with a similar sensitivity and DLs, but with a different susceptibility to matrix effects.

In air-acetylene flame, strong interferences have been observed [due to Al, Si, PO_4^{3-}, SO_4^{2-}, Ca^{2+}, Ca^{2+} + PO_4^{3-}, NaCl, $CaCl_2$, acids (H_3PO_4 > H_2SO_4 > HCl > HNO_3[1461]) etc.] which critically depend on flame stoichiometry, burner height, acids, and releasing agents present.[1460-1464] There is also a slight ionization interference (~10%), unless an ionization buffer is added. The most suitable spectrochemical buffers for Sr in an air-acetylene flame are usually La based,[1458,1459,1461,1462,1465,1466] either La^{3+} alone (0.2 to 1%)[1458,1459,1465,1466] or La^{3+} plus alkali salts,[1460,1462] EDTA, or 8-hydroxyquinoline. Still, the addition of a La buffer should often be accompanied by calibration against matrix-matched standards or even by standard additions.[1462] The acetylene flow and the observation height should be carefully optimized, since the signal, the signal-to-noise ratio, and the extent of interferences all depend strongly on these two parameters.

In N_2O-C_2H_2 flame, an ionization buffer (e.g., 0.1% of K^+ as chloride) is added so as to suppress the severe ionization of Sr. Even in this flame, some interferences, e.g., from

Table 1
STRONTIUM CONCENTRATIONS IN HUMAN BODY FLUIDS AND TISSUES

Sample	Unit	Selected reference values	Ref.
Blood	μg/ℓ	20 ± 2	3
		29	332
		31[a]	13
Bone	μg/g	114 (63 — 281)[b,c]	851
		138.7 ± 9.0[b,d,e]	3
		155.9 ± 14.6[b,d,f]	3
Brain	μg/g	0.08 ± 0.01	3
Dental plaque	μg/g	47.7 ± 8.6 (<0.5 — 1880)	1456
Erythroyctes	μg/ℓ	7.2 ± 0.9	287
Feces	μg/day	1.5[g]	8
Hair	μg/g	4.2 (0.75 — 10.8)	54
		x̃, ♂, 6.7 (4 — 15); x̃, ♀, 13.6 (10 — 20)	290
Kidney	μg/g	0.1 ± 0.02	3
		0.38 ± 0.09[h]	904
Liver	μg/g	0.1 ± 0.03	3
		0.20 ± 0.07[h]	904
Lung	μg/g	0.2 ± 0.02	3
		0.55 ± 0.14[h]	904
Milk	μg/ℓ	20	13
		71 — 182;[i] (17 — 295)	14
Muscle	μg/g	0.05 ± 0.02	3
Nails (finger)	μg/g	(0.43 — 0.86)[h]	16
Plasma or serum	μg/ℓ	30 (10 — 70)	41, 42
		44 ± 5	287
		45	1060, 1116
Saliva	μg/ℓ	11.3 (8.4 — 16);[j] 22 (8 — 63)[k]	671
	ng/g	270 ± 50 (<20 — 3930)	1456
Sweat	mg	0.96[l]	21
Tooth (dentine)	μg/g	70 ± 18[m]	676
		94.3 ± 11.5	23
		180.1	22
Tooth (enamel)	μg/g	65.6 (14 — 450)	26
		67 ± 20 (26 — 132)	1050
		81 ± 11 (26 — 280)	25
		103[n] and 178[o]	51
		111.2 ± 9.9	23
		183.1 ± 15.4 (21 — 1200)	1457
		285.6	22
Urine	mg/ℓ	(<0.01 — 0.03)	1458

Note: Wet weight unless otherwise indicated. Mean ± SD (range in parentheses). x̃, median.

[a] Weighed mean (N = 256).
[b] Ash.
[c] Vertebral.
[d] Rib.
[e] Hard water area (U.K.).
[f] Soft water area (U.K.).
[g] "Reference Man".
[h] Dry weight.
[i] Range of mean values.
[j] Rest saliva.
[k] Paraffin stimulated.
[l] 7.5-hr collection at 38°C.
[m] Surface enamel, children.
[n] European.
[o] African.

PO_4^{3-}, Ca^{2+}, and Ca^{2+} + PO_4^{3-}, still persist,[1463] and standard additions or matrix-matched calibration may be needed.

In ETA, the characteristic concentration and DLs are usually below 1 ng/mℓ (1 to 20 pg); the lowest figures appear to be 0.01 ng/mℓ and 1 pg.[59] In view of the pronounced ability of Sr to form carbides, the use of pyro-coated tubes or *in situ* pyro-coating[295] appears to be advantageous. It is also *essential:* (1) to use Ar as a purge gas instead of N_2;[61,651] (2) to lower the emission noise by applying a higher lamp current, a narrower bandpass, and a shorter slitheight; (3) to check/compensate for any background by using a tungsten halide lamp (or Ni 460.6-[302] or Cr 461.6-nm lines[629]); and (4) to appropriately control the matrix effects.

Signal depression by Ca^{2+}, Ca^{2+} + PO_4^{3-}, and $HClO_4$,[1467] excess Cl^-[1468] and by excess HNO_3 (>1.5 to 2 *M*)[294,1467] has been observed. Suitable media appear to be NH_4EDTA[1468] and dilute HCl or HNO_3 (1 to 1.5 *M*).[1467]

The extent of the matrix effects should greatly depend on the temperature program. The typical ashing temperatures are about 1100°C, but, depending on the particular matrix, even temperatures up to 1500°C can be used. The atomization stage can be carried out between 2300 and 2700°C and should preferably include higher heating rates and lower temperatures, e.g., 2300 or 2400°C rather than 2600 to 2700°C (the latter being typical with older — slower — models of graphite furnaces).

III. AAS METHODS FOR ANALYSIS OF BIOLOGICAL SAMPLES

The flame AAS has been applied to analyses of biological liquids or digests, either directly or after preconcentration. Direct analyses of bone,[728,1462] diet,[731,1466] feces,[1460,1466] fish,[1462] milk,[1462] serum and urine (from treated patients),[1460] tooth dentine,[22] tooth enamel,[22,1463,1465] and urine[1466] have been described.

After preconcentration, strontium has been determined by flame AAS in blood,[332] bone,[299,300,1464] feces,[299,300] food,[1464] serum and urine (from treated patients),[1459] tissues,[299,300,1464] and urine.[298-300,1458,1460]

Several applications of ET-AAS to milk,[651,711] saliva and dental plaque,[1456] serum,[1116] and tooth enamel[1467] have been described.

There are a few major problems in AAS of Sr in biological samples: (1) to avoid losses due to coprecipitation and/or adsorption of the analyte (e.g., as sulfate or phosphate) during ashing and/or storage of samples; (2) to effectively preconcentrate Sr from (certain) samples prior to its flame AAS determination; and (3) to adequately account for matrix interferences encountered in both flame and ET-AAS.

The grinding of some biological specimens may introduce intolerable contamination from the abrasion products; e.g., the contamination rates from agate and corundum have been evaluated to be 0.21 and 0.5 μg Sr/min of grinding, respectively.[624]

In ashing step, the use of H_2SO_4-containing mixtures should be avoided; all other common dry- and wet-ashing techniques appear to be satisfactory.[91] Several variants of dry-ashing procedures, between 400 and 550°C, completed by evaporation of the ashes with HNO_3 or HCl and dissolution in dilute HCl or HNO_3, have been applied with apparent success.[298-300,728,731,1460,1462,1466] Ashing at somewhat higher temperatures — about 500 to 550°C — appears to be better.

Tooth enamel can be digested by single-acid treatment, e.g., by HCl[1467] or $HClO_4$[1463,1465] or HNO_3.[22]

Some early flame AAS procedures involved a major preconcentration step: coprecipitation with $La(OH)_3$ (from urine)[1458] or with CaC_2O_4 (from serum and urine);[1459,1460] ion exchange (from ashed urine[298] and from ashed bone, food, and tissues[1464]); and extraction (e.g., with

Table 2
SELECTED AAS PROCEDURES FOR Sr DETERMINATION IN BIOLOGICAL SAMPLES

Sample	Brief procedure outline	Ref.
Bone, fish, milk	Dry ashing at 450°C for 18 hr, and 1 g of ashes evaporated to dryness with 2 mℓ of HCl and taken up in 10 mℓ of 1 *M* HCl containing 1% w/v of La^{3+}; air-C_2H_2 flame; standards containing HCl, La^{3+}, K^+, and Na^+ (standard addition check advisable)	1462
Diet, feces, urine	Dry ashing at 500°C, ashes dissolved in 3 *M* HCl; and $LaCl_3$ added (10 mg La_2O_3/mℓ); air-C_2H_2 flame; standard addition calibration	1466
Feces	Dry ashing at 500°C; ashes evaporated with HCl, and dissolved in a buffer solution containing 1% w/v $LaCl_3$, 0.1% w/v NaCl, 0.02% w/v KCl, and 0.1% v/v HCl; air-C_2H_2 flame; standards in buffer solution	1460
Milk	1 + 1 dilution with water; ET-AAS: ramp dried, ashed at 1100°C for 2 min, and atomized at 2700°C (standard addition check advisable)	651, 711
Serum, urine	1:10 dilution with a buffer solution containing 1% w/v $LaCl_3$, 0.1% w/v NaCl, 0.02% w/v KCl, and 0.1% v/v HCl; air-C_2H_2 flame; standards in diluent; applicable to Sr-treated patients	1460
Serum	ET-AAS: ramp dried for 2 min, ramp ashed at 1100°C, and atomized at 2450°C (dilution and serum-based standards advisable)	1116
Tooth enamel	1-mg sample dissolved in 0.5 mℓ of 1 *M* HCl, and diluted to 2 mℓ with water. ET-AAS: 100, 900, and 2600°C; standard addition calibration	1467
	0.1 g of enamel dissolved in 1 mℓ of $HClO_4$, 20 mg of NaCl added, and diluted to 10 mℓ; air-C_2H_2 flame; standard addition calibration	1463
Tooth enamel (biopsies)[a]	A piece of vinyl tape with a 3-mm i.d. hole placed on the lingual surface of a tooth, 10 μℓ of 2.5 *M* $HClO_4$ in 20% glycerol dropped, and after etching for 45 sec, the etchant removed onto a 3-mm disc of filter, leached with 2 mℓ of water, and diluted 1 + 1 with 2% $LaCl_3$ solution; air-C_2H_2 flame AAS determination of Sr and Ca; the enamel weight calculated by assuming 37.1% Ca contents (ca. 16 μm of surface enamel etched out)	1467
Urine	To 1—5 mℓ of urine added 2 mℓ of 2% w/v La^{3+} solution and 2 mℓ of a solution containing 10% w/v Na_2CO_3 and 5% w/v NaOH; set aside for 15 min, and centrifuged; the ppt dissolved in 5 mℓ of 1.5 *M* HCl and diluted to 10 mℓ with water; air-C_2H_2 flame; standards prepared similarly	1458

[a] Caution: this technique has been applied only to extracted teeth.

TTA-MIBK,[332] di(2-ethylhexyl)phosphoric acid-toluene, etc.[76]). None of these preconcentration techniques appears to be straightforward.

The preconcentration step can be omitted in analyses of samples containing more than 1 μg/g Sr (in flame atomization); moreover, all biological samples can be analyzed by ET-AAS involving only dilution or digestion.

Analyses of serum and urine samples from Sr-gluconate-treated subjects (micrograms-per-milliliter levels) can be performed by simplified procedures, e.g., by flame AAS of tenfold diluted samples, in presence of a suitable La buffer.[1460]

Finally, the importance of checking every newly adopted procedure by standard additions should be stressed once again.

Selected AAS procedures for the determination of Sr in biological samples are briefly outlined in Table 2.

IV. CONCLUSION

AAS, ICP-ES, and NAA are relevant techniques. Both flame and ET-AAS are prone to interferences. Flame AAS is of limited use (bone, feces, food, hair, teeth); a N_2O-C_2H_2 flame, with ionization suppression and matrix-matched calibration, is advisable.

ET-AAS suffers from carbide formation, emission noise, and complicated calibration. Pyro-coated tubes and high heating rates are beneficial.

Partial losses during storage and digestion are probable. Preconcentration procedures are time consuming and not too reliable.

Chapter 30

TELLURIUM

I. DETERMINATION OF TELLURIUM IN BIOLOGICAL SAMPLES

Many published ''normal'' concentrations of Te in biological samples would seem to be far too high. Recent data suggest that the Te levels in biological liquids should normally be ≤1 ng/mℓ; concentrations in the lower nanograms-per-gram range should be expected for most other biomaterials (Table 1).

There are few sensitive methods for the determination of nanogram amounts of Te: ET- and HG-AAS, some plasma-source AES techniques[95,326] SSMS,[3,25,55,132] and NAA. The reported DLs of some procedures seem to have been well above the normal Te levels in samples obtained from unexposed individuals: in blood (by direct ET-AAS, DL 3 ng/mℓ;[1469] by NAA, DL 2 ng/mℓ[1469]), in urine (by ICP-ES after an ion-exchange preconcentration, DL 0.4 ng/mℓ[326]), in plasma (by SSMS, DL 30 ng/mℓ), and in teeth (by SSMS, DL 30 ng/g[25]). Tellurium contents of food[141] and water[228] have also been below the DLs of the respective procedures. Therefore, normal Te concentrations can be determined only after preconcentration.[95,1470]

Simpler AAS[315,337,1471] or spectrophotometric[414] procedures can be used only in assays of specimens from industrially exposed persons[414] or treated animals.[315,337,1471]

II. ATOMIC ABSORPTION SPECTROMETRY OF TELLURIUM

AAS is a convenient technique for nano- and microgram amounts of Te. In a fuel-lean air-acetylene flame, determinations are selective but not sensitive; the characteristic concentration is between 0.2 and 0.7 μg/mℓ, and the DLs are down to 0.02 to 0.07 μg/mℓ.[56-59,294] EDL is a recommended light source vs. HCL, owing to its higher intensity (5 ×) and lower DLs (2 ×).[138] A narrow bandpass (0.2 nm) is recommended to separate the resonance line at 214.3 nm from the adjacent line at 214.7 nm. The background contribution at this low wavelength is substantial and should be simultaneously corrected. Acid-matched standard solutions are as a rule required.

The Sampling Boat® technique gives DLs of 10 ng/mℓ (10 ng absolute);[913] it is now practically replaced by ET-AAS.

In ETA, the characteristic concentration is as low as 0.07 to 1 ng/mℓ (7 to 20 pg), and DLs are down to 0.03 to 0.1 ng/mℓ (~10 pg).[59-61,294]

Dilute mineral acids, e.g., 0.5 to 2% v/v of HCl, HNO_3, and H_2SO_4, would seem to be tolerable.[228] There are major volatization losses (between 300 and 800°C; depending on the matrix and other experimental factors),[223,228,1469] as well as matrix interferences and background absorption. These adverse effects can be combated by: (1) suitable analyte/matrix modification;[223,1469] (2) carefully optimized ashing and atomization stages (ashing up to 1000 to 1100°C is tolerated in presence of Ag, Pd, or Pt,[1469] while atomization temperatures are between 2000 and 2900°C, depending on heating rates, matrix modifiers, and other factors);[60,228] (3) efficient background correction; and (4) matrix-matched calibration.

The HG-AAS of Te is sensitive and fairly selective. $NaBH_4$ is exclusively used as a reductant,[139-141,155,232,812] the hydride, TeH_2, being introduced into a heated quartz tube[139,140] or in an Ar-H_2-entrained air flame.[141,155,812] DLs are between 0.02 and 2 ng/mℓ (0.5 to 5 ng absolute),[139,140,812] i.e., they can stand comparison with two other methods involving TeH_2 generation: AFS (0.08 ng)[144] and ICP-ES (1 ng/mℓ).[142,145]

Several factors should be considered in HG-AAS of Te:

Table 1
TELLURIUM CONCENTRATIONS IN HUMAN BODY FLUIDS AND TISSUES

Sample	Unit	Selected reference values	Ref.
Blood	μg/ℓ	(0.15 — 0.3)	95
Kidney	μg/g	(<0.05 — 0.33)	106
Liver	μg/g	(<0.005 — 0.12)	106
Lung	μg/g	(<0.005 — 0.025)	106
Tooth (enamel)	μg/g	<0.03	25

Note: Wet weight unless otherwise indicated. Mean ± SD (range in parentheses).

1. The optimum atomization temperature is between 900 and 1000°C.[139,1472]
2. Only Te(IV) is reduced to TeH_2;[139,140,232] any Te(VI) has to be prereduced to Te(IV) by heating with (1 + 1) HCl for a few minutes.[139,1472]
3. The hydride-collection time is important.[139,812]
4. The HCl concentration can be between 0.2 and 6 *N*; other acids may also be used, if their concentration in sample and standard solutions is matched (e.g., 1 to 4 *N* H_2SO_4 or $HClO_4$;[142] 5% v/v H_2SO_4 — 30% v/v HCl;[141] 10% v/v HCl — 4% v/v HNO_3[139]). HNO_3 depresses Te absorbance.
5. Some ions in excess may interfere.[149,155,1472] Smith,[155] in 1 *M* HCl, found a very strong depression (>50%) by a 100-fold excess of Ag, Au, Cd, Co, Cu, Fe, Ge, In, Ni, Pb, Pd, Pt, Re, Rh, Ru, Se, Sn, and one less so (10 to 50%) by the same excess of As, Bi, Ir, Mo, Sb, Si, and W. Thompson et al.,[149] in 5 *M* HCl, found interference by 100 μg/mℓ of Cu, Hg, and Pb, and by 1000 μg/mℓ of Cd, Co, Cr(VI), Mo(VI), Ni, V(V), Zn, CNS^-, and I^-; Fe(III) but not Fe(II) interfered slightly. Coprecipitation with $La(OH)_3$ has been proposed for separating Te and other hydride-forming elements from these potential interferents [except for Fe(III) which is also coprecipitated with $La(OH)_3$].[149] Standard addition calibration/check is encouraged.

III. AAS METHODS FOR ANALYSIS OF BIOLOGICAL SAMPLES

Tellurium has been determined in digested tissues from exposed animals by flame AAS;[315,337] air-acetylene flame, three-slot burner, and background correction have been involved. As low as 30 ng Te has been determined by the Sampling Boat® technique but, due to severe matrix effects, a standard addition calibration was required.[1471]

Direct techniques of HG-AAS[141] and ET-AAS[1469] have proved insufficiently sensitive in efforts to analyze food, fish and tissue digests,[141] and blood,[1469] respectively. The Te levels in some environmental water samples have also been below the DLs of the direct ET-AAS (<1 ng/mℓ)[228] as well as the direct HG-AAS [<0.02 ng/mℓ of Te(IV) and <0.05 ng/mℓ of Te(IV) + Te(VI)].[139]

Thus, most Te assays should include effective and reliable steps of sample decomposition and analyte preconcentration. Unfortunately, plain dry ashing as well as some wet-ashing procedures are prone to volatization losses of the analyte.

Apparently LTA was applied successfully to lyophilized tissues from exposed rats containing gross levels of Te (10 to 190 μg/g, wet weight), and ashes were then dissolved in aqua regia.[315] Van Montfort et al.[95] used faster (20 min!) digestion of blood with microwave-excited oxygen in a closed apparatus with a liquid-N_2 cooled finger (after Kaiser et al.[168]); the condensate was subsequently dissolved under reflux in 0.1 *M* HCl.

Wet-digestion procedures with HNO_3 + $HClO_4$ (for exposed rat organs)[1471] or with HNO_3 + $HClO_4$ + H_2SO_4 (for food, fish, tissues;[141] maize,[1470] etc.) should be satisfactory, provided that slow, temperature-controlled heating, as well as reflux conditions and safety precautions, are assured. Weibust et al.[1469] used bomb decomposition of garlic powder with H_2SO_4 and HCl.

The preconcentration step can be accomplished by liquid-liquid extraction (often as halide or dithiocarbamate),[76,95,1470,1473,1474] HG, coprecipitation [with $La(OH)_3$[149] or with elemental As], and by ion exchange [e.g., on poly(dithiocarbamate) resin from acidified urine[326]]. Organic chelate extracts of Te can be directly injected and analyzed by ET-AAS,[1473] while halide extracts require back extraction in a suitable aqueous phase.[1470]

Sensitivity and selectivity of HG-AAS can be improved by freezing the evolved hydride(s) with subsequent release on heating. This approach may also be useful in speciation studies. Pronounced differences in the behavior of diverse Te species [e.g., Te(IV), Te(VI), alkyl-tellurium compounds, etc.] are to be expected during hydride(s) evolution, stripping-out of solution, transfer to the atomization cell, and atomization pattern.

Finally, a few important notes have to be added to the general characterization of the ETA of Te. The behavior of inorganic and organically bound Te during ashing/atomization is quite different, as recently shown by Weibust et al.[1469] These authors thoroughly studied the thermal stabilizing of the ^{127m}Te isotope in graphite-furnace atomization: 15 elements (Ag, Cd, Cu, Ni, Pd, Pt, Zn, etc.) stabilized the inorganic Te, but only three of these (Ag, Pd, Pt) were effective with the metabolized ^{127m}Te in blood specimens (from treated rats). Without thermal stabilization the maximum ashing temperatures have been only 300 and 400°C for inorganic and organically bound Te, respectively, while with stabilizers added they attained 900 to 1100°C.[1469] The increased volatility of the organically bound Te in presence of Cu, Zn, and KI is an important and unexpected finding.[1469]

A direct ET-AAS procedure for whole blood Te was proposed, based on (1 + 1 + 2) dilution of blood with aqueous 1000 μg/mℓ Pd^{2+} and aqueous 1000 μg/mℓ Ni^{2+}; 50-μℓ aliquots were subsequently injected into a Perkin-Elmer® HGA-500 graphite furnace, and a six-step heating program was run; 120°C (5-sec ramp + 20-sec hold), 150°C (5 + 5 sec), 300°C (5 + 20 sec), 600°C (40 + 10 sec), 1100°C (10 + 20 sec), and 2000°C (0 + 5 sec). The DL, however, 3 ng Te/mℓ, was higher than the normal values of Te in blood from unexposed individuals.[1469]

IV. CONCLUSION

Both nonflame techniques — ET- and HG-AAS — are potentially applicable to nanogram-Te levels in biological digests. Serious methodological problems of these assays — analyte losses, preconcentration efficiency, background and matrix effects, etc. — are awaiting solution.

Chapter 31

THALLIUM

I. DETERMINATION OF THALLIUM IN BIOLOGICAL SAMPLES

The normal levels of Tl are very low: ≤1 ng/mℓ in biological liquids and <10 ng/g in most other biologicals (Table 1).

DPASV and ET-AAS are the most sensitive and convenient analytical techniques for nano- and subnanogram amounts of Tl; the DLs of both techniques are below 0.1 ng/mℓ (down to a few picograms absolute) and they are most suitable for routine work as well.

ASV/DPASV is widely used in analyses of blood,[124,125,1478] feces,[125] nails,[133] serum or plasma,[124,125,321,1478] skin,[133] urine,[124,125,1478] etc. It seems to be directly applicable to samples of hemolyzed blood, plasma, and urine,[124,125,1478,1479] thus eliminating the time-consuming ashing procedures and providing a rapid qualitative screening tool in toxicology and forensic cases.[124,125]

SSMS[3,25,55,325] has DLs as low as 3 ng/mℓ in plasma[55] and 4 ng/g in soft tissues;[3] lower Tl levels in blood, bone, hair, nails, teeth, tissues, and urine have been determined by MS after ashing and extraction.[1475]

In AES procedures,[130,411,1480] an ashing and preconcentration step is required, but even then the DLs are higher than with similar flame AAS procedures.

RNAA is not a very convenient technique for trace Tl.[3,45] Spectrophotometric procedures for Tl are time consuming and not sufficiently sensitive.

II. ATOMIC ABSORPTION SPECTROMETRY OF THALLIUM

The flame AAS of Tl is relatively selective, but not sensitive enough. The characteristic concentration is between 0.1 and 0.64 μg/mℓ, and the DLs range from 0.009 to 0.03 μg/mℓ.[56-59,294] EDL light source provides higher intensity and lower DLs than HCL by factors of 9 and 2.0 to 3.8, respectively.[138]

In an air-acetylene flame, slight ionization and anionic effects can be observed; these can be diminished by employing a stoichiometric flame and a three-slot burner, as well as acid-matched standards and sample solutions containing 0.1% K as chloride. Some background contribution may be expected at the resonance line of 276.8 nm, and therefore a background check is advised. The observation height should not be too low.

Two flame accessories, the Delves cup[415] and Sampling Boat®,[913] with DLs of 10 and 1 ng/mℓ (1 ng absolute), are available. These atomizers are prone to serious matrix effects; matrix-matched or standard addition calibration and background correction are therefore needed.

In ETA, sensitivity is improved by about three orders of magnitude, but due to volatility of some Tl species, the determination is not very selective. The characteristic concentration and DLs are about 0.1 ng/mℓ (ranging between 0.005 and 1.5 ng/mℓ), i.e., typically several picograms absolute (down to 1.5 pg).[59,61,294]

The graphite furnace programming, the background correction efficiency, and the calibration mode are of great importance in ET-AAS of Tl. Some Tl species are lost during the ashing stage at temperatures between 400 and 750°C; premature volatization of the analyte when the atomization stage begins should also be expected (especially in the presence of excess HCl, chlorides, and other halides). It is therefore advisable to carefully optimize the ashing and atomization temperatures, heating rates, and the gas-flow rate. It is difficult to eliminate the inorganic matrix constituents before the atmonization stage, as their volatility

Table 1
THALLIUM CONCENTRATIONS IN HUMAN BODY FLUIDS AND TISSUES

Sample	Unit	Selected reference values	Ref.
Blood	μg/ℓ	0.5	3, 1475
		3[a] (Most below 5)[a]	989
Bone	ng/g	2	1475
Brain	ng/g	<1	3
Feces	μg/day	1[b]	8
Hair	ng/g	10	895
		(4.8—15.8)	1475
Kidney	ng/g	<3	3
		(1.4—4.1)	1475
Liver	ng/g	(0.56—2.85)	1475
		~9	3
Lung	ng/g	(0.36—29.5)	1475
Nails	ng/g	(0.7—4.9)	1475
Plasma or serum	μg/ℓ	<2.6	55
Tooth (dentine)	ng/g	4.7	1475
Tooth (enamel)	ng/g	<40	25
Urine	μg/ℓ	0.4 (0.1—1)	1476
		0.5[b]	8
		(0.07—1.15)[c]	1475
	μg/day	(0.2—0.8)	1477

Note: Wet weight unless otherwise indicated. Mean ± SD (range in parentheses).

[a] Children.
[b] "Reference Man".
[c] ng/g.

is similar to or lower than the volatility of the analyte. In order to reduce the background contribution, the atomization temperatures should be kept low, e.g., between 1300 and 2200°C, but the heating rate should be high.

The versatile purge-gas programming may also facilitate determinations: introducing air or O_2 during the ashing stage;[582] using "gas stop" or "miniflow" of argon during atomization; or applying a diffusion H_2 flame around the Varian Techtron® CRA-63 or CRA-90 graphite furnaces.

The L'vov platform,[229] by analogy with other volatile elements,[85,454-457] might prove very useful in Tl assays; the same should refer to STPF,[553] aerosol deposition,[1481] and atomization under pressure.[1481]

Severe matrix effects from HCl, NaCl, $HClO_4$, urine matrix, etc. have been observed;[1476,1482,1483] suitable media in ET-AAS proved to be 1% H_2SO_4,[553,1482] 1 to 10% HNO_3,[438,920,1482] 1% La^{3+} in 10% HNO_3,[438] and aqueous 7% $LiNO_3$ (to eliminate NaCl interference).[1483] In many cases, calibration by standard additions[438,920,1476,1482,1484] or at least by acid-matched and matrix-matched standards is justified.

III. AAS METHODS FOR ANALYSIS OF BIOLOGICAL SAMPLES

AAS has been widely applied to Tl determinations in biological materials. The published procedures are classified in Table 2. Some extraction methods for preconcentration of trace Tl from biological liquids or digests are depicted in Table 3, and several selected procedures

Table 2
AAS TECHNIQUES FOR Tl DETERMINATION IN BIOLOGICAL MATERIALS

Technique	Biological materials
Flame AAS after dilution	Urine[1480,1490]
Flame AAS after digestion	Tissues[190,319,463]
Flame AAS after liquid-liquid extraction	Blood,[7,332,471,472,1485,1487] hair,[472] serum,[1487] tissues,[190,472,1485] urine,[7,472,1485,1487,1489] urine (speciation)[1489]
Delves cup after dilution or digestion	Plasma and urine[1491]
Delves cup after liquid-liquid extraction	Blood[989,1488]
Sampling Boat® after dilution or digestion	Blood,[1485] urine[1485]
Sampling Boat® after liquid-liquid extraction	Urine[496]
Direct ET-AAS with or without dilution	Blood,[920] hair,[1492] urine,[438,920,1476,1482]
ET-AAS after digestion	Blood,[438] hair,[1484] tissue[920]
ET-AAS after liquid-liquid extraction	Blood,[7] hair,[7] tissues,[7] urine[7,529]

Table 3
EXTRACTION SYSTEMS FOR Tl FROM BIOLOGICAL SAMPLES/DIGESTS

Extraction system	Matrix and conditions	Ref.
APDC-MIBK	Hemolyzed blood	989
	Ashed tissues; pH 7	190
	Hemolyzed or ashed blood; pH 6	1488
	Wet-digested hair; pH 4; R 90.5%	1486
APDC + SDDC in MIBK	Wet-digested hair; pH 5; R 98%	1486
APDC + SDDC in $CHCl_3$	Urine and HNO_3-treated blood or tissues; pH 4; tartrate added	130
Cupferron-MIBK	Wet-digested blood; 2 *N* HCl; triple extraction	332
Dithizone-MIBK	Urine; pH 8	496
Dithizone-CCl_4	Urine (pH 9—10) or wet-ashed hair (pH 9)	1486
HBr-Br_2-ether	Biological digests (hair, nails, tissues, urine, etc.)	1475
	Biological digests (blood, feces, serum, urine)	1487
HBr-Br_2-MIBK	Biological digests (blood, tissues, urine)	1485
Br_2-MIBK	Urine (inorganic Tl); sulfosalicylic acid added	1489
HHDC in toluene or HHDC in MIBK	Urine (pH 7); wet-digested hair (pH 3—5)	1486
SDDC-MIBK	Urine (pH 7—9), R 92%; wet-digested hair (pH 4), R 99%	1486
	Wet-ashed tissues; pH 5—6	1485
	TCA deproteinized blood/urine; pH 6.5—7.0	7, 472, 1485
	Wet-digested hair/tissues; pH 6.5—7.0	7, 472
	Hemolyzed blood	471
	pH-adjusted urine; pH 7	7, 472, 529

are briefly outlined in Table 4. However, some of these procedures are not sensitive enough to be applied to unexposed individuals, although most of them are quite useful in chronic or acute exposure cases as well as in forensic toxicology.[463]

In analyses of solid biological specimens, thallium is put into solution mostly by wet-digestion procedures. All common acid mixtures should be useful, the choice being determined by the next analytical step. In general, the final media of dilute H_2SO_4 or HNO_3 are

Table 4
SELECTED AAS PROCEDURES FOR Tl DETERMINATION IN BIOLOGICAL SAMPLES

Samples	Brief procedure outline	Ref.
Blood	SDDC-MIBK extraction from deproteinized blood at pH 6.5—7.0 for 10 min; flame[a] or ET-AAS: 100, 300, and 2400°C;[b] standards extracted similarly	7, 472, 1485
	LTA, and ashes taken up in 10% v/v HNO_3 containing 1% w/v La^{3+}; ET-AAS: 100, 470, and 2000°C[b]	438
	APDC-MIBK[989,1488] or SDDC-MIBK extraction[471] from hemolyzed (with Triton® X-100) blood, and centrifugation; standards on spiked blood extracted similarly; flame[471,a] or Delves cup[989,1488] or ET-AAS[b]	
Blood, feces, serum, urine	HNO_3-$HClO_4$-H_2SO_4 digestion;[c] saturated Br_2 water added; $TlBr_3$ extracted with ether; evaporated to dryness, and taken up in dilute acid; flame AAS[a]	1487
Hair	0.2-mg sample digested with 0.2 mℓ of HNO_3 at 60°C for 1 hr, 0.1 mℓ of 30% H_2O_2 added,[c] left for 30 min, and diluted to 1 mℓ; ET-AAS (ashing up to 400°C);[a,b] standard addition calibration	1484
Hair, tissues	HNO_3-H_2SO_4 digestion; SDDC-MIBK extraction at pH 6.5—7; ET-AAS: 100, 300, and 2400°C[b]	7
Liver	Tissue boiled with (1 + 1) HNO_3-HCl, and diluted; flame AAS;[a,b] acid-matched standards	319
Tissues	HNO_3-H_2SO_4 digestion, and finally HNO_3-$HClO_4$ added;[c] flame AAS;[a,b] acid-matched standards	190
	Dry ashing at 500°C for 4—5 hr in presence of $Mg(CH_3COO)_2$, and ashes dissolved in 2 *N* HCl; flame AAS directly[a,b] or after extraction with APDC-MIBK at pH 7	190
Urine	1 + 9 dilution with 1% v/v H_2SO_4; ET-AAS: 100, 700, and 2200° C;[a,b] standard addition calibration	1482
	9 + 1 dilution with HNO_3; ET-AAS: 100, 470, and 2000°C;[a,b] standard addition calibration	
	Direct flame AAS;[a,b] standard addition calibration (pulse nebulization and a three-slot burner advisable)	1480
	APDC-MIBK extraction at pH 6.5—7.5 for 10 min, and centrifugation; flame,[472,1485,1489,a] or ET-AAS;[7,529b] standard extracted similarly	
Urine (speciation)	3 mℓ of urine + 1 drop of Br_2 water + 2 drops of 10% w/v sulfosalicylic acid + 3 mℓ of water-saturated MIBK (inorganic Tl extracted); the aqueous layer extracted at pH 7—8 with SDDC-MIBK (organic Tl species extracted)	1489

[a] Only toxic levels determined.
[b] Background correction.
[c] Safety precautions mandatory.

most suitable for following direct injection into graphite tube, whereas $HClO_4$ or HCl are not. In flame AAS any acid or acidic mixture is tolerable. If a liquid-liquid extraction step is involved, then HNO_3 and $HClO_4$ should be avoided in final digests.

Wet-digestion procedures with HNO_3 and H_2O_2,[1484] HNO_3 and H_2SO_4,[1485] HNO_3 and $HClO_4$,[472] and HNO_3, H_2SO_4, and $HClO_4$[190,332,1475,1486,1487] would seem successful. However, they require predigestion with excess HNO_3 and gradual, temperature-controlled heating for the smooth, safe, and complete oxidation of organic matter.[91,266]

Thallium can be released from the organic matrix by treatment with HNO_3 alone,[130] or with HNO_3 + HCl,[319] as well as by TCA precipitation of blood[7,472,1485] or urine[1485] proteins.

Dry-ashing procedures are prone to volatization and retention losses and are not recommended for Tl.[91]

The ashing at 500 to 550°C in presence of $Mg(CH_3COO)_2$ as an ashing aid,[190] and LTA in excited oxygen,[438,1488] proved to be safe and useful for trace Tl in biological samples.

Liquid-liquid extraction is an established, reliable, and rapid technique for Tl preconcentration from biological digests,[7,190,332,472,1475,1485-1488] HNO_3-treated specimens,[130] pH-adjusted urine,[7,472,529,1486,1489] hemolyzed blood,[471,989,1488] or else from TCA supernatants after blood/urine protein precipitation[7,472,1485] (Table 3, see also Cresser[76]).

Thallium is often extracted as dithiocarbamate or as Tl(III) bromide; the latter extracts cannot be directly introduced into a graphite-tube furnace and require a back-extraction step.

Stoeppler et al.[1486] compared the efficiency of several extraction systems including APDC, SSDC, APDC + SDDC, HHDC, and dithizone, from urine and their digests containing a ^{204}Tl tracer; extractions with HHDC in toluene and with dithizone in CCl_4 were most effective.

Organic extracts can be analyzed by flame, Delves cup, Sampling Boat®, or graphite furnace; many early flame AAS procedures are readily adaptable to ET-AAS.

In ET-AAS measurements on diluted or digested biological samples, one should expect serious matrix effects; the ashing parameters are critical; simultaneous and effective background correction is needed;[1439] and standard addition check should always be made. In urine analyses, a combination of matrix modification and calibration by standard additions[438,451,1476,1482] is typical. Possible differences in the ashing and atomization pattern of inorganic and organically bound thallium should not be neglected.

IV. CONCLUSION

DPASV and ET-AAS are reliable methods at nanograms-per-gram levels of Tl. Extractions as dithiocarbamate, Tl(III) bromide, or dithizonate may be used, hexamethylenedithiocarbamate extraction being the most effective. ET-AAS is very sensitive, but prone to volatization losses, background absorption, and matrix effect; hence, versatile programming, high heating rates, L'vov platform (STPF) atomization, thermal stabilization of the analyte, and (often) matrix-matched calibration are required. In a higher concentration region, flame AAS and Delves cup technique may also be used.

Chapter 32

TIN

I. DETERMINATION OF TIN IN BIOLOGICAL SAMPLES

Most early "normal" levels of Sn in biological materials appear to be erroneously high and unreliable, thus revealing blunders in sampling or analysis or both. Tin concentrations should be <0.01 μg/mℓ in body fluids and <1 μg/g in most other biomaterials; only canned food/beverages and feces may contain ⩾1 μg Sn/g (Table 1).

The determination of submicrograms-per-gram concentrations of Sn is by no means an easy analytical task. The accuracy and precision of these assays can be impaired by sampling errors, wet-chemistry steps, preconcentration procedures, reagent blanks, and interfering concomitants; the existence of various environmental tin species also creates problems.

Due to limited sensitivity and/or matrix effects, most, if not all, tin assays require a major precipitation/preconcentration step.[39,129,131,192,326,659,1495-1500] Among the most sensitive instrumental methods for trace tin are ET-AAS (DL <0.3 ng/mℓ in aqueous solutions), SSMS[16,25,54,55,132,275] (DL 4 ng/g in soft tissues[3]), ASV/DPASV[124,125] (DL 2 ng/mℓ in aqueous solutions[123]), AC polarography,[1501] and RNAA[45,108,110,1493,1499,1500,1502] (DL 2 ng/g in blood and dried milk[110,1493]).

Less sensitive are the AES techniques: ICP-ES[35,326] with DL down to 3 ng/mℓ in aqueous solutions,[29] MIP-AES,[131] and the arc-source AES,[26,37,39,41,42,44,288,320,473,904] as well as XRF[128,129,292,293,1502] flame emission (SnH band),[1494] and molecular spectrophotometry.[133,1496,1497,1503] Some of the reported procedures (e.g., References 37, 41, 42, 133, 292, 293, and 326) appear to be insufficiently sensitive.

II. ATOMIC ABSORPTION SPECTROMETRY OF TIN

The flame AAS of Sn is inadequately sensitive due to low volatility and high dissociation energy of tin carbides and oxides. There is a choice between three spectral lines: 224.6, 286.3, and 235.5 nm, the first one being about 1.5- or 2-fold more sensitive, as well as between several flames: hydrogen flames (Ar-H_2-entrained air or fuel-rich premixed H_2-air flame), fuel-rich (red) N_2O-C_2H_2, and luminous air-acetylene flame. Hydrogen flames provide the best sensitivity and lowest DLs (i.e., characteristic concentration and DL of 0.3 to 0.6 and 0.01 to 0.03 μg/mℓ, respectively.[56,57]), but their application is confined to HG-AAS because of numerous severe matrix effects from acids, ions, organic solvents, etc.

The fuel-rich N_2O_2-C_2H_2 flame and 235.5 nm (a non-resonance!) line are recommended for their selectivity; although sensitivity is impaired by a factor of 2 or 3, few interferences are encountered with this flame. Characteristic concentration and DL are around 0.7 to 1.2 and 0.07 to 0.1 μg/mℓ, respectively.[56,58,59]

The worst option is using a fuel-rich, luminous air-acetylene flame, with a three-slot burner; while its sensitivity is approximately the same as with the N_2O-C_2H_2 flame,[294] it is marred by marked interferences, noise levels, and sensitivity drift.

The EDL source provides higher intensity (7 ×) and lower DLs (2 to 10 ×) than the HCL.[138,1504]

At 224.6-nm line, calibration curves are typically bent. Background correction and matrix-matched standards are advisable.

The ET-AAS of Sn is relatively sensitive but not too selective. The characteristic concentration ranges between 0.04 and a few nanograms per milliliter (4 to 100 pg), and DLs are 0.015 to 0.3 ng/mℓ (down to a few picograms).[59-61,294,551,552,1505] Several important factors should be considered in ET-AAS of Sn:

Table 1
TIN CONCENTRATIONS IN HUMAN BODY FLUIDS AND TISSUES

Sample	Unit	Selected reference values	Ref.
Blood	μg/ℓ	<2	1493
Bone	μg/g	0.1 (0.08—1.4)[a]	97
		0.8	37
Brain	μg/g	0.06 ± 0.01	3
Hair	μg/g	0.76 (0.39 — 1.5)	54
		♂, 0.54;[b] ♀, 1.17;[b] (0.048 — 12.0)	54
			288
Kidney	μg/g	0.2 ± 0.04	3
		0.72 ± 0.25[c]	904
		1.95 ± 1.25 (<0.3 — 6.87)	105, 106
Liver	μg/g	(0.08 — 0.32)	108
		0.13 ± 0.55 (<0.3 — 3.18)	105, 106
		0.4 ± 0.08	3
		0.42 ± 0.16	275
Lung	μg/g	0.8 ± 0.2	3
		1.14 ± 0.42 (<0.3 — 2.15)	105, 106
		1.5 ± 0.80[c]	904
Milk	ng/g	≤2 — 3[c]	1493
Muscle	μg/g	0.07 ± 0.01	3
Nails (finger)	μg/g	(2.3 — 43);[c,d] (2.5 — 14)[c,e]	16
Plasma	μg/ℓ	<4[f]	55
Sweat	mg/day	2.2 ?	21
Tooth (enamel)	μg/g	0.21 ± 0.04 (0.03 — 0.92)	25
		x̃, <2; 1.1 (<2 — 93)	26
Urine	μg/ℓ	1.0 (0.56 — 1.6)	1494

Note: Wet-weight unless otherwise indicated. Mean ± SD (range in parentheses). x̃, median.

[a] Femur.
[b] Geometric mean.
[c] Dry weight.
[d] Adults.
[e] Children.
[f] N = 1.

1. Pyrolytically-coated tubes[60] and *in situ* pyro-coating[72] are advantageous, ensuring both lower atomization temperatures (e.g., 2200 vs. 2400°C in "maximum power" heating mode[60]) and higher sensitivity (e.g., 2.5 ×[72]).
2. High heating rates in the atomization stage are essential,[60,1506] so as to: (a) decrease the optimum atomization temperatures to 2200 to 2400°C;[60] (b) extend the useful lifetime of the tubes; (c) reduce fractional vaporization of different tin species; and (d) diminish background and matrix effects.
3. Interruption of the purge gas flow during the atomization stage is favorable[1503,1504,1506] in view of the marked accuracy[1504,1506] and sensitivity improvement (up to 10 ×[1504]).
4. Various tin species exhibit very pronounced differences in behavior during ashing and atomization, resulting in matrix-dependent losses on ashing as well as a quite different atomization pattern. This fact cannot be neglected in analyses of undigested biological or environmental samples which may contain organotins,[1494] complexed tin, etc., as well as interfering concomitants. Vickrey et al.[932,1505] found the sensitivity of inorganic Sn(IV) (as chloride) to differ drastically (by a factor of 1.8 to 36.7!) from that of organotin compounds. Even if inorganic tin is determined, the presence of acids,[1504,1507,1508] some ions,[72,1504,1507,1508,1509] carbide-forming elements,[72,1504] and other

matrix constituents will alter both ashing and atomization behavior of the analyte. Ashing temperatures between 750 and 1500° (for inorganic tin)[1504,1506] and generally below 800°C (in organic matter)[1502,1503,1506] seem tolerable. It is desirable to convert tin to an oxide rather than a sulfide[1506,1509] in order to thermally stabilize the analyte (tin dioxide and sulfide sublime at 1400 and 800°C, respectively[1504]). H_2SO_4 and $HClO_4$ media are inappropriate, due to severe suppression of tin absorbance,[1504,1508] while additions of HNO_3 (e.g., 5% v/v, ashing temperatures up to 1000°C),[1504,1507] NH_4NO_3 (10% w/v),[1506] aqueous NH_3,[57,294] 10% ascorbic acid,[1508] 0.2% w/v of La^{3+},[1507] and 0.1% w/v of Ca^{2+} (as nitrate)[1509] have proved useful as matrix/analyte modifiers. Matrix effects also depend on tube age and pretreatment ("history"), and are not necessarily eliminated by peak-area calibration.[1504]

5. Impregnation of graphite tubes with carbides of Zr,[932,1502,1504,1505,1510] Ta,[1504,1506,1511] W,[1502,1504,1505] Mo,[1504,1505] or V[1505] is a simple, inexpensive, effective, and promising approach likely to improve graphite-tube performance. It offers a number of important advantages: leveling-off (not completely but to a great extent[1505]) the response of different organotin compounds,[932,1505] pronounced sensitivity enhancement (by a factor of 2 to 4 for inorganic tin[932,1504-1506] and by 1 or 2 orders of magnitude for organotins[932,1505]), marked extension of the tube lifetime (e.g., 5 to 8 ×[1504,1511]), long-term response stability,[1505,1511] etc. The first of the listed advantages is of obvious importance in biological/environmental assays and, in particular, in applications of ETA as a chromatographic detector for organotins.[283,932,1505]
6. The use of a hydrogen diffusion flame around the Varian Techtron® CRA-69 or CRA-90 furnaces has improved sensitivity by a factor of 2.5.[552,1509]
7. (Carbide-coated) L'vov platform atomization should be useful.

The HG-AAS is another sensitive technique for nanogram amounts of Sn. Tin hydride (stannane, SnH_4) is evolved from acidic solutions treated with $NaBH_4$, and is introduced into an Ar-H_2 diffusion flame,[812,1512] or into a flame-[150,933,1513] or furnace-heated silica tube,[1514,1515] or into a N_2-H_2 Beckman burner combined with a long silica absorption cell.[1516] The DLs are between 0.1 and 1 ng/mℓ (0.5 to 5 ng),[140,150,812,1514,1515] depending on many instrumental and procedural parameters: volumes of the apparatus and solution, flow rate of the purge gas, hydride collection time/technique, pyrolysis temperature, acidity, buffer components, etc. (see also recent reviews[232,242]). Other instrumental techniques for analyses of hydrides seem to give the same or inferior DLs, e.g., ICP-ES (0.1 ng/mℓ[145]), MIP-AES (2 ng/mℓ or 40 ng[131]), and GC (10 ng/mℓ or 50 ng[146]).

Rigin[1517] employed an electrolytic reduction of Sn in 5% KOH medium on lead cathode, and swept the evolved stannane by an Ar flow into a heated (700°C) quartz tube. The DL was as low as 0.02 ng/mℓ in 5 mℓ, and more than 20 elements did not interfere at a 1:10,000 ratio, thus providing a selectivity superior to the $NaBH_4$ reduction technique.[1517]

HG is strongly affected by acidity;[140,155,812,1514,1515] a narrow range of about 0.05 to 0.2 *M* HCl (i.e., approximately 1% v/v of HCl) appears to be the optimum. The nature of acid seems to be of less importance, and some other dilute acids (H_2SO_4, $HClO_4$, CH_3COOH, tartaric acid, etc.) have also been used.[1512,1516,1516a]

There are three interference studies[1514-1516] which are in good agreement: (1) Vijan and Chan[1514] found in 1% v/v HCl a depression by a 20-fold excess per Ni, Sb, As, and Cu, which was eliminated by adding $Na_2C_2O_4$ (600 μg/mℓ) or, more effectively, by coprecipitation with Mn(IV) hydroxyde; (2) Subramanian and Sastri[1515] found in 0.1 *N* HCl a depression by As and Sb (>20 ×), Co, Cu, and Ni (>50 ×), and by Fe (>2500 ×), and recommended the same coprecipitation; and (3) Nakashima[1516] found in 0.6 *N* HCl and at somewhat higher concentration levels a depression by an excess of Ni (2 ×), Se (40 ×), As (50 ×), Cu (60 ×), Co and Sb (100 ×), and Fe (1000 ×).

Table 2
SELECTED AAS PROCEDURES FOR Sn DETERMINATION IN BIOLOGICAL SAMPLES

Sample	Brief procedure outline	Ref.
Canned food/fish, tissues, plankton, etc.	20- to 200-mg sample solubilized with 1 mℓ of TAAH solution (Lumatom®) at 50°C for ⩾3 hr; ET-AAS: 100, 800, and 2860°C; background correction; standard addition calibration	1502
Canned food	20-g samples digested with HNO_3 and H_2SO_4; NH_4Cl and CH_3OH added; N_2O-C_2H_2 flame; standard addition calibration[a]	1519
Canned food (fish, meat, milk, etc.)	HNO_3 digestion (5- to 40-g samples), finally HCl added, and boiled; 100 μg/mℓ K^+ added; N_2O-C_2H_2 flame AAS; standard solutions containing 10% v/v HCl and 100 μg K^+/mℓ[a]	1518
Food, fish, meat, milk, etc.	H_2SO_4-HNO_3-H_2O_2 digestion[b] or dry ashing in presence of $Mg(NO_3)_2$ (for marine samples); HG-AAS from 1% v/v H_2SO_4[c]	1513
Food	HNO_3 digestion under reflux, completed with H_2SO_4, and finally made up in 10% w/v NH_4NO_3; ET-AAS: 100, 750, and 2600°C (gas stop); tantalized graphite tubes; standard addition calibration;[a] extraction option with neocupferron-$CHCl_3$	1506
Food, meat, tissues, viscera, etc.	TAAH solubilization: 1-g sample + 5 mℓ of Soluene® 350 at 65°C for 1.5—3 hr, and diluted with toluene; ET-AAS: 100, 700, and 2700°C (gas stop); background correction; standard addition calibration	1505
Food, urine	Wet ashing with HNO_3, H_2SO_4, and (finally)[b] H_2O_2; SnI_4 extracted into toluene; ET-AAS	530

[a] Background correction/check advisable.
[b] Safety precautions mandatory.
[c] Standard addition check advisable.

Alkyltins and other organotins may be expected to exhibit an evolution/transport/atomization pattern quite different from that of inorganic tin;[1494] hence, systematic errors may creep into analyses of undigested/untreated samples.[1494,1512,1520] These behavior differences can, of course, be utilized in speciation studies of tin, e.g., by means of fractional volatization of the cold-trapped hydrides.[1494,1512]

III. AAS METHODS FOR ANALYSIS OF BIOLOGICAL SAMPLES

The flame AAS can be used only in analyses of canned food and beverages[1495,1518,1519] and feces. The HG-AAS has been applied to canned fruit juices (after dilution or HCl treatment),[1514,1516,1517] fish,[1516a] wet-ashed food,[1513] and to environmental waters (speciation of organotins[1512]). ET-AAS has been applied to digested blood,[530] bone,[97] food,[530,1506] and urine,[530] as well as to TAAH-homogenized samples of tissues[1503] and canned food/fish.[1502,1503] It, too, is a very sensitive detector for organotins,[283,284,932,1505,1520] provided the graphite tube is adequately pretreated.[932,1505] Several selected procedures are briefly outlined in Table 2.

Contamination is the first serious problem encountered in ultratrace tin assays (nanograms of Sn per gram of blood, urine, milk, nails, hair, etc.). Sources of extraneous tin can be sampling instruments,[622] improperly chosen or cleaned plastics (which may contain organotin stabilizers), airborne dust, conservants and reagents used (e.g., even the dilute HCl,[933] $SnCl_2$ which is a common reagent in AA labs (!), etc.), as well as the ubiquitous tin-coated surfaces, solders, brass, etc. Many early analyses seem to have incorporated substantial positive errors due to contamination.

Concerning the preparatory step, both volatization (as organotin compounds, halides, etc.) and retention losses (as sparingly soluble residues: SnO_2, H_2SnO_3, etc., as well as adsorption) may be expected. Iyengar et al.[1455] found significant, tissue-dependent losses of metabolized ^{113}Sn in certain rat tissues during oven drying at 80 (up to −5%: muscle), 105 (up to −5%: kidney, liver, and muscle), and 120°C (up to −10%: blood, heart, liver, ovary, fur, and uterus; up to −15%: brain, kidney, lung, and muscle). No losses were found in feces (up to 120°C), nor in all the specimens studied on freeze drying up to 72 hr under pressure of 0.05 Torr.[1455]

Creasson et al.[288] recommended wetting of hair samples with H_2SO_4 followed by ashing at 550°C; the use of H_2SO_4 as an ashing aid has also been mentioned by Gorsuch.[91]

LTA in excited oxygen has given satisfaction with liver samples.[275]

Most wet-oxidation mixtures contain H_2SO_4: H_2SO_4 and excess HNO_3 (for bone,[97] canned food/beverages,[1495-1497,1506,1519] fish,[1497] etc.[91]); H_2SO_4 and H_2O_2 (for canned food/ beverages[91,1496,1497]); H_2SO_4, HNO_3, and (finally) H_2O_2 (for blood, food, tissues, urine, etc.[129,530,1495,1499,1500]); and H_2SO_4, HNO_3, and $HClO_4$ (for blood,[131,332] canned food/beverages,[1495,1496] etc.). Temperature-controlled heating and reflux conditions (say, in long-necked quartz Kjeldahl flasks[1499,1500]) are essential.[1499,1500,1506,1523] Pressure vessels (in HNO_3 + H_2SO_4 digestions, with caution!) are also useful.[97]

Acid extraction of tin from food samples with HCl (e.g., 6 to 10 *M* HCl, under mild heating and reflux conditions) or with HNO_3-HCl[1518] is also practiced in food analysis.[1495,1518]

TAAH solubilization with Soluene® 350[1503] or Lumatom®[1502] is outlined in Table 2; although these procedures look feasible, it would be safer to analyze tissue homogenates or untreated biological liquids in impregnated graphite tubes (e.g., Zr[932,1505] or Ta treated[1506]), in order to level off the response of different tin species.

Trace tin can be preconcentrated from (completely!) digested biological samples by liquid-liquid extraction,[39,76,332,530,659,1496,1506] evolution of SnH_4 and/or alkyltin hydrides,[131,242,1494,1512-1516,1520] distillation as volatile $SnBr_4$,[192] coprecipitation with dibenzyldithiocarbamate[129] or with hydroxides of La(III),[1507,1516a] Fe(III), Mn(IV),[1514,1515] etc.,[1495] ion exchange on poly(dithiocarbamate) resin,[326] and chromatography (in tin speciation studies[283,284,932,1520]). Extraction procedures[530,1506] and coprecipitation with $La(OH)_3$[1507] are favorably combined with ETA, whereas coprecipitation with $Fe(OH)_3$ or MnO_2·aq. is better coupled with HG-AAS.[1514,1515]

Many extraction procedures for Sn are compiled and discussed by Cresser.[76] Some extractions have been applied to biological digests; it is important that tin be converted to a definite oxidation state: Sn(II) or Sn(IV). Extractions with cupferron-MIBK, from 2 *M* HCl (ashed blood[332]); neocupferron (i.e., N-nitroso-α-naphtyl hydroxylamine, ammonium salt)-$CHCl_3$, from 5% v/v H_2SO_4 (ashed food[1506]); N-benzoyl-N-phenylhydroxylamine-toluene;[1500] Sn(IV)-KI-H_2SO_4-toluene (from digested blood, food, tissues,[1498-1500] urine,[530] beverages,[1496] etc.[1495]); Sn(II)-dithiocarbamate extractions;[39,659] etc. have been described.

In flame AAS, organic extracts or diluted/digested food, beverages, etc. are readily analyzed in a red N_2O-C_2H_2 flame vs. matrix-matched standards and with simultaneous background correction. Additions of KCl[1518] and NH_4Cl[1519] have proved useful. Standard addition checks are advisable.

In ET-AAS, the roles of graphite-tube pretreatment, age, "history", and temperature programming are very important and critical. Carbide-impregnated tubes/platforms, carefully optimized ashing temperatures, "gas-stop" atomization, fast heating rates, and efficient background correction are all essentials. Introducing O_2 or air during the ashing step (with caution! See Beaty et al.[296,582]) may prove useful. Acids such as H_2SO_4, $HClO_4$, and HCl should be avoided. Acid- and matrix-matched calibrations are typical.

In HG procedures, the HCl (or $HClO_4$ or H_2SO_4) concentration in final solutions is critical. Signal depression from excess of Cu, Fe, Ni, and As is to be expected with some samples. Standard addition calibration is recommended. Some batches of HCl[933] and $NaBH_4$[155,1512] may introduce intolerable blank levels. A simple electrolytic procedure for precleaning the $NaBH_4$ solution has been described by Hodge et al.:[1512] two pure carbon rods have been inserted into a $NaBH_4$ solution (50 mℓ, 4% w/v), and tin impurities have been plated on the cathode for 3 min at 3 V while bubbling He through the solution.

IV. CONCLUSION

Determination of nanograms-per-gram concentrations of tin is a difficult analytical task. RNAA appears to be the most reliable method. Flameless AAS techniques require an efficient preconcentration step — extraction, stannane (SnH_4) generation, coprecipitation, etc. Volatization and retention losses may be encountered in most ashing techniques; suitable are LTA, ashing in O_2 bomb decompositions, and wet digestions in presence of H_2SO_4 and under strict temperature control.

Hydride evolution is liable to acid and matrix effects. The same refers to ET-AAS, in which many important parameters have to be controlled (e.g., heating rates and temperatures, graphite surface pretreatment, etc.). Matrix-matched or standard addition calibration is often needed.

Flame AAS is used in analyses of canned food/beverages (micrograms-per-milliliter levels), but is not devoid of problems, either.

Contamination control is essential, especially in tin assays in body fluids.

Chapter 33

VANADIUM

I. DETERMINATION OF VANADIUM IN BIOLOGICAL SAMPLES

Most previous "normal" values for V in biological samples are artificially raised, probably due to contamination during sampling, storage, and handling of specimens. It is now obvious that the concentrations of V are very low: <1 ng/mℓ in body fluids and well below 0.1 μg/g in most other biomaterials (Table 1). The determination of these levels is far from easy, and reliable analytical results for V in biological materials have been obtained only recently.[1326,1521,1522,1528,1529,1531]

Few analytical techniques provide DLs below 1 ng/g: ICP-ES (DL down to 0.06 ng/mℓ[29] but, typically, nanograms per milliliter),[35] RNAA,[45] ET-AAS (DL down to 0.02 ng/mℓ), and SSMS (DL 0.4 ng/g in soft tissues[3]).[3,25,55,132,275] For higher V levels, the arc-source AES[14,26,36,38-42,44,288,904] and some catalytic procedures[577,1299,1532,1533] can be used. It should be noted, however, that many early procedures,[26,36,44,55,133,1533] are not adequately sensitive, and many early analyses have been performed on samples containing substantial extraneous addition of V.

RNAA has been applied to blood,[1299,1493,1521-1523] bone,[1521] diet,[1493] erythrocytes,[1521] hair,[1521] milk,[110,1521] serum or plasma,[1522,1525,1528,1529] teeth,[1521,1530] tissues,[1296,1521,1522,1528] urine,[1521,1522] etc.[45,1521,1522,1525] Some RNAA procedures call for a preirradiation ashing[1528] or separation.[1522] Only the specimens relatively high in V, e.g., hair[11,290] and nails,[15,290] can be analyzed by INAA.

II. ATOMIC ABSORPTION SPECTROMETRY OF VANADIUM

The flame AAS of V is insufficiently sensitive and prone to pronounced matrix effects. In a reducing N_2O-C_2H_2 flame, under thoroughly optimized conditions (acetylene flow rate, observation height, etc.), the characteristic concentration and DLs range from 0.6 to 2 and 0.02 to 0.07 μg/mℓ, respectively.[56-59,294] The three lines of V triplet, 318.5/318.4/318.3 nm, are measured simultaneously. Numerous interference effects from other concomitants have been observed: enhancement by organic solvents, H_3PO_4, HF, $HClO_4$, Al, Ca, Co, Cr, Fe, Mo, Ti, F^-, I^-, etc; and depression by H_2SO_4. Suitable spectrochemical buffers are aqueous $AlCl_3$ (0.1 to 0.5%), H_3PO_4, or NH_4F. Matrix-matched calibration is advisable.

ET-AAS of V deserves special attention as one of the most sensitive techniques for nanogram amounts of V which moreover is not as straightforward and as sensitive for V as it is for many other elements.

The characteristic concentrations and DLs of V range broadly between 0.15 and 30 ng/mℓ (15 to 400 pg) and between 0.02 and a few ng/mℓ (2 to 360 pg), respectively.[57,59-61,294,551,552,564]

There are several important factors that determine the sensitivity and accuracy of ET-AAS assays:

1. The formation of refractory vanadium carbides with the graphite as well as with organic-matrix residues should be considered, and — as far as possible — prevented. The use of pyro-coated tubes is advantageous: sensitivity is improved by a factor of 2 to 5,[60,1531,1534,1535] somewhat lower atomization temperatures can be used, and the memory effects are less pronounced. *In situ* pyro-coating by adding 1 to 10% CH_4 to the Ar purge gas also improves sensitivity, by a factor of 2 to 5.7.[72,295,1302,1535] The extent of

Table 1
VANADIUM CONCENTRATIONS IN HUMAN BODY FLUIDS AND TISSUES

Sample	Unit	Selected reference values	Ref.
Blood	μg/ℓ	<0.1	1493
		<0.3	1521
		0.77 (Pool)	1522
		<1 (0.4—2)	1523
		<2.5	1524
		3.6 (0.80—7.1)	1525
		5.8 (1.2—25) (Japan)	94, 1526, 1527
Bile	ng/g	(0.55—1.85)	1521
Bone	ng/g	(0.8—8.3)	1521
Brain	μg/g	0.03 ± 0.008	3
Feces	μg/g	(0.141—2.21)[a]	1521
Hair	μg/g	x̃, 0.01 (0.0081—0.48) (Japan)	11
		0.040 ± 0.020 (0.012—0.087) (Yugoslavia)	1521
		x̃, 0.0062; (0.005—0.564) (Canada)	10
		x̃, ♂, 0.07 (0.01—0.36) (Kenya)	290
		x̃, ♀, 0.11 (0.04—0.44) (Kenya)	290
		♂, 0.18;[b] ♀ 0.19;[b] (0.009—2.2) (New York)	288
Kidney	μg/g	(0.0026—0.0033)[c]	1521
		x̃,0.12; 0.11 ± 0.06 (Japan)	12
		0.67 ± 0.35[a] (Japan)	904
Liver	ng/g	(2.53—13.4)	1528
		4.5 and 7.5[d]	1521
		(<7—19)	108
		40 ± 10	3
		44 ± 14	275
		x̃, 110; 110 ± 80 (Japan)	12
		370 ± 80[a] (Japan)	904
Lung	μg/g	x̃, 0.030 (0.019—0.140)	1521
		x̃, 0.09; 0.13 ± 0.08 (0.08—0.24) (Japan)	12
		0.1 ± 0.02 (U.K.)	3
		<0.4[a] (Japan)	904
Milk	μg/ℓ	0.1—0.2	1521
		~0.8[a,e]	110
		3—6[f] (0—21)	14
Muscle	ng/g	(0.45—0.62)[c]	1521
Nails	μg/g	♂, x̃, 0.02;[a] 0.04 ± 0.05[a]	15
		♀, x̃, 0.05;[a] 0.07 ± 0.07[a]	15
		(0.004—0.209)[a]	15
Plasma or serum	μg/ℓ	♂, (0.029—0.939)	1528, 1529
		♀, 0.033 ± 0.012 (0.017—0.053)	1528, 1529
Tooth (enamel)	ng/g	(<2—5.1)	1521
		x̃, 3.5; 3.7 ± 1.5 (1.0—8.3)	1530
		17 (10—30)	25
Urine	μg/ℓ	(<0.2—0.3)	1521
		0.44 (0.3—0.7)	1531
		1.34	1532
		(0.5—2)	1524
		(2—4)	1522
	μg/g creatinine	1.6 (0.45—3.1)	1523

Note: Net weight unless otherwise indicated. Mean ± SD (range in parentheses). x̃, median.

[a] Dry weight.

[b] Geometric mean.

Table 1 (cont.)
VANADIUM CONCENTRATIONS IN HUMAN BODY FLUIDS AND TISSUES

[c] N = 3.
[d] N = 2.
[e] ng/g (ppb).
[f] Range of mean values.

soaking and spreading of the sample solutions into the graphite is also important; an autosampler is recommended.

2. The impregnation of graphite tubes with other carbide-forming elements (W, Si, La, Zr, Ta, and Nb) proved unsatisfactory; sensitivity was impaired by factors of 2.7 for W and 1.05 to 1.4 for the rest of the elements studied.[1535] This effect has been explained by the possible formation of ternary compounds between the impregnating element, V, and graphite.[1535]
3. The purge gas should preferably be high-purity argon. The impurities of oxygen have been shown to significantly lower the sensitivity ($V + O \rightarrow VO$), and the beneficial effect of the added methane has been attributed to the reaction: $O + CH_4 \rightleftharpoons 2H_2 + CO$.[1535]
4. The temperature programming includes relatively high ashing temperatures (e.g., 1100 to 1800°C, in order to eliminate the interfering matrix constituents) and atomization at 2400 to 2900°C with as high as possible a heating rate. Temperature-controlled, high-speed heating is essential for improving the sensitivity, accuracy, precision, and long-term signal stability, as well as to extend the lifetime of the graphite tube.
5. The linear working range is very restricted, often to only 20 times the DL.[511,1536,1537]
6. Background correction is usually needed.
7. Matrix effects are expected, and the calibration should often be by matrix-matched standards or by standard additions. In older models of graphite-tube furnaces which employ slow heating rates of the atomization stage, interferences have been observed by organic matrix residues,[67,294] acids (excess HNO_3, HCl, H_2SO_4, H_3PO_4,[809,1537] salts (KCl, NaCl, $MgCl_2$[1537]), and metal ions (Fe, Ca, Mo, Al, Ti, La, W, etc.[294,809,1324,1535,1537]). The presence of dilute acids (e.g., less than 1 to 2 *M* HCl, H_2SO_4, or HNO_3 can be tolerated provided that similarly acid-matched standard solutions are used. The aqueous NH_3 should be a useful additive as well.[1324]
8. Sensitivity drift may appear[1537] due to etching of pyro-coating and modification of graphite-tube surface, as well as to matrix residue accumulation, and other adverse effects. Therefore, frequent recalibration may be needed.

III. AAS METHODS FOR ANALYSIS OF BIOLOGICAL SAMPLES

The literature on ET-AAS of V in biological samples is scant, reflecting difficulties of V assays around and below the nanograms-per-gram levels. Progress has been achieved only recently, thanks to improved graphite furnaces (faster heating rates, temperature-controlled heating, pyro-coatings, etc.), as well as to efficient preconcentration procedures developed by Japanese scientists.[1526,1527,1534,1538,1539]

ET-AAS has been applied to analyses of blood,[1526,1527,1531,1534,1536] urine,[1531,1534,1538] water,[809,1536] and other biologicals[1527,1536,1539] (Table 2).

Besides the problems of ET-AAS measurements discussed under the preceding heading, there are two problems of the preparation step which are of great concern: (1) the contamination and blank levels arising in sampling, storage, and handling of specimens, and (2) the efficiency and reliability of the preconcentration step.

Table 2
SELECTED AAS PROCEDURES FOR V DETERMINATION IN BIOLOGICAL SAMPLES

Sample	Brief procedure outline	Ref.
Blood, food, hair, tissues, urine	Sample (10 mℓ of blood or 1—10 g of tissue or 50 mℓ or urine digested with HNO_3 and $HClO_4$;[a] evaporated to 10 mℓ, diluted to 20 mℓ, and oxidized to V(V) with $KMnO_4$; vanadium (V) extracted with 1 mℓ of 0.1% N-cinnamoyl-N-(2,3-xylyl)hydroxylamine in CCl_4 from 6 *M* HCl for 3 min; ET-AAS with pyro-coated tubes: 100°C for 20 sec, 1850°C for 20 sec, and 2800°C for 10 sec; background correction	1534, 1538
Fish, oysters, shellfish, etc.	LTA of 10-g samples, ashes evaporated with dilute HCl, and dissolved in 10 mℓ of 0.5 *M* HCl; CRA-63 furnace programming (in volts or scale readings): drying at 5 V for 40 sec, ashing at 4 V for 20 sec, and ramp atomization with ramp rate 4 and cut-off voltage 9.5 V	1327
Food, meat, milk, tissues, etc.	5 to 20-g samples ashed at 450°C for 16 hr, 2 mℓ of HNO_3 added, dried, and re-ashed at 450°C for 2 hr, and ashes dissolved in 2 mℓ of 4 *M* HNO_3 and dilute to 5 mℓ. ET-AAS: 125°C for 60 sec, 1400°C for 60 sec, and 2750°C for 15 sec; background correction	1326, 1536
Urine	20-mℓ sample evaporated to dryness, 10 mℓ of HNO_3 added, and refluxed at 120—130°C for 24 hr; cupferron-MIBK extraction at pH 2.0—2.5; ET-AAS with pyro-coated tubes: 100 (10 + 5 sec), 200 (65 + 5 sec), 1500 (5 + 5 sec), 1500 (1 + 5 sec, miniflow), 2700 (0 + 3 sec, gasstop), and 2700°C for 3 sec. Extracts stable for 2 hr	1531

[a] Safety precautions mandatory.

Uncontrolled contamination of samples is likely to occur since the levels of V in biological specimens (e.g., 0.1 to 10 ng/g) are much lower, by four to six orders of magnitude, than in the lithosphere (about 100,000 ng/g!). Considerable extraneous additions of V can be introduced from stainless steel (some steel instruments contain as much as 10,000,000 ng V/g!), airborne particles,[108] adherent dirt (to hair, nails, skin, etc.), and containers and labware materials (glass,[1522] glazed porcelain, polyethylene,[163] agate,[624] corundum,[624] etc.). Let us cite just a few examples.

Van Grieken et al.[624] found pronounced and quite intolerable contamination during grinding procedures, e.g., 0.5 μg V/min in agate and 0.8 μg V/min in corundum.[624] Cleaner homogenation techniques are obviously needed for solid specimens of teeth, bone, tissues, etc.[532,533]

Lievens et al.[108] measured the dust fallout in a dust-free and in a normal laboratory; it differed by a factor of 123 (0.3 vs. 37 pg/cm²/day).

Karin et al.[163] found as much as 0.84 μg V/g in polyethylene bottles; 0.74 μg/mℓ was leached with 8 *M* HCl for 3 days.

Some samples of hair,[10,623] nails,[15,623] teeth,[1530] etc. require thorough cleaning procedures, including, e.g., washing under sonification and brisk washing in dilute H_2O_2[15,623] or brief etching (of teeth) in 0.5 *M* $HClO_4$.[1530]

Some sampling and handling procedures[623,1521,1528,1529] even appear difficult to follow in ordinary laboratories and in routine practice. Cornelis et al.[1528,1529] obtained the lowest published normal values for serum vanadium by a scrupulous sampling and handling procedure: the blood was taken with the aid of a plastic cannula trocar (Vygon®) flushed with 40 mℓ of blood(!) before sampling in an ultra-clean Spectrosil® quartz tube; the serum was then lyophilized, ashed at 450°C, and analyzed by RNAA.

All procedures for the oxidation of organic matter would seem to be applicable,[91] the choice depending on the next step (extraction or directly introducing digests into the graphite tube) as well as on the blank levels introduced by one or another wet-digestion procedure.

LTA in excited oxygen is a clean and successful decomposition technique.[132,275,626,1327] Dry ashing should involve ashing aids — wetting samples with HNO_3[39,74,1326,1528,1529,1536] or with H_2SO_4[91,288] — so that lower ashing temperatures (e.g., only 450°C) can be used. Cornelis et al.[1528,1529] recommended to keep the ashing temperature ~450°C, so as to avoid serious etching of quartz ware and associated adverse effects occurring at higher temperatures.

The pressure decomposition in PTFE-lined autoclaves[1525] is clean and rapid, although the organic matter is not completely oxidized[1288,1289] and safety precautions are to be taken.

In wet-decomposition techniques, specially purified reagents and precleaned labware of quartz or PTFE are involved. Mixtures of HNO_3 and $HClO_4$ (caution!),[1526,1534,1538] HNO_3 and H_2SO_4,[1522] HNO_3 and H_2SO_4, completed with H_2O_2,[170,1521] and HNO_3, H_2SO_4, and $HClO_4$[91] have been successfully applied in RNAA or ET-AAS procedures.

Solvent extraction[76,1531] (Table 3) is mostly used for preconcentration of trace V from digested biological samples, though ion exchange of V(IV) on Chelex® 100 has also been applied.[412,1525] Prior to extraction, the analyte should be quantitatively converted into the desired oxidation state — V(V) or V(IV) or (rarer) to V(III). Extractions with cupferron (N-nitrosophenylhydroxylamine)[1528,1529,1531] analogous reagents[1521,1526,1527,1530,1534,1538,1539] are most popular and effective, but dithiocarbamates,[39] or 8-hydroxyquinoline,[1522] or a mixture of the last two reagents,[39] have been used as well.

In ET-AAS, excess acids (more than 1 to 2 *M*) should be avoided, and either evaporation or neutralization with aqueous ammonia[1324] should be attempted. The concentration of the dilute HCl,[1327] or HNO_3,[74,1326,1536] or H_2SO_4[67] should be matched in sample and standard solutions. The adverse effects mentioned earlier — memory, background absorption, restricted linearity of working curves, sensitivity drift, and matrix interferences — are to be expected and checked.

IV. CONCLUSION

NAA is the current technique of choice, but ET-AAS is steadily gaining ground. What matters is to avoid extraneous contamination during sampling and preparation. Essentials in ET-AAS are use of pyro-coated tubes, rapid temperature-controlled heating, and matrix-matched calibration. Any ashing technique is applicable but cleaner procedures are preferable, viz. LTA, pressurized digestions, etc. Chelate extraction is effected with cupferron and analogous reagents.

Table 3
EXTRACTION OF VANADIUM FROM BIOLOGICAL DIGESTS

Extraction system	Matrix and conditions	Ref.
N-benzoyl-N-phenyl-hydroxylamine in toluene	Wet-digested samples (bile, blood, bone, food, hair, milk, teeth, tissues, urine); 5 *M* HCl-1.8 *M* H_2SO_4 medium; oxidized with $KMnO_4$ to V(V); RNAA	1296, 1521
N-cinnamoyl-N(2,3-xylyl)hydroxylamine in CCl_4	Wet-digested samples (blood, food, hair, tissues, urine); oxidized with $KMnO_4$; extraction from 6 *M* HCl; ET-AAS	1534, 1538
N-cinnamoyl-N-phenylhydroxylamine	Biological digests; interference from iron; ET-AAS	1539
N-benzoyl-o-tolyl-hydroxylamine in CCl_4	Wet-digested samples (blood, tissues, etc.), treated with $KMnO_4$ and H_2SO_4 [$\rightarrow$ V(V)]; ET-AAS	1526, 1527, 1539
N-nitrosophenyl-hydroxylamine in $CHCl_3$	Ashed liver/serum; 1 *M* HCl; RNAA	1528, 1529
HHDC + 8-hydroxyquinoline (4:1) in $CHCl_3$ + i-pentanol (3:1)	Ashed food, feces, urine; pH = 4.8—5.0; exhaustive extraction (4—5 ×); part of a multielement scheme; arc-source AES	39
8-Hydroxyquinoline in MIBK	Wet-ashed samples (blood, serum, tissues, urine, etc.); pH = 3—4; RNAA	1522

Chapter 34

ZINC

I. DETERMINATION OF ZINC IN BIOLOGICAL SAMPLES

Normal levels of Zn in biological specimens are relatively high: in general, 0.1 to 10 μg/mℓ and 10 to 200 μg/g in liquid and solid biomaterials, respectively (Table 1). There are a dozen relevant analytical methods; among them flame AAS lends itself best to routine application.

The most sensitive methods are ASV and ET-AAS, both with DLs down to 0.01 ng/mℓ. ASV has been applied to blood,[683] hair,[393] nails,[133] serum,[321,683] teeth,[395] tissues,[133] urine,[124,125] etc. It is relatively simple, accessible, inexpensive, as well as a multielement (for say, Cu, Cd, Pb, and Zn) technique, but calls for a complete release of Zn from its chemical bonds with the matrix and for a somewhat larger sample size (vs., e.g., AAS or ICP-ES). These and perhaps a few more drawbacks may be assigned to polarographic and spectrophotometric procedures.[40,414,1545]

While ICP-ES is gaining ground in simultaneous multielement assays (with a DL of a few nanograms Zn per milliliter[29,35]), the arc-source AES[14,38,41-44,320,904] is regarded as a classic. ICP-ES has been applied to blood,[30,689] bone,[392] diet,[690] milk,[35] serum,[30,289,580,688,689] tissues,[289,348,411,412,518,580,636] etc.[29,348]

Dried undigested biological specimens are simultaneously analyzed for many elements by INAA,[45,121] XRF,[128] PIXE,[128] and SSMS.[3,16,25,52,54,55,132,275,409]

INAA assays on blood,[686] CSF,[621] erythrocytes,[631,840] fish,[45,853] hair,[11,48,98,99,102,104,118,184,290,408,635] milk,[182] nails,[290,560] placenta,[377] platelets,[119] serum/plasma,[120,631,840] serum protein fractions,[120] teeth,[51,665,676] tissues,[12,631,686] etc.[45,121] have been described. RNAA is also well documented,[45,104,108,110,111,116,287,633,665,670]

XRF[128] applications to blood,[3,128,691] bone,[851] hair,[128,293,408,1546] nails,[692] tissues,[3,128,292,636] urine,[695] etc. have been described.

PIXE[97,128,195,407,408,696-698,852] provides excellent absolute DLs (e.g., 0.1 pg[852]) in small samples (biopsies,[407] etc.[128]) and/or in studying Zn gradients in single hairs,[128,195,408] tooth enamel,[97,852] bone, etc.

A few laboratories are able to perform exceedingly accurate analyses by IDMS[410] and IPAA.[104]

Flame AFS may be adopted with diluted plasma (1:1000![404]), blood, urine, tissue digests, etc.; minor modifications of existing AAS apparatus are required.

II. ATOMIC ABSORPTION SPECTROMETRY OF ZINC

The flame technique is sensitive and selective. Characteristic concentration and DLs are 0.007 to 0.02 and 0.0008 to 0.002 μg/mℓ, respectively.[56-59,294]

EDL is recommended vs. HCL for higher intensity (10 ×) and lower DL (1.3 to 1.7 and 3.5 × in flame and ET-AAS, respectively[138]).

Background absorption should always be checked. Only bulk-matrix effects, caused by excess salts (>1%) and unmatched viscosity and/or acidity of sample and standard solutions, have been observed; they may, however, be a source of minor inaccuracies, depending on performance of a particular piece of equipment, instrumental conditions, calibration mode, etc.[722,730] A fuel-lean air-acetylene flame and a not-too-low burner height are recommended. Any dilute mineral acid is tolerable (<10% v/v). A three-slot burner and a pulse-nebulization attachment (injection or dipping versions) are very useful with small sample aliquots (20 to 200 μg/ℓ).[704,705,722]

Table 1
ZINC CONCENTRATIONS IN HUMAN BODY FLUIDS AND TISSUES

Sample	Unit	Selected reference values	Ref.
Blood	µg/mℓ	♀, (3.8—6.8); ♂, (4.1—7.9)	661
		(3.42—7.94)	1540
		5.7 ± 0.7[a]	178
		4 — 9;[b] (2.6—11.2)	96
		6.7 ± 0.05	3
Bone	µg/g	27 ± 2	268
		x̄, 95 (92—102)[c]	97
		x̄, 99.8; 94.0 (49.9—129)[d]	103
		100[e]	367
		217 (155—299)[f,g]	851
Brain	µg/g	10.5 ± 0.7;[h] 12.9 ± 0.7[i]	1541
		12 (6—22)	662
		x̄, 35.2; 39.1 ± 15.4 (19.7—63.5)[d]	103
		34.3;[d,h] 72.9[d,i]	663
CSF	µg/mℓ	(0.020—0.100)	839
		(0.09—0.31)	621
		0.170 ± 0.042	1542
Erythrocytes	µg/g	10.6 ± 0.8 (9.4—11.8)	665
		11.15 ± 1.83	840
		35.05 ± 3.4 (31.2—40.6)[d]	631
Feces	mg/day	2.8 (4.5—158)	96
		(5.1—10.3)	838
		11[j]	8
Hair	µg/g	♂, 107;[k] ♀, 112[k] (20.1—313) (New York)	288
		x̄, ♂, 110; ♀, 126; (88—201) (Kenya)	290
		♂, 120.3; ♀, 141.2; 132.6 ± 34.3 (53.1—282) (U.K.)	368
		x̄, 138 ± 44 (62—430) (India)	101
		141.9—154.3[b] (U.S.A.)	180
		x̄, 150 and 159 (U.S.A.)	9
		x̄, 142 — 181[l] (47—486) (Austria)	99
		x̄, ♂, 150 (108—190); x̄, ♀, 167 (120—357) (Canada)	10
		x̄, 170 (76—550) (Japan)	11
		x̄, 174; 173 ± 40 (92—255)[d] (U.K.)	103
		181 — 215[m]	100
		♂, 200; ♀, 215 (G.F.R.)	367
		(40—485)[d] (Canada)	104
		(70—490) (U.S.S.R.)	102
		208 ± 137 (<2—824) (Italy)	98
		♂, 232.1; ♀, 233.6 (Nepal)	666
Hair (pubic)	µg/g	156 ± 50 (64—285)	841
		♀, 178 ± 56[n]	734
Kidney	µg/g	x̄, 25; 31 ± 6.3 (18—47)	105, 106
		29.4 ± 12.8	739
		37.4 ± 5.9	3
		150 ± 72[d]	904
		x̄, 190;[d] 188 ± 82.6 (63.6—320)[d]	103
Kidney cortex	µg/g	33.7	107
		48.1 ± 4.9	3
		91 ± 22 (64—126);[d,o] 218 ± 89 (119—448)[d,p]	372
Liver	µg/g	40 ± 3	268
		x̄, 41.0; 46 ± 6.9 (30—99)	105, 106
		41.7 ± 20.0	739
		59 (53—66)	108
		63.8 ± 19.1	275
		67 ± 20 (44—92)	633

Table 1 (continued)
ZINC CONCENTRATIONS IN HUMAN BODY FLUIDS AND TISSUES

Sample	Unit	Selected reference values	Ref.
		x̃, 156;[d] 169 ± 58.6 (41.7—298)[d]	103
		194 ± 92 (57—312)[d]	319
		210 ± 89[d]	904
Lung	μg/g	(8.9—10.9)	109
		x̃, 10.7;14.3 ± 3.5 (6.4—24)	105, 106
		13.7 ± 0.9	3
		16.6 ± 12.9	739
		x̃, 47.5;[d] 50.2 ± 14.3 (21.0—77.6)	103
		77 ± 22[d]	904
Milk	μg/mℓ	1.34(0.7—4.33)	736
		1.5 (1.4—1.7)	182
		1.66 ± 0.52	667
		1.3—2.0[b] (1.0—3.0)	14
		(0.14—3.95)	668
		(0.316—5.300)	13
Nails	μg/g	73 ± 8[q]	133
Nails (finger)	μg/g	(62—360)[d]	16
		108 ± 25	1543
		x̃, 139;[d] 151 ± 50 (93—292)	103
		♂, 178 (132—391);[d] ♀ 220 (130—360)[d]	738
Nails (thumb)	μg/g	♂, 153 ± 68 (58—272)[d]	560
		♀,184 ± 75 (85—300)[d]	560
Plasma/serum	μg/mℓ	0.85 ± 0.07 (0.80—0.99)[a]	633
		0.877 ± 0.143	1544
		0.91 ± 0.14 (0.65—1.23)	665
		0.95 ± 0.44	111
		♀, 1.03 ± 0.14; ♂, 1.22 ± 0.20	840
Skin	μg/g	15.6 ± 9.0	739
		25.0 (4.1—61.1)[d]	103
		x̃, 39.7;[d] 40.0 ± 10.1[d]	112
		13.5 ± 8.1;[d,r] 79.4 ± 32.1[d,s]	565
Sweat	μg/mℓ	♀, 0.507 ± 0.151; ♂ 0.960 ± 0.425	381
		♂, 0.55 ± 0.48 (0.13—1.46)	675
		♀, 1.25 ± 0.77 (0.53—2.62)	675
		1.15 ± 0.30	1545
Tooth (whole)	μg/g	(111—227)	1099
Tooth (dentine)	μg/g	126 ± 21;[t] 181 ± 155[u]	24
		172.8 ± 11.8	23
		199 ± 78[d]	676
		(210—260)[d]	845
Tooth (enamel)	μg/g	145 ± 12;[u] 180 ± 49[t]	24
		199 ± 14 (91—400)	25
		231 (129—1197)	26
		238;[v] 349[w]	51
		263.4 ± 14.8	23
		x̃, 339;[d] 366 ± 181 (58—992)[d]	103
Urine	μg/ℓ	480 ± 320	483
		(400—1000)	27
	μg/day	353 ± 23 (141—779)	663
		♀, 414; ♂, 585; (276—817)	1540
		300—700;[b] (77—1155)	96
		520 ± 300 (140—1200)	111

Note: Wet weight unless otherwise indicated. Mean ± SD (range in parentheses). x̃, median.

Table 1 (continued)
ZINC CONCENTRATIONS IN HUMAN BODY FLUIDS AND TISSUES

[a] μg/g (ppm).
[b] Range of mean values.
[c] Femur.
[d] Dry weight.
[e] Rib.
[f] Ash.
[g] Vertebral.
[h] White matter.
[i] Grey matter.
[j] "Reference Man".
[k] Geometric mean.
[l] Range of medians.
[m] Variations over the same head.
[n] Before delivery.
[o] Newborn.
[p] Adults.
[q] Finger and toenails.
[r] Dermis.
[s] Epidermis.
[t] Permanent.
[u] Primary.
[v] European.
[w] African.

The Delves cup[415,912] and Sampling Boat®[913] accessories help decrease DLs to 0.05 (5 pg) and 0.03 ng/mℓ (30 pg), respectively, but suffer from matrix effects, transient background absorption, and complicated calibration.

In ETA, zinc determinations are very sensitive, but may not be as accurate as expected. Characteristic concentration and DLs are down to 0.01 ng/mℓ (0.05 to 3 pg absolute, often <0.1 pg).[57,59-61,294,551,552] The practical DLs, however, may be impaired by one or two orders of magnitude, due to blank levels/fluctuations and sporadic contamination.

Owing to volatility of zinc (chloride[1547]) species, ashing temperatures have to be restricted to 400 to 800°C (often ≤500°C). Hence, most matrix constituents cannot be expelled from the tube during preatomization heating and give rise to pronounced interferences, background absorption, and sensitivity loss. Strong interferences by salts and acids have been reported.[57,1542,1547] e.g., NaCl, $CaCl_2$, $MgCl_2$, $FeCl_3$, PO_3^{4-}, $HClO_4$, etc. Thermal stabilization of the analyte by additions of HNO_3,[27,438] H_3PO_4 and its NH_4 salts, $(NH_4)_2SO_4$,[454,455] NH_3,[1547] $AgNO_3$,[1547] $La(NO_3)_3$,[438] etc. has been observed. Of utmost importance is the temperature programming: ashing ramp and temperature, rate of heating during atomization (higher rates advisable[60,655]), as well as including a clean-out stage in each heating cycle.

Pyro-coated[60,1548,1549] or tantalized[1511] graphite tubes, L'vov platform atomization,[85,229,445,455] and gas-stop/miniflow conditions have proved to be advantageous. Matrix-matched calibration is often mandatory.[421,1548,1550,1551]

III. AAS METHODS FOR ANALYSIS OF BIOLOGICAL SAMPLES

AAS applications are ample (Table 2). Obviously, most analyses can be performed by *flame* AAS, which is simple, accurate, rapid, inexpensive, and amenable to automated, microscale operation (pulse-nebulization mode).[701,704,705] Preconcentration and ETA would rarely be justified, say, with some specimens which are low in Zn (CSF, saliva, etc.) and/or are of limited size (separated blood[1548] or serum protein[785,786,1549] fractions, etc.).

While AAS assays of Zn are generally straightforward, two sources of analytical inac-

Table 2
AAS TECHNIQUES FOR Zn DETERMINATION IN BIOLOGICAL MATERIALS

Technique	Biological materials
Direct flame AAS with or without dilution	Blood,[1540,1552] CSF,[664,709,1553] erythrocytes,[1552] milk,[710,712] serum or plasma[27,539,664,714,718-723,749,872,1540,1544,1552-1560] sweat,[381,675] synovial fluid,[758] urine[7,27,339,483,539,664,714,1540,1553,1561]
Flame AAS after TCA deproteinization	Erythrocytes,[1561-1563] milk,[1564] serum or plasma,[7,680,718,719,722,723,725,1556,1561-1563,1565,1566] urine[1563]
Flame AAS after digestion	Blood,[726,727] bone,[90,268,728] brain,[662,680,740,791,1541] erythrocytes,[727,1567] feces,[248,729] fish,[442,458,475,535,730,792,853,1568] food,[250,591,729-731] hair,[31,180,228,368,370,426,460,461,557,662,666,732-735] liver,[462,517,791,793,853] liver (biopsies),[515] meat,[250,517,729,730] milk,[250,374,538,591,729,730,736,745] nails,[692,737,738] semen,[1553] serum,[727] teeth,[24,50,481,545] tissues[7,43,190,250,268,442,517,535,538,541,545,718,739]
Flame AAS after TAAH solubilization	Blood,[521] hair,[464] plasma,[741] tissues,[464,521,741,1568] urine[521]
Flame AAS after acid extraction	Feces,[729,1569] food,[729,1569] hair,[466] liver,[319,467] milk[729]
Flame AAS after liquid-liquid extraction	Blood,[332,473,482,526,530] bone,[976] feces,[39] fish,[334,474,494] food,[39,530,659,690,1570] serum,[482,640,657,719] teeth,[976] tissues[39,473,474,482,530,1561]
Pulse-nebulization flame AAS	Plasma protein fractions,[751] serum,[699-704,756] tissues[704,760]
Delves cup	Blood[487]
Loop method	Serum[757]
Direct ET-AAS with or without dilution	Blood,[421,500,686,762,920,1561] erythrocytes,[1548] neutrophils,[1548] lymphocytes,[1548] saliva,[379,1571] serum,[27,762,1550,1551,1565] urine[27,438,920,1550]
ET-AAS after digestion	Blood,[438,686,727] erythrocytes,[727] hair,[426] nails,[1543] liver biopsies,[515] tissues,[373] tooth enamel,[84] tooth surface enamel biopsies[1050]
ET-AAS after TAAH solubilization	Plasma[610] tissues[610]
Direct ET-AAS (solid sampling)	Bone,[1099] ''Bovine Liver'' (NBS SRM),[454] fish,[655] hair,[778] liver,[452,454] mussels,[86] nails,[1543] ''Oyster Tissue'' (NBS SRM),[455] teeth,[1099] tissues[655]
Miscellaneous AAS techniques (and procedures requiring special apparatus or separation procedures)	Blood fractions,[1548] CSF,[1542] fish,[655] leukocytes,[158] liver,[445,452] (rat) liver fractions,[793] milk (Zn-binding study),[1564,1572,1573] plasma,[785] plasma (Zn-binding study),[752] plasma/serum protein fractions,[751,753,755,785-787,1549,1574] saliva protein fractions,[379,1575,1576] tissues[373,417]

curacies deserve attention: (1) minor but persistent contamination and reagent blanks; and (2) optimization of instrumental parameters and calibration mode.

There are numerous sources of extraneous Zn additions: rubber (stoppers,[1578] gloves, etc.); some batches of blood storage tubes;[420,421,1579,1580] needles (during sampling of CSF[621]); heparin;[656] hemolysis (in serum/plasma and CSF assays[656,1542,1560,1579]); cells, food, debris, and stimulants (in saliva collection[1571]); hemodialysis (elevated serum Zn[1557,1581]); hair cosmetic treatments (shampoos,[9] permanent wave lotions,[734] etc.); airborne dust;[1560] porcelain crucibles;[1570] ordinary grinding procedures (in agate, alumina, corundum, etc.[624]); and reagents, water, analytical utensils, pipet tips, etc. These potential sources of errors are controlled by clean sampling[106,1544,1582] and grinding/homogenation[426,532,533,944] procedures as well as appropriately precleaned labware,[1542,1544] It might be better to analyze plasma instead of serum,[656,751,1560] both of which are now proved to be identical in their Zn content.[1551,1583] Every batch of evacuated blood-collection tubes should be examined by leaching two or three tubes with pooled preanalyzed plasma.[420] It is noteworthy that blood specimens have undergone significant contamination during storage in commercially cleaned glass and plastic containers within 120 hr of storage at all storage temperatures (-70, -10, 4, and 22°C).[421]

Some small or spot samples may not be truly representative. Although this problem is not as severe with Zn as with many other pollutants, it should be taken into consideration while obtaining such specimens and/or interpreting their Zn content. This would seem to refer to bone, food, feces,[96] saliva,[1571,1577] nails,[1543] certain tissues (kidney,[372] brain[663,740]), autopsy tissues which may have undergone post-mortem changes,[789] aged urine specimens containing Zn-enriched sediments,[165] plasma/serum specimens obtained at different hours of the day[1540,1583,1584] or in different body positions,[620,1583] analyses of single hairs,[101,1546] as well as to sweat specimens obtained in different conditions and with different sampling techniques.[381]

For long-term storage, specimens are kept frozen (at $-20°C$) in precleaned plastic ware; no changes have been observed with serum/plasma (under storage from 2 weeks[1580] to 1 year[1554]), saliva (at least for 5 days), urine, etc.

Washing of hair[10,31,48,732-735,815] and nails[290,692,738] prior to analysis still appears to be a matter of controversy. There are extreme approaches, from "no-wash"[393,625] to severe treatments with chelating agents or dilute acids. It has been proved that endogenous Zn is liable to leaching from hair, and, therefore, treatments with dilute acids,[466,735] acidified (pH <6) detergents,[735] and aqueous EDTA[734,815] should be rejected. Measurable extraction by detergents takes place, as well,[692,735] while organic solvents appear ineffective (with nails[692]). The IAEA procedure (acetone-water ($3\times$)-acetone)[48] or water wash under sonification[290] would be the best current choice with hair, while nails need mechanical scraping, as well.[290,692]

All drying techniques — lyophilization,[166,167,536] oven drying at 110 to 120°C,[166,423] and microwave-assisted (rapid) drying[535] — are applicable to biomaterials prior to Zn assays.

A few direct ET-AAS applications to solid biological samples — milligram amounts of bone, nails, teeth, tissues, etc. (Table 2) — have been described. While these direct assays look attractive, they involve major problems (nonrepresentative sampling, matrix effects, background, matrix-residue accumulation, complicated calibration, etc.). A less sensitive line, Zn, 307.6, nm, may be used. Peak-area calibration,[86,1099,1543] standard additions,[86,454,455] and L'vov platform[454,455] have all proved useful. Nevertheless, typical Zn assays in solid biomaterials are better performed after the analyte is put into solution by digestion, acid extraction, or TAAH solubilization.

All common wet-digestion procedures are applicable, provided clean ware and reagents are used and certain safety precautions (in working with $HClO_4$, H_2O_2, and pressurized vessels) are observed. Apparently successful are digestions with HNO_3 and H_2SO_4;[7,97,458,730,737] HNO_3 and $HClO_4$;[248,392,541,732,734,738,740,1541,1567] HNO_3, $HClO_4$, and H_2SO_4;[190,332,442] HNO_3 and (finally) H_2O_2;[530,853] H_2SO_4 and H_2O_2;[368,793,1570] HNO_3, H_2SO_4, and (finally) H_2O_2;[170,462,475,690] etc. Rapid digestion procedures with pressurized vessels[90,458,494,518,545,1109] or in microwave oven[277] have been described.

Inasmuch as Zn is easily put into solution, there are simplified preparation techniques which involve a single-acid treatment or incomplete digestion (e.g., with HNO_3 and HCl: food, feces, meat, and tissues,[729] liver,[319] etc.). Many single-acid procedures are documented: HNO_3 (blood,[686,726] bone,[90,728] feces,[729] food,[729] hair,[31,425,460,466,557,666,735] kidney cortex,[373] meat,[517,729] milk,[729] nails,[1543] liver,[515,686,1109] teeth,[545] tooth enamel,[24,50] and tissues[467,517,518,545,718,729]); HCl (diet,[1569] feces,[1569] tooth enamel[84]); $HClO_3$ (brain, hair[662]); and $HClO_4$(CSF[1542]). Final acidity should approximately be known so as to match the standard solutions (say, in 5% v/v of H_2SO_4[730] or in any other dilute acid [$<10\%$ v/v)].

Ashing in O_2[133,288,725] and LTA[169,275,438,462,626] are clean decomposition techniques which quantitatively recover Zn; only Behne and Matamba[167] found small (-11%) but significant ($p<0.001$) losses of serum zinc on LTA vs. lyophilization and INAA.

Dry ashing is less satisfactory and its use would be rarely justified (say, with larger and/or fatty samples, milk, bone, teeth). Volatization losses are not probable up to 500°C, but partial retention of Zn on crucible walls or in silica and other insoluble residues may be encountered. Thorough treatment of the ashes with aqua regia[370,538,1570] or evaporation with HNO_3 or HCl[334,374,442,538,739] is practiced so as to improve Zn recovery. Sometimes, ashing aids are added [$Mg(CH_3COO)_2$,[190] $Mg(NO_3)_2$,[442] $CaCO_3$,[1570] K_2SO_4,[538] HNO_3,[442,538,739] H_2SO_4,[91] etc.], but they may also give rise to intolerable blanks.

Solubilization with detergent solutions is useful in analyses of hair,[464] dried milk,[712] tissues,[464,521,610,741,1568] etc.,[521,610] but not fish meal.[1568] TAAH solutions (aqueous or in organic solvents, e.g. Lumatom®, Soluene®, etc.) are often used; standard addition calibration is advisable.

Preconcentration is rarely if ever needed in Zn assays, as the analyte levels are relatively high. Zn(II) is extracted from digested biomaterials,[39,332,334,473,474,482,494,530,659,690,976,1570] pH-adjusted urine,[1561] TCA supernatants of deproteinized blood[526] or serum,[640,657,719] as well as from milk (in a Zn-binding study[1564])with dithiocarbamates (at pH 6 to 8, with APDC,[334,526,1570] DDDC,[690] SDDC,[494] HHDC,[39,640,657,659,719] etc.), dithizone (at pH 9 to 9.5, citrate added),[474,530] 8-hydroxyquinoline (at pH 8),[665] 1-(2-pyridylazo)-2-naphthol (pH ~7),[1561] diphenylcarbazone-pyridine (pH 5.5 to 7.8),[473,482] as well as ion-association Zn(II)-chloride[332] or -bromide[976] complexes.

Other preconcentration techniques are also documented — ion exchange,[412,1542] coprecipitation,[446] etc. — but their use would seem somewhat out of date.

Most liquid samples can actually be analyzed following simple dilution. Serum is diluted with water or dilute acid (e.g., 0.1 *M* HCl or HNO_3)[1540] or else with mixed water-organic solvents (e.g., 10% v/v propanol[714] or 6% v/v butanol[664,702,1544,1556]). It is essential to use viscosity-matched standards unless at least eight- to tenfold dilution is made.[57,702,704,723,1556,1558] Lesser dilution ratios are often practiced — 1 + 1[700,701,718,1555,1581] or 1 + 4[27,1554,1560,1567,1583] — so as to measure higher absorbance signals. Viscosity of standards should be increased by adding 0.03% polyvinylalcohol,[700,701] or 3% dextran,[1555] or else in 3% bovine albumin[718](with 1 + 1 diluted sera), or in 5% glycerol (with 1 + 4 diluted sera).[27,721,1554,1560,1567] Some authors use even more complex standard solutions, say, with added electrolytes: NaCl, or NaCl + KCl, or even NaCl + KCl + $CaCl_2$ + $MgCl_2$.[721] Use of such composite standards does not necessarily improve accuracy.[722] The point is that the effect of viscosity depends on the diameter[1559] and length of the nebulizer capillary, the effect of solution surface tension may also be involved,[703] together with some iffy factors like nebulizer adjustment and age, spray chamber construction, observation height, fuel:oxidant ratio, etc. It is recommended to check every adopted procedure by standard additions and, if possible, use higher dilution factors. Pulse-nebulization technique (injection or dipping) is very useful and can be run automatically.[58,704,705] It is applicable to serum analyses (Tables 2 and 3), separated serum protein fractions,[751,787] urine, CSF, and tissue digests.[417,760,761] TCA precipitation of serum proteins is hardly needed (moreover, it is likely to involve inaccuracies[722,878,1556,1565]).

The main problem in all flame AAS assays is to account for the numerous minor, but persistent sources of errors. Peak-area[704] and standard additions calibration can improve precision and accuracy to a certain degree. Another option is to use matrix-matched standards, viz. pooled serum or urine which have been stripped out of Zn and then spiked with known additions. Such low-Zn pools can be prepared by batch sorption on Chelex® 100 resin[722,1551] or else by dialysis[1548,1550] and may be stored deep frozen. They have proved very useful in serum analyses by ET-AAS.[1548,1550,1551]

IV. CONCLUSION

Flame AAS is the best routine method for the determination of Zn in biological samples. Viscosity- or acid-matched standards are required in analyses of diluted sera or biological digests, respectively. Background check is always needed. Pulse-nebulization AAS is in most cases preferable to ET-AAS. Minor sources of inaccuracy persist in flame technique

Table 3
SELECTED AAS PROCEDURES FOR Zn DETERMINATION IN BIOLOGICAL SAMPLES

Sample	Brief procedure outline	Ref.
Blood, erythrocytes	20-fold dilution with 0.05% Triton® X-100; flame AAS	1552
Blood	1:160 dilution with water; ET-AAS: 110, 500, and 2300°C; standard addition calibration; likewise Reference 500	421
	400-fold dilution with water; ET-AAS: 100, 840, and 2200°C	1561
Bone	HNO_3 digestion (open vessel or autoclave[a]); flame AAS; background correction; acid-matched standards	90, 728
"Bovine Liver" (NBS SRM)	2-mg sample (dry weight) digested in a closed teflon® vessel with HNO_3[a], Pulse-nebulization flame AAS; acid-matched standards	760, 761
Brain, muscle, tissues	TAAH solubilization: 0.5—1 mℓ of Soluene® 100 containing 2% w/v of APDC per 100 mg of tissue at 60°C overnight or 24 hr, and dilution with toluene; flame AAS; standard addition calibration; ET-AAS option[610]	741
CSF	1 + 1 dilution with water; flame AAS[b]	709
	1 + 9 dilution with 6% v/v n-butanol; flame AAS; standards and blank containing 0.15 *M* NaCl and 6% v/v n-butanol	1569
Erythrocytes	Specimens lysed with 2 vol of water, and diluted; ET-AAS: 110 (ramp), 400 (ramp), 2100 (miniflow or gas stop), and 2700°C (clean-out). Matrix-matched standards	1548
Feces, food	0.5-g sample mixed with 6 mℓ of 1 *M* HCl and left for 24 hr, 4 mℓ of $CHCl_3$ and 2 mℓ of CH_3OH added, mixed, and centrifuged; supernatant analyzed by flame AAS, standards containing HCl, $CHCl_3$, and CH_3OH	1569
Feces, food, meat, milk	1 to 2-g sample boiled for 30 min with 15 mℓ of a (27:3:20 by vol) mixture of $HCl\text{-}HNO_3\text{-}H_2O$; diluted to 50 mℓ; flame AAS[b]	729
Fish, liver	Digestion with HNO_3 and (finally) H_2O_2;[a] flame AAS[b]	853
Food, (fish, meat, milk, etc.)	$HNO_3\text{-}H_2SO_4$ digestion, flame AAS; standards in ~5% v/v H_2SO_4[b]	730
Food, fish, milk, etc.	HNO_3 digestion in a teflon®-lined autoclave;[a] an aliquot completely digested in an open vessel with $HClO_4$[a] and diluted with 1% $HClO_4$; flame AAS; acid-matched standards	250
Hair	HNO_3 digestion; flame AAS;[b] standards in 10% v/v HNO_3	460
Hair, nails	HNO_3 digestion, completed with $HClO_4$;[a] flame AAS; acid-matched standards	732, 738
Kidney cortex	HNO_3 digestion; Zeeman ET-AAS	373
Liver (biopsies)	5-mg sample (dry weight) digested with 1mℓ of HNO_3 at 80°C for 24 hr slowly evaporated, 2 mℓ of 0.01 *M* HNO_3 added, and vigorously mixed; flame AAS; standards in 0.01 *M* HNO_3	515
Liver, meat, muscle	1-g sample heated with 10 mℓ of (1 + 1) HNO_3 at 80°C, evaporated to ca. 3 mℓ, and diluted to 25 mℓ; flame AAS; acid-matched standards	517
Milk	1 + 3 dilution with detergent (1% saponin or 0.2% Meriten (nonylphenyl polyglycolether) or 0.2% Na dodecylbenzenesulfonate); flame AAS; standard in detergent solution; applicable to evaporated milk[712]	710
Milk (zinc-binding study)	Centrifuged, fat-free sample diluted with Tris buffer, and gel filtrated on columns of Sepharose® 2B[1573] and/or Sephadex® G75[1572,1573]	1573
Nails	$HNO_3\text{-}H_2SO_4$ digestion; flame AAS; standards in H_2SO_4	737
Plasma	1 + 4 dilution with water; flame AAS; standards in aqueous 5% v/v glycerol; a three-slot burner, fuel-rich flame, and optimized burner height essential	1554, 1560
Plasma/serum	100-fold dilution with water; ET-AAS: ramp drying to 95—100°C, ramp ashing to 400—450°C, atomization at 2100—2400°C (gas stop advisable); standards on dialyzed plasma pool	1548, 1550

Table 3 (continued)
SELECTED AAS PROCEDURES FOR Zn DETERMINATION IN BIOLOGICAL SAMPLES

Sample	Brief procedure outline	Ref.
Plasma/serum	200-fold dilution with water; ET-AAS: 78, 103, 500, and 2231°C; standards on low-Zn serum (pretreated with Chelex® 100 for 24 hr)	1551
Saliva, saliva protein fractions	2- to 4-fold dilution; ET-AAS: 100—120° 400—600, and 2100—2300, and clean-out at 2500—2600°C; background correction; standard addition calibration; saliva protein fractions[379] separated by gel filtration on Sephadex® G-150	379, 1571
Serum	1 + 9 dilution with water; flame AAS; standards in 1% v/v HNO_3[27,1558] or in 14 m*M* NaCl-0.5 m*M* KCl[1556]	27, 1556, 1558
	Pulse-nebulization of 10-fold diluted sera; automated dipping method or manual injection; background correction; peak-area calibration; aqueous standard solutions	704
	1 + 1 dilution with water; pulse-nebulization flame AAS (50- or 100-μℓ injections); standard solutions in 0.03% w/v polyvinylalcohol (Merck-Schuchardt); background correction, alternatively, calibration vs. control sera	700, 701
Serum (speciation)	An eq vol of 20% w/w polyethylene glycol 6000 added to serum; albumin-bound Zn determined in the supernatant by flame AAS; the α_2-macroglobulin-bound Zn (in ppt) calculated by difference between total Zn and albumin-bound Zn	1574
Serum protein fractions	Gel filtration on Sephadex® G-100; ET-AAS	786
Urine	1 + 4 dilution; flame AAS; three-slot burner; standards on pooled or synthetic urine; background correction	7, 27
	100-fold dilution with water; ET-AAS; 95 (10 + 50 sec), 450 (15 + 15 sec), and 2400°C; background correction; standard addition calibration	1550

[a] Safety precautions mandatory.
[b] Background and standard addition checks advisable.
[c] Good ventilation essential.

and call for optimization of instrumental parameters, appropriate calibration, and good quality control. There are many sources of external contamination. Simple (wet) pretreatment procedures are advisable, e.g., single-acid digestions.

ET-AAS entails major problems: background, matrix effects, premature volatization, sporadic contamination, etc. Matrix-matched calibration, matrix modification, and versatile programming are needed. L'vov platform atomization and, more generally, the STPF concept are promising.

REFERENCES

1. **Allain, P. and Mauras, Y.,** Determination of Al in blood, urine, and water by ICP-ES, *Anal. Chem.*, 51, 2089, 1979.
2. **Alfrey, A. C., Le Gendre, G. R., and Kaehny, W. D.,** The dialysis encephalopathy syndrome. Possible Al intoxication, *New Engl. J. Med.*, 294, 184, 1976.
3. **Hamilton, E. I., Minski, M. J., and Cleary, J. J.,** Concentration and distribution of some stable elements in healthy human tissues from the United Kingdom. An environmental study, *Sci. Total Environ.*, 1, 341, 1973.
4. **Crapper, D. R., Krishnan, S. S., and Quittkat, S.,** Al, neurofibrillary degeneration and Alzheimer's disease, *Brain*, 99, 67, 1976.
5. **Clavel, J. P., Jandon, M. C., and Galli, A.,** Estimation of Al in biological liquids by AAS in a graphite oven, *Ann. Biol. Clin.*, 36, 33, 1978.
6. **Pegon, Y.,** Determination of Al in biological liquids by flameless AA, *Anal. Chim. Acta*, 101, 385, 1978.
7. **Berman, E.,** *Toxic Metals and Their Analysis*, Heyden, London, 1980.
8. International Commission on Radiological Protection (ICRP), *Report of the Task Group on Reference Man*, Pergamon Press, Oxford, 1975.
9. **Gordus, A.,** Factors affecting the trace-metal content of human hair, *J. Radioanal. Chem.*, 15, 229, 1973.
10. **Ryan, D. E., Holzbecher, J., and Stuart, D. C.,** Trace elements in scalp-hair of persons with multiple sclerosis and of normal individuals, *Clin. Chem.*, 24, 1996, 1978.
11. **Takeuchi, T., Hayashi, T., Takada, J., Koyama, M., Kozuka, H., Tzuji, H., Kusaka, Y., Ohmori, S., Shinogi, M., Aoki, A., Katayama, K., and Tomiyama, T.,** Survey of trace elements in hair of normal Japanese, in *Nuclear Activation Techniques in the Life Sciences 1978*, IAEA, Vienna, 1979, 545.
12. **Yukawa, M., Amano, K., Suzuki-Yasumoto, M., and Terai, M.,** Distribution of trace elements in the human body determined by NAA, *Arch. Environ. Health*, 35, 36, 1980.
13. **Iyengar, G. V., Kollmer, W. E., and Bowen, H. J. M.,** *The Elemental Concentration of Human Tissues and Fluids*, Verlag Chemie, New York, 1978.
14. **Angelieva, R. and Syarova, D.,** A study of the mineral content of human milk, *Khig. Zdraveopaz.*, 11, 376, 1968.
15. **Masironi, R., Koirtyohann, S. R., Pierce, J. O., and Schanschula, R. G.,** Calcium content of river water, trace elements concentrations in toenails, and blood pressure in village population in New Guinea, *Sci. Total Environ.*, 6, 41, 1976.
16. **Harrison, W. W. and Clemena, G. G.,** Survey analysis of trace elements in human fingernails by SSMS, *Clin. Chim. Acta*, 36, 485, 1972.
17. **Alderman, F. R. and Gitelman, H. J.,** Improved electrothermal determination of Al in serum by AAS, *Clin. Chem.*, 26, 258, 1980.
18. **Oster, O.,** Al content of human serum determined by AAS with a graphite furnace, *Clin. Chim. Acta*, 114, 53, 1981.
19. **Kaehny, W. D., Hegg, A. P., and Alfrey, A. C.,** Gastrointestinal absorption of Al from Al-containing antacids, *New Engl. J. Med.*, 296, 1389, 1977.
20. **Wawschinek, O.,** The determination of Al in human plasma, 8th Int. Microchem. Symp., Graz, Austria, August 25 to 30, 1980, 30.
21. **Consolazio, C. F., Nelson, R. A., Matoush, L. R. O., Hughes, R. C., and Urone, P.,** *Trace Mineral Losses in Sweat*, Rep. No. 284, Fitzsimmons General Hospital, Denver, 1964.
22. **Derise, N. L. and Ritchey, S. J.,** Mineral composition of normal human enamel and dentine and the relation of composition to dental caries. II. Microminerals, *J. Dent. Res.*, 53, 853, 1974.
23. **Retieff, D. H., Cleaton-Jones, P. E., Turkstra, J., and De Wet, W. J.,** The quantitative analysis of 16 elements in normal human enamel and dentine by NAA and high-resolution gamma spectrometry, *Arch. Oral Biol.*, 16, 1257, 1971.
24. **Lakomaa, E.-L. and Rytömaa, I.,** Mineral composition of enamel and dentin of primary and permanent teeth in Finland, *Scand. J. Dent. Res.*, 85, 89, 1977.
25. **Losee, F. L., Cutress, T. W., and Brown, R.,** Natural elements of the periodic table in human dental enamel, *Caries Res.*, 8, 123, 1973.
26. **Losee, F. L., Curzon, M. E. J., and Little, M. F.,** Trace element concentrations in human enamel, *Arch. Oral Biol.*, 19, 467, 1974.
27. **Pozzoli, L., Minoia, C., and Angeler, S.,** Metodi di analisi per la determinazione dei metalli nei liquidi biologici, *Aurora Leg. Cart. di C. Ge*, Pavia, 1976.
28. **Valentin, H., Preusser, P., and Schaller, K.-H.,** The analysis of Al in serum and urine for monitoring of exposed persons, *Int. Arch. Occup. Environ. Health*, 38, 1, 1976.
29. **Barnes, R. M.,** Recent advances in emission spectroscopy: ICP discharges for spectrochemical analysis, *CRC Crit. Rev. Anal. Chem.*, 7, 203, 1978.

30. **Kniseley, R. N., Fassel, V. A., and Butler, C. C.,** Application of ICP excitation sources to the determination of trace metals in μL volumes of biological fluids, *Clin. Chem.,* 19, 807, 1973.
31. **Assarian, G. S. and Oberleas, D.,** Effect of washing procedures on trace-element content of hair, *Clin. Chem.,* 23, 1771, 1977.
32. **Greenfield, S. and Smith, P. B.,** The determination of trace metals in μL samples by plasma torch excitation with special reference to oil, organic compounds, and blood samples, *Anal. Chim. Acta,* 59, 341, 1972.
33. **Lichte, F. E., Hopper, S., and Osborn, T. W.,** Determination of Si and Al in biological matrices by ICP-ES, *Anal. Chem.,* 52, 120, 1980.
34. **Schramel, P., Wolf, A., and Klose, B.-J.,** Determination of Al in blood serum by ICP spectroscopy, *J. Clin. Chem. Clin. Biochem.,* 18, 591, 1980.
35. **Schramel, P., Klose, B.-J., and Hasse, S.,** Efficiency of ICP-ES for the determination of trace elements in bio-medical and environmental samples, *Fresenius Z. Anal. Chem.,* 310, 209, 1982.
36. **Karyakin, A. V. and Gribovskaya, I. F.,** *Emission Spectroscopic Analysis of Biological Samples,* Khimia, Moscow,1979.
37. **Kehoe, R. A., Cholak, J., and Story, R. V.,** Spectrochemical study of the normal ranges of concentration of certain trace metals in biological materials, *J. Nutr.,* 19, 579, 1940.
38. **Scoblin, A. P. and Belous, A. M.,** *Microelements in Bone Tissue,* Medizina, Moscow, 1968.
39. **Pokrovskaya, E. I., Puzanova, O. G., and Tereshtschenko, A. P.,** Determination of microelements in human diet and excreta by use of extraction, *Kosm. Biol. Med.,* 5, 66, 1972.
40. **Murthy, G. K.,** Trace elements in milk, *CRC Crit. Rev. Environ. Control,* 4, 1, 1974.
41. **Niedermeier, W.,** Analytical emission spectroscopy in biomedical research, in *Applied Atomic Spectroscopy,* Vol. 2, Grove, E. L., Ed., Plenum Press, New York, 1978, 219.
42. **Niedermeier, W., Griggs, J. K., and Johnson, R. S.,** Emission spectrometric determination of trace elements in biological fluids, *Appl. Spectrosc.,* 25, 53, 1971.
43. **Aleksandrov, S., Konstantinova, A., and Tsalev, D. L.,** Comparative investigation of the spectrochemical, AA, and NAA for the determination of Cu, Zn, Cd, Mn, Pb, and Al in biological sample, *God. Sofii. Univ. Khim. Fac.,* 69, 153,1974/1975.
44. **Tipton, I. H., Cook, M. J., Steiner, R. L., Boye, C. A., Perry, H. M., Jr., and Schroeder, H. A.,** Trace elements in human tissue. I. Methods, *Health. Phys.,* 9, 89, 1963.
45. **Bowen, H. J. M.,** Application of activation techniques to biological analysis, *CRC Crit. Rev. Anal. Chem.,* 10, 127, 1980.
46. **Garmestani, K., Blotcky, A. J., and Rack, E. P.,** Comparison between NAA and graphite furnace AAS for trace Al determination in biological material, *Anal. Chem.,* 50, 144, 1978.
47. **Tanner, J. T. and Friedman, M. H.,** NAA for trace elements in foods, *J. Radioanal. Chem.,* 37, 529, 1977.
48. **Ryabukhin, Yu. S., Ed.,** *Activation Analysis of Hair as an Indicator of Contamination of Man by Environmental Trace Element Pollutants,* IAEA, Vienna, 1978.
49. **Versieck, J. and Cornelis, R.,** Normal levels of trace elements in human blood plasma or serum, *Anal. Chim. Acta,* 116, 217, 1980.
50. **Derise, N. L., Ritchey, S. J., and Furr, A. K.,** Mineral composition of normal human enamel and dentin and the relation of composition to dental caries. I. Macrominerals and comparison of methods of analyses, *J. Dent. Res.,* 53, 847, 1974.
51. **Turkstra, J.,** The multi-element analysis of teeth and other biological materials by instrumental neutron activation, in *The Analysis of Biological Materials,* Butler, L. R. P., Ed., Pergamon Press, Oxford, 1979, 49.
52. **Gooddy, W., Williams, T. R., and Nicholas, D.,** SSMS in investigation of neurological disease. I. Multielement analysis in blood and CSF, *Brain,* 97, 327, 1974.
53. **Gregory, N. L.,** Interfering ions in the elemental analysis of biological samples by mass spectrometry, *Anal. Chem.,* 44, 231, 1972.
54. **Yurachek, J. P., Clemena, G. G., and Harrison, W. W.,** Analysis of human hair by SSMS, *Anal. Chem.,* 41, 1666, 1969.
55. **Wostenholme, W. A.,** Analysis of dried blood plasma by SSMS, *Nature (London),* 203, 1284, 1964.
56. *Analytical Methods for Atomic Absorption Spectrophotometry,* Perkin-Elmer, Norwalk, Conn., 1976.
57. **Price, W. J.,** *Spectrochemical Analysis by Atomic Absorption,* Heyden & Son, London, 1979.
58. *AA-975 Automated Multi-Element Atomic Absorption Systems,* Varian Techtron Pty., Victoria, Australia, 1981.
59. *Analytical Values for AA Spectrophotometers 157/257/551/951,* Instrumentation Laboratory, Wilmington, Mass., 1980.
60. **Fernandez, F. J. and Iannarone, J.,** Operating paramters for the HGA-2200 graphite furnace, *At. Absorpt. Newsl.,* 17, 117, 1978.
61. *GTA-95 Graphite Tube Atomization System,* Varian Techtron, Victoria, Australia, 1981.

61a. **Persson, J.-A., Frech, W., and Cedergren, A.,** Investigation of reactions involved in flameless AA procedures. V. An experimental study of factors influencing the determination of Al, *Anal. Chim. Acta,* 92, 95, 1977.

62. **Matsusaki, K., Yoshino, T., and Yamamoto, Y.,** A method for removal of chloride interference in the determination of Al by AAS with graphite furnace,*Talanta,* 26, 377, 1979.

63. **King, S. W., Wills, M. R., and Savory, J.,** Electrothermal AAS determination of Al in blood serum, *Anal. Chem. Acta,* 128, 221, 1981.

64. **Ranisteano-Bourdon, S., Prouillet, F., and Bourdon, R.,** Estimation of Al and Ga in biological fluids, *Ann. Biol. Clin.,* 36, 39, 1978.

65. **Julshamn, K., Andersen, K.-J., Willassen, Y., and Braekkan, O. R.,** A routine method for the determination of Al in human tissue samples using standard additions and graphite furnace AAS, *Anal. Biochem.,* 88, 552, 1978.

66. **Julshamn, K.,** Inhibition of response by $HClO_4$ in flameless AA, *At. Absorpt. Newsl.,* 16, 149, 1977.

67. **Schramel, P.,** Determination of 8 metals in the international biological standard by flameless AAS, *Anal. Chim. Acta,* 67, 69, 1973.

68. **Toda, W., Lux, J., and Van Loon, J. C.,** Determination of Al in solutions from gel filtration of human serum by electrothermal AAS, *Anal. Lett.,* 13, 1105, 1980.

69. **Le Gendre, G. R. and Alfrey, A. C.,** Measuring picogram amounts of Al in biological tissue by flameless AAA of a chelate, *Clin. Chem.,* 22, 53, 1976.

70. **Kovalchik, T. M., Kaehny, W. D., Hegg, A. P., Jackson, J. T., and Alfrey, A. C.,** Al kinetics during hemodialysis, *J. Lab. Clin. Med.,* 92, 712, 1978.

71. **Gorsky, J. E. and Dietz, A. A.,** Determination of Al in biological samples by AAS with a graphite furnace, *Clin. Chem.,* 24, 1485, 1978.

72. **Thompson, K. C., Gooden, R. G., and Thomerson, D. R.,** A method for the formation of pyrolytic graphite coatings and enhancement by Ca addition technique for graphite rod flameless AAS, *Anal. Chim. Acta,* 74, 289, 1974.

73. **Krishnan, S. S., Gillespie, K. A., and Crapper, D. R.,** Determination of Al in biological material by AAS, *Anal. Chem.,* 44, 1469, 1972.

74. **Krishnan, S. S., Quittkat, S., and Crapper, D. R.,** AAA for traces of Al and V in biological tissue. A critical evaluation of the graphite furnace atomizer, *Can. J. Spectrosc.,* 1, 25, 1976.

75. **Katsumura, K., Ishizaki, M., Sesamoto, K., Ueno, S., and Hosogai, Y.,** Al determination in foods by AAA, *Ibaraki-ken Eisei Kenkyusho Nempo,* 12, 25, 1974.

76. **Cresser, M. S.,** *Solvent Extraction in Flame Spectroscopic Analysis,* Butterworths, London, 1976.

77. **Langmyhr, F. J. and Tsalev, D. L.,** AAS determination of Al in whole blood, *Anal. Chim. Acta,* 92, 79, 1977.

78. **Alder, J. F., Alger, D., Samuel, A. J., and West, T. S.,** Design and development of a multichannel AA spectrometer for simultaneous determination of trace metals in hair, *Anal. Chim. Acta,* 87, 301, 1976.

79. **Alder, J. F., Samuel, A. J., and West, T. S.,** Single element determination of trace metals in hair by carbon-furnace AAS, *Anal. Chim. Acta,* 87, 313, 1976.

80. **Alder, J. F., Samuel, A. J., and West, T. S.,** The anatomical and longitudinal variations of trace element concentration in human hair, *Anal. Chim. Acta,* 92, 217, 1977.

80a. **Yokel, R. A.,** Hair as an indicator of excessive Al exposure, *Clin. Chem.,* 28, 662, 1982.

80b. **Alder, J. F. and Batoreu, M. C. C.,** Ion-exchange resin beads as solid standards for the electrothermal AAS determination of metals in hair, *Anal. Chim. Acta,* 135, 229, 1982.

81. **Fuchs, C., Brasche, M., Paschen, K., Nordbeck, H., and Quellhorst, E.,** Determination of Al in serum by flameless AA, *Clin. Chim. Acta,* 52, 71, 1974.

82. **Smeyers-Verbeke, J., Verbeelen, D., and Massart, D. L.,** Te determination of Al in biological fluids by means of grahite furnace AAS, *Clin. Chim. Acta.,* 108, 67, 1980.

83. **Dolinšek, F., Štupar, J., and Špenko, M.,** Determination of Al in dental enamel by the carbon cup AA method, *Analyst,* 100, 884, 1975.

84. **Reda, A., Srinvasan, B. N., and Brudevold, F.,** Trace elements in microsamples of human enamel, *J. Dent. Res.,* 52, 126, 1973.

85. **Grégoire, D. C. and Chakrabarti, C. L.,** Atomization from a platform in graphite furnace AAS, *Anal. Chem.,* 49, 2018, 1977.

86. **Lord, D. A., McLaren, J. W., and Wheeler, R. C.,** Determination of trace metals in fresh water mussels by AAS with direct solid sample injection, *Anal. Chem.,* 49, 257, 1977.

87. **Schramel, P.,** The application of peak integration in flameless AAS, *Anal. Chim. Acta,* 72, 414, 1974.

88. **McDermot, J. K. and Whitehill, I.,** Determination of Al in biological tissues by flameless AAS, *Anal. Chim. Acta,* 85, 195, 1976.

89. **Thornton, D. J., Todd, D. M., Lott, J. A., and Liss, L.,** Stability of serum Al concentration during storage, Paper 508, Jt. Meet. AACC and CSCC, Boston, July 20 to 25, 1980.

90. **Tsalev, D. L.,** unpublished data, 1982.

91. **Gorsuch, T. T.,** *The Destruction of Organic Matter,* Pergamon Press, Oxford, 1970.
92. **Julshamn, K., and Andersen, K.-J.,** A study on the digestion of human muscle biopsies for trace metal analysis using an organic tissue solubizer, *Anal. Biochem.,* 98, 315, 1979.
93. **Ishizaki, M., Kataoka, F., Murakami, R., Fujiki, M., and Yamaguchi, S.,** Determination of Sb in blood by flameless AAS using carbon tube atomizer, *Sangyo Igaku,* 19, 510, 1977.
94. **Ishizaki, M., Ueno, S., Oyamada, N., Kataoka, F., Murakami, R., Kubota, K., Katsumura, K., and Noda, M.,** Determination of 13 elements (As, Cd, Co, Cu, Fe, Hg, Mn, Ni, Pb, Sb, Se, V, Zn) in human blood, *Ibaraki-ken Eisei Kenkyusho Nempo,* 15, 71, 1977.
95. **Van Montfort, P. F. E., Agterdenbos, J., and Jutte, B. A. H. G.,** Determination of Sb and Te in human blood by microwave-induced emission spectrometry, *Anal. Chem.,* 51, 1553, 1979.
96. **Clemente, G. F., Cigna Rossi, L., and Santaroni, G. P.,** Trace element intake and excretion in the Italian population, *J. Radioanal. Chem.,* 37, 549, 1977.
97. **Lindh, U., Brune, D., Nordberg, G., and Wester, P.-O.,** Levels of Sb, As, Cd, Cu, Pb, Hg, Se, Ag, Sn, and Zn in bone tissue of industrially exposed workers, *Sci. Total Environ.,* 16, 109, 1980.
98. **Clemente, G. F., Cigna Rossi, L., and Santaroni, G. P.,** Trace element composition of hair in the Italian population, in *Nuclear Activation Techniques in the Life Sciences 1978,* IAEA, Vienna, 1979, 527.
99. **Lanzel, E.,** Activation analysis of environmental metals in human head hair, *J. Radioanal. Chem.,* 58, 347, 1980.
100. **Obrusnik, I., Gislason, J., Maes, D., McMillan, D. K., D'Auria, J., and Pate, B. D.,** The variation of trace element concentrations in single human head hairs, *J. Radioanal. Chem.,* 15, 115, 1973.
101. **Arunachalam, J., Gangadharan, S., and Yegnasubramanian, S.,** Elemental data on human hair sampled from Indian student population and their interpretation for studies in environmental exposure, in *Nuclear Activation Techniques in the Life Sciences 1978,* IAEA, Vienna, 1979, 499.
102. **Korobenkova, M. M., Pelekis, L. L., and Tsirkunova, I. E.,** INAA study of Se, Hg, Sc, Cr, Co, Fe, Zn, and Sb in hair of Byelorussian inhabitants, in *Nuclear Analytical Methods in Environmental Control,* Gidrometeoizdat, Leningrad, 1980, 188.
103. **Liebscher, K. and Smith, H.,** Essential and nonessential trace elements. A method of determining whether an element is essential or nonessential in human tissue, *Arch. Environ. Health,* 17, 881, 1968.
104. **Jervis, R. E., Tiefenbach, B., and Chattopadhyay, A.,** Scalp hair as a monitor of population exposure to environmental pollutants, *J. Radioanal. Chem.,* 37, 751, 1977.
105. **Brune, D., Nordberg, G. F., Wester, P. O., and Bivered, B.,** Accumulation of heavy metals in tissues of industrially exposed workers, in *Nuclear Activation Techniques in the Life Sciences 1978,* IAEA, Vienna, 1979, 643.
106. **Brune, D., Nordberg, G., and Wester, P. O.,** Distribution of 23 elements in the kidney, liver and lungs of workers from a smeltery and refinery in North Sweden exposed to a number of elements and of a control group, *Sci. Total Environ.,* 16, 13, 1980.
107. **Plantin, L.-O.,** Trace elements in cardiovascular diseases. Contribution to the joint WHO/IAEA project, in *Nuclear Activation Techniques in the Life Sciences 1978,* IAEA, Vienna, 1979, 321.
108. **Lievens, P., Versieck, J., Cornelis, R., and Hoste, J.,** The distribution of trace elements in normal human liver determined by semi-automated RNAA, *J. Radioanal. Chem.,* 37, 483, 1977.
109. **Molokhia, M. M. and Smith, H.,** Trace elements in the lung, *Arch. Environ. Health,* 15, 745, 1967.
110. **Byrne, A. R., Kosta, L., Ravnik, V., Štupar, J., and Hudnik, V.,** A study of certain trace elements in milk, in *Nuclear Activation Techniques in the Life Sciences 1978,* IAEA, Vienna, 1979, 255.
111. **Wester, P. O.,** Trace elements in serum and urine from hypertensive patients before and during treatment with chlorthalidone, *Acta Med. Scand.,* 194, 505, 1973.
112. **Molin, L. and Wester, P. O.,** The estimated daily loss of trace elements from normal skin by desquamation, *Scand. J. Clin. Lab. Invest.,* 36, 679, 1976.
113. **Schroeder, H. A. and Nason, A. P.,** Trace-element analysis in clinical chemistry, *Clin. Chem.,* 17, 461, 1971.
114. **Taylor, P. J.,** Acute intoxication from $SbCl_3$, *Br. J. Ind. Med.,* 23, 318, 1966.
115. **Smith, B. M. and Griffiths, M. B.,** Determination of Pb and Sb in urine by AAS with electrothermal atomization, *Analyst,* 107, 253, 1982.
116. **Schlenz, R.,** Dietary intake of 25 elements by man estimated by NAA, *J. Radioanal Chem.,* 37, 539, 1977.
117. **Olmez, I., Gulovali, M. C., Gordon, G. E., and Henkin, R. I.,** Trace elements in human saliva, *Trace Subst. Environ. Health,* 12, 231, 1978.
118. **Al-Shahristani, H., Shihab, K. M., and Jalil, M.,** Distribution and significance of trace element pollutants in hair of the Iraqi population, in *Nuclear Activation Techniques in the Life Sciences 1978,* IAEA, Vienna, 1979, 515.

119. **Kiem, J., Iyengar, G. V., Borberg, H., Kasperek, K., Siegers, M., Feinendeden, L. A., and Gross, R.,** Sampling and sample preparation of platelets for trace analysis and determination of certain selected bulk and trace elements in normal human platelets by means of NAA, in *Nuclear Activation Techniques in the Life Sciences 1978,* IAEA, Vienna, 1979, 143.
120. **Behne, D., Brätter, P., Gawlik, D., Rösick, U., and Schmelzer, W.,** Combination of protein separation methods and NAA in the determination of protein-bound trace element fractions, in *Nuclear Activation Techniques in the Life Sciences 1978,* IAEA, Vienna, 1979, 117.
121. **Bagdavadze, N. B.,** in *Application of Activation Analysis in Biology and Medicine,* Andronikashvili, E., Ed., Mezniereba, Tbilissi, 1977, 78.
122. **Valente, I., Minski, M. J., and Bowen, H. J. M.,** Rapid determination of Sb in biological and environmental samples using instrumental NAA, *J. Radioanal. Chem.,* 45, 417, 1978.
123. **Morrison, G. H.,** Elemental trace analysis of biological materials, *CRC Crit. Rev. Anal. Chem.,* 8, 287, 1979.
124. **Franke, J. P. and de Zeeuw, R. A.,** DPASV as a rapid screening technique for heavy metal intoxications, *Arch. Toxicol.,* 37, 47, 1976.
125. **Franke, J. P.,** Potentials of ASV for the Toxicological Analysis of Heavy Metals, D.Sc. thesis, State University, Groningen, 1978.
126. **Woidich, H. and Pfannhauser, W.,** Determination of Sb in biological materials and environmental samples by AAS, *Nahrung,* 24, 367, 1980.
127. **Dmitriev, M. T. and Grigor'eva, F. M.,** Determination of Sb in the atmosphere and biomedia by X-ray fluorescence, *Gig. Sanit.,* 10, 52, 1978.
128. **Valković, V.,** *Analysis of Biological Material for Trace Elements Using X-Ray Spectroscopy,* CRC Press, Boca Raton, Fla., 1980.
129. **Seltner, H. D., Linder, H. R., and Schreiber, B.,** Behavior of different oxidation states of As, Sb, Se, and Sn using dithiocarbamates for their separation from environmental, food and drug samples, *Int. J. Environ. Anal. Chem.,* 10, 7, 1981.
130. **Blacklock, E. C. and Sadler, P. A.,** A rapid screening method for heavy metals in biological materials by emission spectroscopy, *Clin. Chim. Acta,* 113, 87, 1981.
131. **Robbins, W. B., Caruso, J. A., and Fricke, F. L.,** Determination of Ge, As, Se, Sn, and Sb in complex samples by hydride generation-microwave-induced plasma AES, *Analyst,* 104, 35, 1979.
132. **Evans, C. A., Jr. and Morrison, G. H.,** Trace element survey analysis of biological materials by SSMS, *Anal. Chem.,* 40, 869, 1968.
133. **Morsches, B. and Tölg, G.,** Contributions to the determination of trace elements in limited amounts of biological material. II. Separation scheme and determination methods, *Fresenius Z. Anal. Chem.,* 250, 81, 1970.
134. **DiGennaro, C. and Muttoni, E.,** Determination of Sb in dry yeast by AAS, *Riv. Zootec. Vet.,* 2, 97, 1979.
135. **Kneip, T. J.,** Analytical method for antimony in air and urine, *Health Lab. Sci.,* 13, 90, 1976.
136. **Collett, D. L., Fleming, D. E., and Taylor, G. A.,** Determination of Sb by stibine generation and AAS using a flame-heated silica furnace, *Analyst,* 103, 1074, 1978.
137. **Ward, R. J., Black, C. D. V., and Watson, G. J.,** Determination of Sb in biological materials by electrothermal AAS, *Clin. Chim. Acta,* 99, 143, 1979.
138. **Barnett, W. B., Kollmer, J. W., and De Nuzzo, S. M.,** The application of electrodeless discharge lamps in atomic absorption, *At. Absorpt. Newsl.,* 15, 33, 1976.
139. **Sinemus, H. W., Melcher, M., and Welz, B.,** *Wertigkeitseinflusse bei der Bestimmung von Sb, As, Se, Te, und Bi mit der Hydrid-AAS-Technik am Beispiel Bodenseewasser,* Bodenseewerk Perkin-Elmer & Co., GmbH, Überlingen, 1980.
140. **Thompson, K. C. and Thomerson, D. R.,** AA studies on the determination of Sb, As, Bi, Ge, Pb, Se, Te, and Sn by utilizing the generation of covalent hydrides, *Analyst,* 99, 595, 1974.
141. **Fiorino, J. A., Jones, J. W., and Capar, S. G.,** Sequential determination of As, Se, Sb, and Te in foods via rapid hydride evolution and AAS, *Anal. Chem.,* 48, 120, 1976.
142. **Thompson, M., Pahlavanpour, B., Walton, S. J., and Kirkbright, G. F.,** Simultaneous determination of trace concentrations of As, Sb, Bi, Se, and Te in aqueous solutions by introduction of the gaseous hydrides into an ICP source in emission spectrometry. I. Preliminary study, *Analyst,* 103, 568, 1978.
143. **Tsujii, K.,** Differential determination of Sb(III) and Sb(V) by non-dispersive AFS using a $NaBH_4$ reduction technique, *Anal. Lett.,* 14, 181, 1981.
144. **Thompson, K. C.,** Atomic fluorescence determination of Sb, As, Se, and Te using the hydride generation technique, *Analyst,* 100, 307, 1075.
145. *ARL Model 341 Hydride Generator,* Applied Research Laboratories, Sunland, Calif., 1982.
146. **Kadeg, R. D. and Christian, G. D.,** Gas chromatographic determination of selected group IV to VI element hydrides, *Anal. Chim. Acta,* 88, 117, 1977.

147. **Scogerboe, R. K. and Bejmuk, A. P.,** Simultaneous determination of As, Ge, and Sb by gas chromatography after hydride generation, *Anal. Chim. Acta,* 94, 297, 1977.
148. **Andreae, M. O., Asmodé, J.-F., Foster, P., and Luc Van't dack,** Determination of Sb(III), Sb(V), and methylantimony species in natural waters by AAS with hydride generation, *Anal. Chem.,* 53, 1766, 1981.
149. **Thompson, M., Pahlavanpour, B., Walton, S. J., and Kirkbright, G. F.,** Simultaneous determination of trace concentrations of As, Sb, Bi, Se, and Te in aqueous solutions by introduction of gaseous hydrides into an ICP source for emission spectrometry. II. Interference study, *Analyst,* 103, 705, 1978.
150. **Hon, P. K., Lau, O. W., Cheung, W. C., and Wong, M. C.,** The AAS determination of As, Bi, Pb, Sb, Se, and Sn with a flame-heated silica T-tube after hydride generation, *Anal. Chim. Acta,* 115, 355, 1980.
151. **Nakashima, S.,** Selective determination of Sb(III) and Sb(V) by AAS following stibine generation, *Analyst,* 105, 732, 1980.
152. **Yamamoto, M., Urata, K., and Yamamoto, Y.,** Differential determination of Sb(III) and Sb(V) by hydride generation AAS, *Anal. Lett.,* 14, 21, 1981.
153. **Yamamoto, M., Urata, K., Murashige, K., and Yamamoto, Y.,** Differential determination of As(III) and As(V), and Sb(III) and Sb(V) by hydride generation AAS, and its application to the determination of these species in sea water, *Spectrochim. Acta,* 36B, 671, 1981.
154. Analytical Methods Committee, Determination of small amounts of Sb in organic matter, *Analyst,* 105, 66, 1980.
155. **Smith, A. E.,** Interferences in the determination of elements that form volatile hydrides with $NaBH_4$ using AAS and the argon-hydrogen flame, *Analyst,* 100, 300, 1975.
156. **Morita, K. and Mishima, M.,** AAS determination of Sb and Se in whole blood by MIBK extraction, *Bunseki Kagaku,* 30, 170, 1981.
157. **Kataoka, F., Ishizaki, M., Ueno, S., Oyamada, N., Murakami, R., Kubota, K., and Katsumara, K.,** Determination of Sb in biological materials by flameless AAS using carbon tube atomizer, *Ibaraki-ken Eisei Kenkyusho Nempo,* 15, 65, 1977.
158. **Dornemann, A. and Kleist, H.,** Determination of nanotraces of Sb in biological samples, *Fresenius Z. Anal. Chem.,* 294, 402, 1979.
159. **Kneip, T. J.,** As, Se, and Sb in urine and air. Analytical method for hydride generation and AAS, *Health Lab. Sci.,* 14, 53, 1977.
160. **Lajunen, L. H. J., Häyrynen, H., Yrjänheikki, E., and Hakala, E.,** Trace element analysis of Sb by AAS using the hydride generation method, Euroanalysis IV, Helsinki/Espoo, August 23 to 28, 1981, 137.
161. **Cornelis, R.,** NAA of hair. Failure of a mission, *J. Radioanal. Chem.,* 15, 305, 1973.
162. **Bate, L. C.,** Adsorption and elution of trace elements on human hair, *Int. J. Appl. Radiat. Isot.,* 17, 417, 1966.
163. **Karin, R. W., Buono, J. A., and Fasching, J. L.,** Removal of trace elemental impurities from polyethylene by nitric acid, *Anal. Chem.,* 47, 2296, 1975.
164. **Bajo, S. and Suter, U.,** Wet ashing of organic matter for the determination of Sb, *Anal. Chem.,* 54, 49, 1982.
165. **Cornelis, R., Speecke, A., and Hoste, J.,** NAA for bulk and trace elements in urine, *Anal. Chim. Acta,* 78, 317, 1975.
166. **Iyengar, G. V., Kasperek, K., and Feinendegen, L. E.,** Retention of the metabolized trace elements in biological tissues following different drying procedures. I. Sb, Co, I, Hg, Se, and Zn in rat tissues, *Sci. Total Environ.,* 10, 1, 1978.
167. **Behne, D. and Matamba, P. A.,** Drying and ashing of biological samples in the trace element determination by NAA, *Fresenius Z. Anal. Chem.,* 274, 195, 1975.
168. **Kaiser, G., Tschöpel, P., and Tölg, G.,** Decomposition with activated oxygen in the determination of extremely low contents of trace elements in organic materials, *Fresenius Z. Anal. Chem.,* 253, 177, 1971.
169. **Gleit, C. E. and Holland, W. D.,** Use of electrically excited oxygen for the low temperature decomposition of organic substances, *Anal. Chem.,* 34, 1454, 1962.
170. **Schreiber, B. and Linder, H. R.,** Apparatus for the programmed wet-decomposition of organic samples, *Fresenius Z. Anal. Chem.,* 298, 404, 1979.
171. **Ward, R. J., Black, C. D. V., and Watson, G. J.,** The determination of free and liposome-entrapped Sb in biological samples and its application to the chemotherapy of experimental leishmaniasis, in *Drug Measurement and Drug Effects in Laboratory Health Science,* Siest, G. and Young, D. S., Eds., S. Karger, Basel, 1980, 111.
172. **Subramanian, K. S. and Meranger, J. C.,** Determination of As(III), As(V), Sb(III), Sb(V), Se(IV), and Se(VI) by extraction with APDC-MIBK and electrothermal AAS, *Anal. Chim. Acta,* 124, 131, 1981.
173. **Kamada, T. and Yamamoto, Y.,** Selective determination of Sb(III) and Sb(V) with APDC, SDDC, and dithizone by AAS with a carbon tube atomizer, *Talanta,* 24, 330, 1977.
174. **Slovák, Z. and Dočekal, B.,** Sorption of As, Sb, and Bi on glycomethacrylate gels with bound thiol groups for direct sampling in electrothermal AAS, *Anal. Chim. Acta,* 117, 293, 1980.

175. **Kagey, B. T., Bumgarner, J. E., and Creason, J. P.,** Arsenic levels in maternal-fetal tissue sets, *Trace Subst. Environ. Health,* 11, 252, 1977.
176. **Heydorn, K.,** Environmental variations of As levels in human blood determined by NAA, *Clin. Chim. Acta,* 28, 349, 1970.
177. **Ishizaki, M. and Kataoka, F.,** Determination of As in blood by flameless AAS using a carbon tube atomizer, *Sangyo Igaku,* 19, 136, 1977.
178. **Brune, D., Samsahl, K., and Wester, P. O.,** A comparison between the amounts of As, Au, Br, Cu, Fe, Mo, Se, and Zn in normal and uremic human whole blood by means of NAA, *Clin. Chim. Acta,* 13, 285, 1966.
179. **Cross, J. D., Dale, I. M., Leslie, A. C. D., and Smith, H.,** Industrial exposure to As, *J. Radioanal. Chem.,* 48, 197, 1979.
180. **Hammer, D. I., Finklea, J. F., Hendricks, R. H., Shy, C. M., and Horton, R. J. M.,** Hair trace metal levels and environmental exposure, *Am. J. Epidemiol.,* 93, 84, 1971.
181. **Wester, P. O., Brune, D., and Nordberg, G.,** As and Se in lung, liver and kidney tissue from dead smelter workers, *Br. J. Ind. Med.,* 38, 179, 1981.
182. **Grimanis, A. P., Vassilaki-Grimani, M., Alexion, D., and Papadatos, C.,** Determination of seven trace elements in human milk, powdered cow's milk and infant foods by NAA, in *Nuclear Activation Techniques in the Life Sciences 1978,* IAEA, Vienna, 1979, 241.
183. **Peter, F., Growcock, G., and Strune, G.,** Determination of As in urine by AAS with electrothermal atomization, *Anal. Chim. Acta,* 104, 177, 1979.
184. **Brown, A. C. and Crounse, R. G., Eds.,** *Hair, Trace Elements, and Human Illness,* Praeger Publishers, New York, 1980.
185. **Iyengar, G. V.,** *Elemental Composition of Human Milk,* IAEA Report RC/75-2A, IAEA, Vienna, 1977.
186. **Heydorn, K., Damsgaard, E., Larsen, N. A., and Nielsen, B.,** Sources of variability of trace element concentrations in human serum, in *Nuclear Activation Techniques in the Life Sciences 1978,* IAEA, Vienna, 1979, 129.
187. **Johnson, C. A., Lewin, J. F., and Fleming, P. A.,** The determination of some "toxic" metals in human liver as a guide to normal levels in New Zealand. II. As, Hg, and Se, *Anal. Chim. Acta,* 82, 79, 1976.
188. **Baselt, R. C.,** *Analytical Procedures for Therapeutic Drug Monitoring and Emergency Toxicology,* Biomedical Publications, Davis, Calif. 1980.
189. **Puttemans, F., Vernaillen, M., and Massart, D. L.,** Densitometric evaluation of the As content of biological materials using the method of Gutzeit, *Anal. Lett.,* 13, 635, 1980.
190. **Stahr, H. M., Ed.,** *Analytical Toxicology Methods Manual,* Iowa State University Press, Ames, 1977.
191. **Smith, T. J., Crecelius, E. A., and Reading, J. C.,** Airborne As exposure and excretion of methylated As compounds, *Environ. Health Perspect.,* 19, 89, 1977.
192. **Nixon, D. E.,** The Determination of Ultratrace Quantities of the Toxic Metals in Biomedical and Environmental Samples, Ph.D. thesis, Iowa State Univeristy, Ames, 1976.
193. **Braman, R. S. and Foreback, C. C.,** Methylated forms of As in the environment, *Science,* 182, 1247, 1973.
194. **Crecelius, E. A.,** Changes in the chemical speciation of As following ingestion by man, *Environ. Health Perspect.,* 19, 147, 1977.
195. **Cookson, J. A. and Pilling, F. D.,** Trace element distribution across the diameter of human hair, *Phys. Med. Biol.,* 20, 1015, 1975.
195a. **Davis, P. H., Dulude, G. R., Griffin, R. M., Matson, W. R., and Zink, E. W.,** Determination of total As at ng level by high-speed ASV, *Anal. Chem.,* 50, 137, 1978.
196. **Franke, J. P. and de Zeeuw, R. A.,** Direct determination of As in urine by DPASV, *Pharm. Weekbl. Sci. Ed.,* 3, 166, 1981.
197. **Buchet, J. P., Lauwerys, R., and Roels, H.,** Comparison of several methods for determination of As compounds in water and in urine. Their application for the study of As metabolism and for monitoring of workers exposed to As, *Int. Arch. Occup. Environ. Health,* 46, 11, 1980.
198. **Howard, A. G. and Arbab-Zavar, M. H.,** Determination of "inorganic" As(III) and As(V), "methylarsenic" and "dimethylarsinic" species by selective hydride evolution AAS, *Analyst,* 106, 213, 1981.
199. **Aggett, J. and Aspell, A. C.,** The determination of As(III) and the total As by AAS, *Analyst,* 101, 341, 1976.
200. **Ishizaki, M., Fujiki, M., and Yamaguchi, S.,** Studies on the analytical method for As in biological materials, and on the accumulation and localization of As in rats. Determination of total As and hydrolytic As in urine by flameless AAS using a carbon tube atomizer, *Sangyo Igaku,* 21, 234, 1979.
201. **Fitchett, A. W., Daughtrey, E. H., Jr., and Mushak, P.,** Quantitative measurements of inorganic and organic As by flameless AAS, *Anal. Chim. Acta,* 79, 93, 1975.
202. **Odanaka, Y., Matano, O., and Goto, S.,** Biomethylation of inorganic As by the rat and some laboratory animals, *Bull. Environ. Contam. Toxiol.,* 24, 452, 1980.

203. **Odanaka, Y., Matano, O., and Goto, S.,** Enhancing and depressing effects of various co-existing reagents on determination of inorganic and methylated arsenicals in environmental materials by graphite furnace AAS, *Bunseki Kagaku,* 28, 517, 1979.
204. **Ricci, G. R., Shepard, L. S., Colovos, G., and Hester, N. E.,** Ion chromatography with AAS detection for determination of organic and inorganic As species, *Anal. Chem.,* 53, 610, 1981.
205. **Lauwerys, R. R., Buchet, J. P., and Roels, H.,** Determination of trace levels of As in human biological materials, *Arch. Toxicol.,* 41, 239, 1979.
206. **Iverson, D. G., Anderson, M. A., Holm, T. R., and Stanforth, R.,** An evaluation of column chromatography and flameless AAS for As speciation as applied to aquatic systems, *Environ. Sci. Technol.,* 13, 1491, 1979.
207. **Mushak, P., Dessauer, K., and Walls, E. L.,** Flameless AA and gas-liquid chromatographic studies in arsenic bioanalysis, *Environ. Health Perspect.,* 19, 5, 1977.
208. **Yasui, A., Tsutsumi, C., and Toda, S.,** Selective determination of inorganic As(III), (V), and organic arsenic in biological materials by solvent extraction-AAS, *Agric. Biol. Chem.,* 42, 2139, 1978.
209. **Shaikh, A. U. and Tallman, D. E.,** Species-specific analysis for ng quantities of As in natural waters by arsine generation followed by graphite furnace AAS, *Anal. Chim. Acta,* 98, 251, 1978.
210. **Zingaro, R. Z.,** How certain trace elements behave, *Environ. Sci. Technol.,* 13, 282, 1979.
211. **Hinners, T. A.,** As speciation: limitations with direct hydride analysis, *Analyst,* 105, 751, 1980.
212. **Edmonds, J. S. and Francesconi, K. A.,** Estimation of methylated arsenicals by vapor generation AAS, *Anal. Chem.,* 48, 2019, 1976.
213. **Edmonds, J. S. and Francesconi, K. A.,** Methylated As from marine fauna, *Nature (London),* 265, 436, 1977.
214. **Uthus, E. O., Collings, M. E., Cornatzer, W. E., and Nielsen, F. H.,** Determination of total As in biological samples by arsine generation and AAS, *Anal. Chem.,* 53, 2221, 1981.
215. **Maher, W. A.,** Determination of inorganic and methylated As species in marine organisms and sediments, *Anal. Chim. Acta,* 126, 157, 1981.
216. **Morita, M., Uehiro, T., and Fuwa, K.,** Determination of As compounds in biological samples by liquid chromatography with inductively coupled argon plasma — AES detection, *Anal. Chem.,* 53, 1806, 1981.
217. **Diamondstone, B. I. and Burke, R. W.,** Some difficulties encountered in speciation studies of As, *Analyst,* 102, 613, 1977.
218. **Lakso, J. U., Rose, L. J., Peoples, S. A., and Shirachi, D. Y.,** A Colorimetric method for determination of arsenite, arsenate, methanearsonic and dimethylarsinic acid in biological and environmental samples, *J. Agric. Food. Chem.,* 27, 1229, 1979.
219. **Peoples, S. A., Lakso, J., and Lais, T.,** The simultaneous determination of methyl arsenic acid and inorganic arsenic in urine, *Proc. West. Pharmacol. Soc.,* 14, 178, 1971.
220. **Odanaka, Y., Matano, O., and Goto, S.,** Identification of dimethylated arsenic by GC-MS in blood, urine, and feces of rats treated with feric methanearsonate, *J. Agric. Food. Chem.,* 26, 505, 1978.
221. **Beckermann, B.,** Determination of monomethylarsonic acid and DMAA by derivatization with thioglycolic acid methylester and gas/liquid chromatographic separation, *Anal. Chim. Acta,* 135, 77, 1982.
222. **Tam, G. K. H., Charbonneau, S. M., Lacroix, G., and Bryce, F.,** Confirmation of inorganic As and DMAA in urine and plasma of dog by ion-exchangeand thin-layer chromatography, *Bull. Environ. Contam. Toxicol.,* 21, 371, 1979.
223. **Ediger, R. D.,** AAA with the graphite furnace using matrix modification, *At. Absorpt. Newsl.,* 14, 127, 1975.
224. **Chakraborti, D., De Jonghe, W., and Adams, F.,** The determination of As by electrothermal AAS with a graphite furnace. I. Difficulties in the direct determination, *Anal. Chim. Acta,* 119, 331, 1980.
225. **Sanzolone, R. F. and Chao, T. T.,** Matrix modification with Ag for the electrothermal atomization of As and Se, *Anal. Chim. Acta,* 128, 225, 1981.
226. **Pierce, F. D. and Brown, H. R.,** Comparison of inorganic interferences in AAS determination of As and Se, *Anal. Chem.,* 49, 1417, 1977.
227. **Cooksley, M. and Barnett, W. B.,** Matrix modification and the method of additions in flameless AA, *At. Absorpt. Newsl.,* 18, 101, 1979.
228. **Kunselman, G. C. and Huff, E. A.,** The determination of As, Sb, Se, and Te in environmental water samples by flameless AA, *At. Absorpt. Newsl.,* 15, 29, 1976.
229. **L'vov, B. V., Pelieva, L. A., and Sharnopolsky, A. I.,** Reduction of matrix effect in tube-furnace AAA of solutions by means of sample evaporation off graphite platform, *Zh. Prikl. Spectrosk.,* 27, 395, 1977.
230. **Thiex, N.,** Solvent extraction and flameless AA determination of As in biological material, *J. Assoc. Off. Anal. Chem.,* 63, 496, 1980.
231. **Siemer, D. D. and Koteel, P.,** Comparisons of methods of hydride generation AAS As and Se determination, *Anal. Chem.,* 49, 1097, 1977.
232. **Godden, R. G. and Thomerson, D. R.,** Generation of covalent hydrides in AAS, *Analyst,* 105, 1137, 1980.

233. **Rigin, V. I.,** Dispersionless atomic fluorescence determination of As in biological materials, *Zh. Anal. Khim.,* 33, 1966, 1978.
234. **Brooks, R. R., Ryan, D. E., and Zhang, H.,** AAS and other instrumental methods for quantitative measurements of As, *Anal. Chim. Acta,* 131, 1, 1981.
235. **Liddle, J. R., Brooks, R. R., and Reeves, R. D.,** Some parameters affecting hydride generation from As(V) for AAS, *J. Assoc. Off. Anal. Chem.,* 63, 1175, 1980.
236. **Kang, H. K. and Valentine, J. L.,** Acid interference in the determination of As by AAS, *Anal. Chem.,* 49, 1829, 1977.
237. **Yamamoto, Y. and Kumamaru, T.,** Comparative study of Zn tablet and $NaBH_4$ tablet reduction systems in the determination of As, Sb, and Se by AAS via their hydrides, *Fresenius Z. Anal. Chem.,* 281, 353, 1976.
238. **Flanjak, J.,** AAS determination of As and Se in offal and fish by hydride generation, *J. Assoc. Off. Anal. Anal. Chem.,* 61, 1299, 1978.
239. **Agemian, H. and Thomson, R.,** Simple semi-automated AAS method for determination of As and Se in fish tissue, *Analyst,* 105, 902, 1980.
240. **Howard, A. G. and Arabab-Zavar, M. H.,** Sequential spectrophotometric determination of inorganic As(III) and As(V) species, *Analyst,* 105, 338, 1980.
241. **Knechtel, J. R. and Fraser, J. L.,** Preparation of a stable borohydride solution for use in AA studies, *Analyst,* 103, 104, 1978.
242. **Robbins, W. B. and Caruso, J. A.,** Development of hydride-generation methods for atomic-spectroscopic analysis, *Anal. Chem.,* 51, 889A, 1979.
243. **Knudson, K. J. and Christian, G. D.,** A note on the determination of As using $NaBH_4$, *At. Absorpt. Newsl.,* 13, 74, 1974.
244. **Zaprianov, Z. K.,** unpublished data, 1982.
245. **Petrov, I. I.,** AAS Methods for the Determination of As and Cd in Soils and Soil Extracts with a View to Hygienic Control, Ph.D. thesis, Institute of Hygiene and Occupational Health, Sofia, 1980.
246. **Oster, O.,** The Direct Determination of As and Hg in Human Body Fluids with the Hydride AA Technique, Paper No. 3.18 presented at Jt. Congr. Scand. Germ. Soc. Clin. Chem., Hamburg, October 8 to 11, 1980.
247. **Orheim, R. M. and Bovee, H. H.** AA determination of ng quantities of As in biological media, *Anal. Chem.,* 46, 921, 1974.
248. **Hilliard, E. P. and Smith, J. D.,** Minimum sample preparation for the determination of ten elements in pig feces and feeds by AAS and a spectrophotometric procedure for total phosphorus, *Analyst,* 104, 313, 1979.
249. **Ihnat, M. and Thompson, B. K.,** Acid digestion, hydride evolution AAS method for determining As and Se in foods. II. Assessment of collaborative study, *J. Assoc. Off. Anal. Chem.* 63, 814, 1980.
250. **Holak, W.,** Analysis of foods for Pb, Cd, Cu, Zn, As, and Se, using closed system sample digestion: collaborative study, *J. Assoc. Off. Anal. Chem.,* 63, 485, 1980.
251. **Woidich, H. and Pfannhauser, W.,** Determination of As in biological material by AAS, *Fresenius Z. Anal. Chem.,* 276, 61, 1975.
252. **Yasui, A. and Tsutsumi, C.,** Adaptability of wet decomposition method to food samples for the determination of As by arsine generation-AAS, *Bunseki Kagaku,* 26, 809, 1977.
253. **Katagiri, Y., Kawai, M., and Tati, M.,** Determination of As in biological materials by AAS, *Sangyo Igaku,* 15, 67, 1973.
254. **Welz, B. and Melcher, M.,** Determination of As and Se in body fluids and tissues with the hydride-generation method, *Fresenius Z. Anal. Chem.,* 290, 106, 1978.
255. **Melcher, M.,** *Determination of As and Hg in Tissue Samples, Using the MHS Technique,* AA Lab Notes No. 18/E, Bodenseewerk Perkin-Elmer & Co., GmbH, 1978.
256. **Holm, J.,** A simplified digestion method and measuring technique for determining Pb, Cd, and As in animal tissues by AAS, *Fleischwirtschaft,* 58, 864, 1978.
257. **Norin, H. and Vahter, M.,** Rapid method for selective analysis of total urinary metabolites of inorganic As, *Scand. J. Work Environ. Health,* 7, 38, 1981.
258. **Heintges, M. G., Toffaletti, J., and Savory, J.,** Determination of As in Urine by Electrothermal AAS Using a Quartz Cell, Paper 227, *29th Natl. Meet. Am. Assoc. Clin. Chem.,* Chicago, July 17 to 22, 1977.
259. **de Groot, G., Van Dijk, A., and Maes, R. A. A.,** Determination of As in urine by AAS with the hydride generation technique, *Pharm. Weekbl.,* 112, 949, 1977.
260. **Welz, B. and Melcher, M.,** Use of a new antifoaming agent for the determination of As in urine with hydride AA technique, *At. Absorpt. Newsl.,* 18, 121, 1979.
261. **Tam, G. K. H. and Lacroix, G.,** Determination of As in urine and feces by dry ashing, AAS, *Int. J. Environ. Anal. Chem.,* 8, 283, 1980.
262. **Lo, D. B. and Coleman, R. L.,** Determination of As in animal tissues using graphite furnace AAS, *At. Absorpt. Newsl.,* 18, 10, 1979.

263. **Ishizaki, M.,** Determination of total As in biological samples by flameless AAS using a carbon tube atomizer, *Bunseki Kagaku,* 26, 667, 1977.
264. **Freeman, H., Uthe, J. F., and Fleming, B.,** A rapid and precise method for the determination of inorganic and organic As with and without wet ashing using a graphite furnace, *At. Absorpt. Newsl.,* 15, 49, 1976.
265. **Raptis, S. E., Wegscheider, W., and Knapp, G.,** The determination of As at ng/g- and μg/g-levels in organic and biological matrices, *Mikrochim. Acta,* I, 93, 1981.
266. **Bock, R.,** *A Handbook of Decomposition Methods in Analytical Chemistry,* International Textbook Company, Glasgow, 1979.
267. **Yanagi, K. and Ambe, M.,** Determination of As in biological, environmental, and geological materials by arsine evolution-flameless AAS, *Bunseki Kagaku,* 30, 209, 1981.
268. **Kneip, T., Kleinman, M., Bernstein, D., and Raddick, R.,** Trace elements in human tissue, in *The Use of Biological Specimens for the Assessment of Human Exposure to Environmental Pollutants,* Berlin, A., Wolff, A. H., and Hasegawa, Y., Eds., Martinus Nijhoff Publishers, Boston, 1979, 221.
269. **Cox, D. H.,** Arsine evolution-electrothermal AA method for determination of ng levels of total As in urine and water, *J. Anal. Toxicol.,* 4, 207, 1980.
270. **Simon, R. K., Christian, G. D., and Purdy, W. C.,** Comparision of methods for elimination of organic matter in human urine, *Am. J. Clin. Pathol.,* 49, 733, 1968.
271. **Friend, M. T., Smith, C. A., and Wishart, D.,** Ashing and wet oxidation procedures for the determination of some volatile trace metals in foodstuffs and biological materials by AAS, *At. Absorpt. Newsl.,* 16, 46, 1977.
272. **Knapp, G., Sadjadi, B., and Spitzy, H,.** Behavior of volatile elements in the incineration of organic matrices with chloric acid, *Fresenius Z. Anal. Chem.,* 274, 275, 1975.
273. **Scheubeck, E., Gehring, J., and Pickel, M.,** A pressure decomposition device for a rapid treatment of large amounts of biological and organic matter and its use in the determination of trace quantities of heavy metals, *Fresenius Z. Anal. Chem.,* 297, 113, 1979.
274. **Scheubeck, E., Nielsen, A., and Iwantscheff, G.,** Rapid decomposition process for large samples of biological materials and its use in determination of trace quantities of heavy metals, *Fresenius Z. Anal. Chem.,* 294, 398, 1979.
275. **Locke, J., Boase, D. R., and Smalldon,** The quantitative multielement analysis of human liver tissue by SSMS, *Anal. Chim. Acta,* 104, 233, 1979.
276. **Imai, S.,** Loss of elements in low-temperature plasma ashing, *Bunseki Kagaku,* 27, 611, 1978.
277. **Abu-Samra, A., Morris, J. S., and Koirtyohann, S. R.,** Wet ashing of some biological samples in a microwave oven, *Anal. Chem.,* 47, 1475, 1975.
278. **Brooke, P. J. and Evans, W. H.,** Determination of total inorganic As in fish, shellfish and fish products, *Analyst,* 106, 514, 1981.
279. **Kamada, T.,** Selective determination of As(III) and As(V) with APDC, SDDC and dithizone by means of flameless AAS with a carbon-tube atomizer, *Talanta,* 23, 835, 1976.
280. **Chakraborti, D., de Jonghe, W., and Adams, F.,** The determination of As by electrothermal AAS with a graphite furnace. II. Determination of As(III) and As(V) after extraction, *Anal. Chim. Acta,* 120, 121, 1980.
281. **Andreae, M. O.,** Determination of arsenic species in natural waters, *Anal. Chem.,* 49, 820, 1977.
282. **Drasch, G., Meyer, L. V., and Kauert, G.,** Application of the furnace AA method for the detection of As in biological samples by means of the hydride technique, *Fresenius Z. Anal. Chem.,* 304, 141, 1980.
283. **Parris, G. E., Blair, W. R., and Brinckman, F. E.,** Chemical and physical considerations in the use of AA detectors coupled with a gas chromatograph for determination of trace organometalic gases, *Anal. Chem.,* 49, 378, 1977.
284. **Brinckman, F. E., Blair, W. R., Jewett, K. L., and Iverson, W. P.,** Application of a liquid chromatograph coupled with a flameless AA detector for speciation of trace organometalic compounds, *J. Chromatogr. Sci.,* 15, 493, 1977.
285. **Pacey, G. E. and Ford, J. A.,** Arsenic speciation by ion-exchange separation and graphite furnace AAS, *Talanta,* 28, 935, 1981.
286. **Mauras, Y. and Allain, P.,** Determination of Ba in water and biological fluids by emission spectrometry with an ICP, *Anal. Chim. Acta,* 110, 271, 1979.
287. **Olehy, D. A., Schmitt, R. A., and Bathard, W. F.,** NAA of Mg, Ca, Sr, Ba, Mn, Co, Cu, Zn, Na, and K in human erythrocytes and plasma, *J. Nucl. Med.,* 7, 917, 1966.
288. **Creason, J. P., Hinners, T. A., Bumgarner, J. E., and Pinkerton, C.,** Trace elements in hair, as related to exposure in Metropolitan New York, *Clin. Chem.,* 21, 603, 1975.
289. **Dahlquist, R. L. and Knoll, J. W.,** ICP-AES: analysis of biological materials and soils for major, trace, and ultra-trace elements, *Appl. Spectrosc.,* 32, 1, 1978.
290. **Othman, I. and Spyrou, N. M.,** The abundance of some elements in hair and nail from the Machokos District of Kenya, *Sci. Total Environ.,* 16, 267, 1980.
291. **Hadzistelios, I. and Papadopoulou, C.,** Radiochemical microdetermination of Mn, Sr, and Ba by ion-exchange, *Talanta,* 16, 337, 1969.

292. **Forssen, A.,** Inorganic elements in the human body. Occurrence of Ba, Br, Ca, Cd, Cs, Cu, K, Mn, Ni, Sn, Sr, Y, and Zn in the human body, *Ann. Med. Exp. Biol. Fenn.,* 50, 99, 1972.
293. **Forssen, A. and Eramesta, O.,** Inorganic elements in the human body. Ba, Br, Ca, Cd, Cu, K, Ni, Pb, Sn, Sr, Ti, Y, and Zn in hair, *Ann. Acad. Sci. Fenn.,* Ser. A5, 162, 1974.
294. **Welz, B.,** *Atomic Absorption Spectroscopy,* Verlag Chemie, New York, 1976.
295. **Wall, C. D.,** The use of *in situ* pyrolytic coating with the HGA-70 graphite furnace, *At. Absorpt. Newsl.,* 17, 61, 1978.
296. **Beaty, R. D. and Cooksey, M. M.,** The influence of furnace conditions on matrix effects in graphite furnace AA, *At. Absorpt. Newsl.,* 17, 53, 1978.
297. **Cioni, R., Mazzucotelli, A., and Ottonello, G.,** Matrix effects in the flameless AA determination of trace amounts of Ba in silicates, *Anal. Chim. Acta,* 82, 415, 1976.
298. **Tompsett, S. L.,** Determination of Li, Sr, Ba and Au by AAS, *Proc. Assoc. Clin. Biochem.,* 5, 125, 1968.
299. **Sutton, D. C. and Miro, M.,** Sequential Determination of Stable Ba, Sr, Ca, and Mg in Biological Materials, Report HASL-212, U.S. Atomic Energy Commission, Washington, D.C., 1969.
300. **Miro, M. and Sutton, D. C.,** Ba and Sr determination in biological materials by AAS, *Appl. Spectrosc.,* 24, 220, 1970.
301. **Berggren, P.-O.,** The determination of Ba, La and Mg in pancreatic islets by electrothermal AAS, *Anal. Chim. Acta,* 119, 161, 1980.
302. **Dumott, T. C.,** *Atomic Absorption with Electrothermal Atomization,* 2nd ed., Pye Unicam, Cambridge, 1981.
303. **Stiefel, Th., Schulze, K., Zorn, H., and Tölg, G.,** Toxicokinetic and toxicodynamic studies of Be, *Arch. Toxicol.,* 45, 81, 1980.
304. **Kaiser, G., Grallath, E., Tschöpel, P., and Tölg, G.,** Contribution to the optimization of the chelate-gas-chromatographic determination of Be in limited amounts of organic materials, *Fresenius Z. Anal. Chem.,* 259, 257, 1972.
305. **Grewal, D. S. and Kearns, F. X.,** A simple and rapid determination of small amounts of Be in urine by flameless AA, *At. Absorpt. Newsl.,* 16, 131, 1977.
306. **Desai, S. R. and Sudhalatha, K. K.,** Determination of micro-amounts of Be in urine, *Talanta,* 14, 1346, 1967.
307. **Takata, T., Hitosugi, M., Kadowaki, T., Inoue, Y., and Seki, K.,** Determination of Be in blood by fluorimetric analysis, *Sangyo Igaku,* 20, 114, 1978.
308. **Bokowski, D. L.,** Determination of Be by a direct-reading AA spectrophotometer, *Am. Ind. Hyg. Assoc. J.,* 29, 474, 1968.
309. **Stiefel, Th., Schulze, K., Tölg, G., and Zorn, H.,** A combined method for determination of Be in biological matrices by flameless AAS, *Anal. Chim. Acta,* 87, 67, 1976.
310. **Hurlbut, J. A.,** Determination of Be in biological tissues and fluids by flameless AAS, *At. Absorpt. Newsl.,* 17, 121, 1978.
311. **Geladi, P. and Adams, F.,** The determination of Be and Mn in aerosols by AAS with electrothermal atomization, *Anal. Chim. Acta,* 105, 219, 1979.
312. **Owens, J. W. and Gladney, E. S.,** Determination of Be in environmental materials by flameless AAS, *At. Absorpt. Newsl.,* 14, 76, 1975.
313. **Stiefel, Th., Schulze, K., Tölg, G., and Zorn, H.,** Analysis of trace elements distributed in blood. Determination of Be concentrations $\geqslant 0.01$ ng/g in human and animal blood components by preparative electrophoresis and flameless AAS, *Fresenius Z. Anal. Chem.,* 300, 189, 1980.
314. **Nakashima, R.,** Determination of Be in biological materials by flameless AAS, *Bunseki Kagaku,* 27, 185, 1978.
315. **Lockwood, T. H. and Limtiaco, L. P.,** Determination of Be, Cd, and Te in animal tissues using electronically excited oxygen and AAS, *Am. Ind. Hyg. Assoc. J.,* 36, 57, 1975.
316. **Hurlbut, J. A.,** The Direct Determination of Be in Urine by Flameless AAS, Report RFP-2151, U.S. Atomic Energy Commission, Washington, D.C., 1974.
317. **Kauffmann, J.-M., Patriarche, G. J., and Christian, G. D.,** A rapid determination of trace amounts of Bi in urine and blood using DPASV at the hanging Hg electrode, *Anal. Lett.,* 14, 1209, 1981.
318. **Allain, P.,** Determination of Bi in blood, urine and CSF by flameless AAS, *Clin. Chim. Acta,* 64, 281, 1975.
319. **Johnson, C. A.,** The determination of some toxic metals in human liver as a guide to normal levels in New Zealand. I. Determination of Bi, Cd, Cr, Co, Cu, Pb, Mn, Ni, Ag, Tl, and Zn, *Anal. Chim. Acta,* 81, 69, 1976.
320. **Meltzer, L. E., Rutman, J., George, P., Rutman, R., and Kitchell, J. R.,** The urinary excretion pattern of trace metals in Diabetes Mellitus, *Am. J. Med. Sci.,* 244, 282, 1962.
321. **Šinko, J. and Gomišček, S.,** Determination of Pb, Cd, Cu, Tl, Bi, and Zn in blood serum by anodic stripping polarography, *Mikrochim. Acta,* 163, 1972.

322. **Djudzman, R., Van den Eeckhout, E., and De Moerloose, P.,** Determination of Bi by AAS with electrothermal atomization after low-temperature ashing, *Analyst,* 102, 688, 1977.
323. **Rooney, R. C.,** Determination of Bi in blood and urine, *Analyst,* 101, 749, 1976.
324. **Palliere, M. and Garnez, G.,** Technique for the AAS determination of Bi in biological material, *Ann. Pharm. Fr.,* 34, 183, 1976.
325. **Fitchett, A. W., Buck, R. P., and Mushak, P.,** Direct determination of heavy elements in biological media by spark source mass spectrometry, *Anal. Chem.,* 46, 710, 1974.
326. **Barnes, R. M. and Genna, J. S.,** Concentration and spectrochemical determination of trace metals in urine with poly(dithiocarbamate) resin and ICP-AES, *Anal. Chem.,* 51, 1065, 1979.
327. **Palliere, M. and Gernez, G.,** Determination of traces of Bi in blood, *Ann. Pharm. Fr.,* 38, 123, 1980.
328. **Gladney, E. S.,** Matrix modification for the determination of Bi by flameless AA, *At. Absorpt. Newsl.,* 16, 114, 1977.
329. **Frech, W.,** Rapid determination of Bi in steel by flameless AA, *Fresenius Z. Anal. Chem.,* 275, 353, 1975.
330. **Bourdon, R., Galliot, M., and Prouillet, E.,** Estimation of Cu, Pb, Mn, Bi, Cd, and Au in biological fluids by flameless AAS, *Ann. Biol. Clin.,* 32, 413, 1974.
331. **Inui, T., Fudagawa, N., and Kawase, A.,** Extraction and AAS determination of Bi with electrothermal atomization, *Fresenius Z. Anal. Chem.,* 299, 190, 1979.
332. **Delves, H. T., Shepherd, G., and Vinter, P.,** Determination of 11 metals in small samples of blood by sequential solvent extraction and AAS, *Analyst,* 96, 260, 1971.
333. **Suzuki, T., Takeda, M., and Uchiyama, M.,** Determination of Bi and Pb in milk powder by solvent extraction flameless AAS, *Shokuhiu Eiseigaku Zasshi,* 20, 198, 1979.
334. **Outon, A., Creus, J. M., Charro, A., and Simal, J.,** Investigation of metal contamination in shellfish and sea-water of Rias Bajas de Galicia. III. Determination of Pb, Cu, Zn, and Bi, *An. Bromatol.,* 31, 81, 1979.
335. **Hall, R. J. and Faber, T.,** Determination of Bi in body tissues and fluids after administration of controlled doses, *J. Assoc. Off. Anal. Chem.,* 55, 639, 1972.
336. **Aihara, M. and Kuboku, M.,** Determination of Bi or Tl by AAS after extraction with K xanthate-MIBK, *Bunseki Kagaku,* 29, 243, 1980.
337. **Kinser, R. E.,** Determination of Bi and Te in tissues by AAS, *Am. Ind. Hyg. Assoc. J.,* 27, 260, 1966.
338. **Devoto, G.,** Determination of Bi in urine by AAS, *Boll. Soc. Ital. Biol. Sper.,* 44, 1253, 1968.
339. **Willis, J. B.,** Determination of Pb and other heavy metals in urine by AAS, *Anal. Chem.,* 34, 614, 1962.
340. **Tsalev, D. L. and Petrova, V. P.,** Hexamethyleneammonium hexamethylenedithiocarbamate — n-butylacetate as an extractant of Bi, Cd, Cu, Hg, In, Pb, and Pd from acidic media, *Dokl. Bolg. Acad. Nauk,* 32, 911, 1979.
341. **Slovák, Z., Smrž, M., Dočekal, B., and Slováková, S.,** Analytical behavior of hydropholic glycomethacrylate gels with bound thiol groups, *Anal. Chim. Acta,* 111, 243, 1979.
342. **Imbus, H. R., Cholak, J., Miller, L. H., and Sterling, T.,** Boron, Cd, Cr, and Ni in human blood and urine, *Arch. Environ. Health,* 6, 286, 1963.
343. **Fisher, R. S. and Freimuth, H. C.,** Blood boron levels in human infants, *J. Invest. Dermatol.,* 30, 85, 1958.
344. **Monnier, D., Menzinger, C. A., and Marcantonatos, M.,** Direct microdetermination of B in blood by the fluorimetric method using 2-hydroxy-4-methoxy-4′-chlorobenzophenone, *Anal. Chim. Acta,* 60, 233, 1972.
345. **Kaczmarczyk, A., Messer, J. R., and Peirce, C. E.,** Rapid method for determination of B in biological materials, *Anal. Chem.,* 43, 271, 1971.
346. **Daughtrey, E. H., Jr. and Harrison, W. W.,** Analysis of trace levels of B by ion-exchange-hollow cathode emission, *Anal. Chim. Aca,* 72, 225, 1974.
347. **Schramel, P.,** Determination of B in milk and milk products by ICP-ES analysis, *Z. Lebensm.-Unters. Forsch.,* 169, 255, 1979.
348. **Schachter, M. M. and Boyer, K. W.,** Digestion of organic matrices with a single acid for trace determination, *Anal. Chem.,* 52, 360, 1980.
349. **Belcher, R., Ghonaim, S. A., and Townshend, A.,** MECA — a new flame analytical technique. III. The determination of boron, *Anal. Chim. Acta,* 71, 255, 1974.
350. **Bogdanski, S. L., Henden, E., and Townshend, A.,** MECA. XVI. Determination of B, Se, and other elements by using a flame generated within the cavity, *Anal. Chim. Acta,* 116, 93, 1980.
351. **Pickett, E. E. and Franklin, M. L.,** Filter flame photometer for determination of B in plants, *J. Assoc. Off. Anal. Chem.,* 60, 1164, 1977.
352. **Mair, J. W., Jr. and Day, H. G.,** Curcumin method for spectrophotometric determination of B extracted from radiofrequency ashed tissues using 2-ethyl-1,3-hexanediol, *Anal. Chem.,* 44, 2015, 1972.
353. **Kliegel, W.,** *Bor in Biologie, Medizin und Pharmazie,* Springer-Verlag, New York, 1980.

354. **Szydlowski, F. J.,** Boron in natural waters by AAS with electrothermal atomization, *Anal. Chim. Acta,* 106, 121, 1979.
355. **van der Geugten, R. P.,** Determination of B in river water with flameless AAS (graphite furnace technique), *Fresenius Z. Anal. Chem.,* 306, 13, 1981.
356. **Bader, H. and Brandenberger, H.,** Boron determination in biological materials by AAS, *At. Absorpt. Newsl.,* 7, 1, 1968.
357. **Holak, W.,** AA determination of B in Food, *J. Assoc. Off. Anal. Chem.,* 54, 1138, 1971.
358. **Holak, W.,** Collaborative study of the determination of boric acid in foods by AAS, *J. Assoc. Off. Anal. Chem.,* 55, 890, 1972.
359. **Allain, P. and Mauras, Y.,** Microdetermination of Pb and Cd in blood and urine by graphite furnace AAS, *Clin. Chim. Acta,* 91, 41, 1979.
360. **Stoeppler, M. and Brandt, K.,** Contributions to automated trace analysis. V. Determination of Cd in whole blood and urine by electrothermal AAS, *Fresenius Z. Anal. Chem.,* 300, 372, 1980.
361. **Valenta, P., Rützel, H., Nürnberg, H. W., and Stoeppler, M.,** Trace chemistry of toxic metals in biomatrices. II. Voltametric determination of trace content of Cd and other toxic metals in human whole blood, *Fresenius Z. Anal. Chem.,* 285, 25, 1977.
362. **Lagesson, V. and Andrasko, L.,** Direct determination of Pb and Cd in whole blood and urine by flameless AAS, *Clin. Chem.,* 25, 1948, 1979.
363. **Pleban, P. A. and Pearson, K. H.,** Determination of Cd in whole blood and urine by Zeeman AAS, *Clin. Chim. Acta,* 99, 267, 1979.
364. **Elinder, C.-G., Kjellström, T., Lind, B., Molander, M.-L., and Silander, T.,** Cadmium concentrations in human liver, blood, and bile: comparison with a metabolic model, *Environ. Res.,* 17, 236, 1978.
365. **Legotte, P. A., Rosa, W. C., and Sutton, D. C.,** Determination of Cd and Pb in urine and other biological samples by graphite furnace AAS, *Talanta,* 27, 39, 1980.
366. **Tati, M., Katagiri, Y., and Kawai, M.,** Urinary and fecal excretion of Cd in normal Japanese: an approach to non-toxic levels of Cd, in *Effects and Dose-Response Relationships of Toxic Metals,* Elsevier, Amsterdam, 1976, 331.
367. **Anke, M., Schneiderti, J., Grün, M., Groppel, B., and Hennig, A.,** The diagnosis of Mn-, Zn-, and Cu-insufficiency and of Cd-exposure, *Zbl. Pharm. Pharmakother. Laboratoriumsdiagn.,* 117, 688, 1978.
368. **Barlow, P. J. and Kapel, M.,** Metal and S contents of hair in relation to certain mental states, in *Hair, Trace Elements and Human Illness,* Brown, A. C. and Crounse, R. G., Eds., Praeger Publishers, New York, 1980, chap. 7.
369. **Kowal, N. E., Johnson, D. E., Kraemer, D. F., and Pahren, H. R.,** Normal levels of Cd in diet, urine, blood, and tissues of healthy inhabitants of the U.S., *J. Toxicol. Environ. Health,* 5, 995, 1979.
370. **Eads, E. A. and Labdin, C. E.,** Survey of trace elements in human hair, *Environ. Res.,* 6, 247, 1973.
371. **Curry, A. S. and Knott, A. R.,** "Normal" levels of Cd in human liver and kidney in England, *Clin. Chim. Acta,* 30, 115, 1970.
372. **Livingston, H. D.,** Measurement and distribution of Zn, Cd, and Hg in human kidney tissue, *Clin. Chem.,* 18, 67, 1972.
373. **Pleban, P. A., Kerkay, J., and Pearson, K. H.,** Polarized Zeeman-effect flameless AAS of Cd, Cu, Pb, and Mn in human kidney cortex, *Clin. Chem.,* 27, 68, 1981.
374. **Casey, C. E.,** Concentrations of some trace elements in human and cows' milk, *Proc. Univ. Otago Med. Sch.,* 54, 7, 1976.
375. **Gergely, A. and Linder-Szotyori, K.,** Investigation of Cd content of human milk and baby foods in Hungary, in *Kadmium-Symp. 1977,* Bolck, F., Ed., Friedrich-Schiller University, Jena, 1979.
376. **van Hattum, B., De Voogt, P., and Copius Peereboom, J. W.,** An analytical procedure for the determination of Cd in human placentae, *Int. J. Environ. Anal. Chem.,* 10, 121, 1981.
377. **Baglan, R. J., Brill, A. B., Schulert, A., Wilson, D., Larsen, K., Dyer, N., Mansour, M., Schaffner, W., Hoffman, L., and Davies, J.,** Utility of placental tissue as an indicator of trace element exposure to adult and fetus, *Environ. Res.,* 8, 64, 1974.
378. **Raptis, S. and Müller, K.,** A new method for determination of Cd in blood serum, *Clin. Chim. Acta,* 88, 393, 1978.
379. **Langmyhr, F. J., Eyde, B., and Jonsen, J.,** Determination of total content and distribution of Cd, Cu, and Zn in human parotid saliva, *Anal. Chim. Acta,* 107, 211, 1979.
380. **Gozhaya, L. D. and Manova, T. G.,** Spectrochemical determination of traces of Fe, Cu, Mn, Ag, Pd, and Cd in biological fluid (human saliva), *Trudi VNII khim. reaktivov i os. tch. khim. v. (IREA),* Moscow, 1969, 168.
381. **Cohn, J. R. and Emmett, E. A.,** The excretion of trace metals in human sweat, in *Toxicology and Occupational Medicine,* Diechmann, W. B., Organizer, Elsevier/North-Holland, New York, 1979, 319.
382. **Iwakura, M.,** Quantitative analysis of Cd in teeth of inhabitants of Cd-contaminated region, *Koku Eisei Gakkai Zasshi,* 22, 1, 1972.

383. **Kaneko, Y., Inamori, I., and Nishimura, M.,** Zinc, Pb, Cu and Cd in human teeth from different geographical areas in Japan, *Bull. Tokyo Dent. Coll.,* 15, 233, 1974.
384. **Michel, R. G., Hall, M. L., Ottaway, J. M., and Fell, G. S.,** Determination of Cd in blood and urine by flame AFS, *Analyst,* 104, 491, 1979.
385. **Gardiner, P. E., Ottaway, J. M., and Fell, G. S.,** Accuracy of direct determination of Cd in Urine by carbon-furnace AAS, *Talanta,* 26, 841, 1979.
386. **Kovats, A. and Bohm, B.,** Urinary Cd determination by flameless AAS with atomization in a graphite furnace, *Stud. Cercet. Biochim.,* 19, 125, 1976.
387. **Lehnert, G., Schaller, K. H., and Haas, Th.,** AAS determination of Cd in serum and urine, *Z. Klin. Chem. Klin. Biochem.,* 6, 174, 1968.
388. **Perry, E. F., Koirtyohann, S. R., and Perry, H. M., Jr.,** Determination of Cd in blood and urine by graphite furnace AAS, *Clin. Chem.,* 21, 626, 1975.
389. **O'Laughlin, J. W., Hamphill, D. D., and Pierce, J. O.,** *Analytical Methodology for Cd in Biological Matter — A Critical Review,* Int. Lead Zinc Research Organization, Inc., New York, 1976.
390. **Sacchini, A.,** Cadmium in environmental and biological materials: problems related to analytical methods, *Rass. Chim.,* 32, 23, 1980.
391. **Oehme, M. and Lund, W.,** Comparison of digestion procedures for the determination of heavy metals (Cd, Cu, Pb) in blood by ASV, *Fresenius Z. Anal. Chem.,* 298, 260, 1979.
392. **Capar, S. C. and Gould, J. H.,** Lead, fluoride, and other elements in bone meal supplements, *J. Assoc. Off. Anal. Chem.,* 62, 1054, 1979.
392a. **Alt, F.,** Comparative determination of Cd in blood by four different methods, *Fresenius Z. Anal. Chem.,* 308, 137, 1981.
393. **Chittleborough, G. and Steel, B. J.,** The determination of Zn, Cd, Pb and Cu in human hair by DPASV at a hanging Hg drop electrode after nitrate fusion, *Anal. Chim. Acta,* 119, 235, 1980.
394. **Jönsson, H.,** Determination of Pb and Cd in milk with modern analytical methods, *Z. Lebensm.-Unters. Forsch.,* 160, 1, 1976.
395. **Oehme, M., Lund, W., and Jonsen, J.,** The determination of Cu, Pb, Cd, and Zn in human teeth by ASV, *Anal. Chim. Acta,* 100, 389, 1978.
396. **Golimowski, J., Valenta, P., Stoeppler, M., and Nürnberg, H. W.,** A rapid high-performance analytical procedure with simultaneous voltametric determination of toxic trace metals in urine, *Talanta,* 26, 649, 1979.
397. **Lund, W., Thomassen, Y., and Dølve, P.,** Flame-AAA for trace metals after electrochemical preconcentration on a wire filament, *Anal. Chim. Acta,* 93, 53, 1977.
398. **Franke, J. P. and de Zeeuw, R. A.,** Toxic metal analysis by DPASV in clinical and forensic toxicology, *J. Anal. Toxicol.,* 1, 291, 1977.
399. **Jervis, R. E., Tiefenbach, B., and Chattopadhyay, A.,** Determination of trace Cd in biological materials by neutron and photon activation analyses, *Can. J. Chem.,* 52, 3008, 1974.
400. IAEA, *Nuclear Activation Techniques in the Life Sciences 1978,* IAEA, Vienna, 1979.
401. **Biggin, H. C., Chen, N. S., Ettinger, K. V., Fremlin, J. H., Morgan, W. D., Nowotny, R., Chamberlain, M. J., and Harvey, T. C.,** Cd by in vivo NAA, *J. Radioanal. Chem.,* 19, 207, 1974.
402. **Harvey, T. C., Thomas, B. J., McLellan, J. S., and Fremlin, J. H.,** Measurement of liver-Cd concentrations in patients and in industrial workers by NAA, *Lancet,* I, 1269, 1975.
403. **Fell, G. S., Ottaway, J. M., and Hussein, F. E. R.,** Application of blood Cd analysis to industry using atomic fluorescence method, *Br. J. Ind. Med.,* 34, 106, 1977.
404. **Fell, G. S., Hough, D. C., Hussein, F. E. R., and Ottaway, J. M.,** Determination of Zn and Cd in biological samples by AFS, *Proc. Anal. Div. Chem. Soc.,* 13, 271, 1976.
405. **Worrell, G. J., Vickers, T. J., and Williams, F. D.,** A solvent extraction-atomic fluorescence system for the determination of Cd in complex samples, *Anal. Chim. Acta,* 75, 453, 1975.
406. **Ahlgren, L. and Mattsson, S.,** Cd in man measured in vivo by X-ray fluorescence analysis, *Phys. Med. Biol.,* 26, 19, 1981.
407. **Meinel, B., Bode, J. Ch., Koenig, W., and Richter, F. W.,** Study of sample collection and preparation methods for multielement analysis in liver tissue by PIXE, *J. Clin. Chem. Clin. Biochem.,* 17, 15, 1979.
408. **Valkovic, V.,** *Trace Elements in Human Hair,* Carland STPM Press, New York, 1977.
409. **Verbueken, A., Michiels, E., and Van Grieken, R.,** Total analysis of plant material and biological tissue by SSMS, *Fresenius Z. Anal. Chem.,* 309, 300, 1981.
410. **Gramlich, J. W., Wachlan, L. A., Murphy, T. J., and Moore, L. J.,** The determination of Zn, Cd and Pb in biological and environmental materials by IDMS, *Trace Subst. Environ. Health,* 11, 376, 1977.
411. **Locke, J.,** The application of plasma source AES in forensic science, *Anal. Chim. Acta,* 113, 3, 1980.
412. **Jones, J. W., Capar, S. G., and O'Haver, T. C.,** Critical evaluation of a multi-element scheme using plasma emission and hydride evolution AAS for the analysis of plant and animal tissues, *Analyst,* 107, 353, 1982.

413. **Mertz, D. P., Koschnick, R., Wilk, G., and Pfeilsticker, K.,** Untersuchungen über den Stoffwechsel von Spurenelemente beim Menschen, *Z. Klin. Chem. Klin. Biochem.*, 6, 171, 1968.
414. **Gadaskina, I. D., Gadaskina, N. D., and Filov, V. A.,** *Determination of Industrial Inorganic Toxic Substances in the Organism,* Medizina, Leningrad, 1975.
415. **Kerber, J. D. and Fernandez, F. J.,** Determination of trace metals in aqueous solutions with the Delves sampling cup technique, *At. Absorpt. Newsl.*, 10, 78, 1971.
416. **Kanabrocki, E. L., Case, L. F., Fields, T., Graham, G., Miller, E. B., Oester, Y. T., and Kaplan, E.,** Mn and Cu levels in human urine, *J. Nucl. Med.*, 6, 780, 1965.
417. **Uchida, T., Iida, C., and Kojima, I.,** Discrete nebulization in AAS with a long absorption tube, *Anal. Chim. Acta,* 113, 361, 1980.
418. **Lauwerys, R., Buchet, J.-P., Roels, H., Berlin, A., and Smeets, J.,** Intercomparison program of Pb, Hg, and Cd analysis in blood, urine and aqueous solutions, *Clin. Chem.*, 21, 551, 1975.
419. **Paulev, P.-E., Solgaard, P., and Tjell, J. C.,** Interlaboratory comparision of Pb and Cd in blood, urine and aqueous solutions, *Clin. Chem.*, 24, 1797, 1978.
420. **Nackowski, S. B., Putnam, R. D., Robbins, D. A., Varner, M. O., White, L. D., and Nelson, K. W.,** Trace metal contamination of evacuated blood collection tubes, *Am. Ind. Hyg. Assoc. J.*, 38, 503, 1977.
421. **Méranger, J. C., Hollebone, B. R., and Blanchette, G. A.,** Effects of storage times, temperatures and container types on accuracy of AA determination of Cd, Cu, Hg, Pb, and Zn in whole heparinized blood, *J. Anal. Toxicol.*, 5, 33, 1981.
422. **Zydowicz, S., Tytko, S. A., Willis, C. E., and Shamberger, R. J.,** *Cd Analysis by Flameless AA,* Paper 047, 29th Natl. Meet. Am. Assoc. Clin. Chem., Chicago, July 17 to 22, 1977.
423. **Fourie, H. O. and Peisach, M.,** Loss of trace elements during dehydration of marine zoological material, *Analyst,* 102, 193, 1977.
424. Analytical Methods Committee, Determination of small amounts of Cd in organic matter. II. Determination of amounts down to sub-microgram level, *Analyst,* 100, 761, 1975.
425. **Sperling, K.-R. and Bahr, B.,** Determination of heavy metals in sea water and marine organisms by flameless AAS. XII. Correspondence and some possible sources of error in an intercalibration of Cd, *Fresenius Z. Anal. Chem.*, 306, 7, 1981.
426. **Bagliano, G., Benischek, F., and Huber, I.,** A rapid and simple method for the determination of trace metals in hair samples by AAS, *Anal. Chim. Acta,* 123, 45, 1981.
427. **Subramanian, K. S. and Méranger, J. C.,** APDC-MIBK-graphite furnace AA system for some trace metals in drinking water, *Int. J. Environ. Anal. Chem.*, 7, 25, 1979.
428. **Dellien, I. and Persson, L.,** Effect of hydrogen-ion concentration on the extraction of Co, Ni, Cd and Pb with APDC/MIBK: time stability of the extracts, *Talanta,* 26, 1101, 1979.
429. **Petrov, I. I., Tsalev, D. L., and Vassileva, E. T.,** Pulse-nebulization AAS after a preconcentration from acidic media, *Dokl. Bolg. Acad. Nauk,* 34, 679, 1981.
430. **Vesterberg, O. and Bergström, T.,** Determination of Cd in blood by use of AAS with crucibles — and a rational procedure for dry-ashing, *Clin. Chem.*, 23, 555, 1977.
431. **Cernik, A. A.,** A preliminary procedure for the determination of Cd in blood, *At. Absorpt. Newsl.*, 12, 163, 1973.
432. **Issaq, H. J.,** Effects of matrix on the determination of volatile metals in biological samples by flameless AAS with graphite tube atomizer, *Anal. Chem.*, 51, 657, 1979.
433. **Sperling, K.-R. and Bahr, B.,** Determination of heavy metals in sea water and in marine organisms by flameless AAS. XI. Quality criteria for graphite-tubes — a warning, *Fresenius Z. Anal. Chem.*, 299, 206, 1979.
434. **Vesterberg, O. and Wrangskogh, K.,** Determination of Cd in urine by graphite-furnace AAS, *Clin. Chem.*, 24, 681, 1978.
435. **Wright, F. C. and Riner, J. C.,** Determination of Cd in blood and urine with the graphite furnace, *At. Absorpt. Newsl.*, 14, 103, 1975.
436. **Hoenig, M., Vanderstappen, R., and van Hoeyweghen, R.,** Electrothermal atomization of Cd in presence of complex matrix, *Analysis,* 7, 17, 1979.
437. **Wegscheider, W., Knapp, G., and Siptzy, H.,** Statistical investigation of interferences in graphite furnace AAS. II. Cadmium, *Fresenius Z. Anal. Chem.*, 283, 97, 1977.
438. **Machata, G. and Binder, R.** The determination of metallic trace elements Pb, Tl, Zn, and Cd in biological materials by flameless AA, *Z. Rechtsmed.*, 73, 29, 1973.
439. **Tsalev, D. L. and Petrov, I. I.,** Pulse nebulization of chloroform and carbon tetrachloride extracts in flame AAS, *Anal. Chim. Acta,* 111, 155, 1979.
440. **Lund, W., Larsen, B. V., and Gundersen, N.,** The application of electrodeposition techniques to flameless AAS. III. The determination of Cd in urine, *Anal. Chim. Acta,* 81, 319, 1976.
441. **Marletta, G. P.,** A reagent-free determination of nanogram levels of Cd, *Riv. Merceol.*, 18, 97, 1979.

442. **Watling, H. R. and Wardale, I. M,.** Comparison of wet and dry ashing for the analysis of biological materials by AAS, in *The Analysis of Biological Materials,* Butler, L. R. P., Ed., Pergamon Press, Oxford, 1979, 69.
443. **Hwang, J. Y., Mokeler, C. J., and Ullucci, P. A.,** Maximization of sensitivities in Ta-ribbon flameless AAS, *Anal. Chem.,* 44, 2018, 1972.
444. **Ullucci, P. A. and Hwang, J. Y.,** Determination of Cd in biological materials by AA, *Talanta,* 21, 745, 1974.
445. **Talmi, Y.,** Determination of Zn and Cd in environmentally based samples by the use of the radio frequency spectrometric source, *Anal. Chem.* 46, 1005, 1974.
446. **Nichols, J. A. and Woodriff, R.,** Coprecipitation of heavy metals directly in graphite crucibles for furnace AAS, *J. Assoc. Off. Anal. Chem.,* 63, 500, 1980.
447. **Robinson, J. W. and Weiss, S.,** Direct determination of Cd in whole blood using an RF-heated carbon-bed atomizer for AAS, *Spectrosc. Lett.,* 11, 715, 1978.
448. **Robinson, J. W., Wolcott, D. R., Slevin, P. J., and Hindman, G. D.,** Determination of Cd by AA in air, water, sea water, and urine with a RF carbon bed atomizer, *Anal. Chim. Acta,* 66, 13, 1973.
449. **Sumino, K., Yamamoto, R., Hatayama, F., Kitamura, S., and Itoh, H.,** Laser AAS for histochemistry, *Anal. Chem.,* 52, 1064, 1980.
450. **Langmyhr, F. J., and Kjuus, I.,** Direct AAS determination of Cd, Pb, Mn in bone and of Pb in ivory, *Anal. Chim. Acta,* 100, 139, 1978.
451. **Langmyhr, F. J., Sundly, A., and Jonsen, J.,** AAS determination of Cd and Pb in dental material by atomization directly from the solid state, *Anal. Chim. Acta,* 73, 81, 1974.
452. **Nichols, J. A., Jones, R. D., and Woodriff, R.,** Background reduction during direct atomization of solid biological samples in AAS, *Anal. Chem.,* 50, 2071, 1978.
453. **Koizumi, H. and Yasuda, K.,** Determination of Pb, Cd, and Zn using the Zeeman effect in AAS, *Anal. Chem.,* 48, 1178, 1976.
454. **Chakrabarti, C. L., Wan, C. C., and Li, W. C.,** Direct determiation of traces of Cu, Zn, Pb, Co, Fe and Cd in bovine liver by graphite-furnace AAS using the solid sampling and the platform techniques, *Spectrochim. Acta,* 35B, 93, 1980.
455. **Chakrabarti, C. L., Wan, C. C., and Li, W. C.,** AAS determination of Cd, Pb, Zn, Cu, Co and Fe in oyster tissue by direct atomization from the solid state using the graphite platform technique, *Spectrochim. Acta,* 35B, 547, 1980.
456. **Fernandez, F. J., Beaty, M. M., and Barnett, W. B.,** Use of L'vov platform for furnace AA applications, *At. Spectrosc.,* 2, 16, 1981.
457. **Hinderberger, E. J., Kaiser, M. L., and Koirtyohann, S. R.,** Furnace AAA of biological samples using the L'vov platform and matrix modification, *At. Spectrosc.,* 2, 1, 1981.
458. **Paus, P. E.,** Bomb decomposition of biological materials, *At. Absorpt. Newsl.,* 11, 129, 1972.
459. **Oleru, U. G.,** Epidemiological implications of environmental Cd. I. The probable utility of human hair for occupational trace metal (Cd) screening, *Am. Ind. Hyg. Assoc. J.,* 36, 229, 1975.
460. **Sorenson, J. R. J., Melby, E. G., Nord, P. J., and Petering, H. G.,** Interferences in the determination of metallic elements in human hair. An evaluation of Zn, Cu, Pb, and Cd using AAS, *Arch. Environ. Health,* 27, 36, 1973.
461. **Schroeder, H. A. and Nason, A. P.,** Trace elments in human hair, *J. Invest. Dermatol.,* 53, 71, 1969.
462. **Locke, J.,** The determination of eight elements in human liver tissue by flame AAS in sulphuric acid solution, *Anal. Chim. Acta,* 104, 225, 1979.
463. **Carpenter, R. C.,** The determination of Cd, Cu, Pb and Tl in human liver and kidney tissue by flame AAS after enzymatic digestion, *Anal. Chim. Acta,* 125, 209, 1981.
464. **Murthy, L., Menden, E. E., Eller, P. M., and Petering, H. G.,** AA determination of Zn, Cu, Cd, and Pb in tissues solubilized by aqueous tetramethylammonium hydroxide, *Anal. Biochem.,* 53, 365, 1973.
465. **Kaplan, P. D., Blackstone, M., and Richdale, N.,** Direct determination of Cd, Ni, and Zn in rat lungs by AAS, *Arch. Environ. Health,* 27, 387, 1973.
466. **Hinners, T. A., Terrill, W. J., Kent, J. L., and Colucci, A. V.,** Hair-metal binding, *Environ. Health Perspect.,* 8, 191, 1974.
467. **Hinners, T. A.,** AAA of liver without ashing, *Fresenius Z. Anal. Chem.,* 277, 377, 1975.
468. **Westerlund-Helmerson, U.,** Determination of Pb and Cd in blood by a modification of the Hessel method, *At. Absorpt. Newsl.,* 9, 133, 1970.
469. **Iguchi, T., Tati, M., and Miyata, S.,** Determination of Cd in urine and blood by AAS, *Igaku to Seibutsugaku,* 73, 286, 1966.
470. **Yeager, D. W., Cholak, J., and Meiners, B. G.** Determination of Cd in biological and related material by AA, *Am. Ind. Hyg. Assoc. J.,* 34, 450, 1973.
471. **Amore, F.,** Determination of Cd, Pb, Tl, and Ni in blood by AAS, *Anal. Chem.,* 46, 1597, 1974.
472. **Berman, E.,** Determination of Cd, Tl and Hg in biological materials by AA, *At. Absorpt. Newsl.,* 6, 57, 1967.

473. **Brovko, I. A.,** Diphenylcarbazone as a reagent for extraction of Cd, Co, Cu, Mn, Ni, and Zn, *Zh. Anal. Khim.* 35, 2095, 1980.
474. **Ármannsson, H.,** Dithizone extraction and flame AAS for the determination of Cd, Zn, Pb, Cu, Ni, Co and Ag in sea water and biological tissues, *Anal. Chim. Acta,* 110, 21, 1979.
475. **Ágemian, H., Sturtevant, D. P., and Austen, K. D.,** Simultaneous acid extraction of six trace metals from fish tissue by hot-block digestion and determination by AAS, *Analyst,* 105, 125, 1980.
476. **Tsutsumi, C., Koizumi, H., and Yoshikawa, S.,** AAS determination of Pb, Cd, and Cu in foods by simultaneous extraction of the iodides with MIBK, *Bunseki Kagaku,* 25, 150, 1976.
477. **Evans, W. H., Read, J. I., and Lucas, B. E.,** Evaluation of a method for the determination of total Cd, Pb and Ni in foodstuffs using measurement by flame AAS, *Analyst,* 103, 580, 1978.
478. **Matsuzawa, K.,** AAS determination of Cd, Cu and Pb in biological materials, *Nagano-ken Eisei Kogai Kenkyusho Kenkyu Hokoku,* 1, 15, 1979.
479. **Holroyd, P. M. and Snodin, D. J.,** Determination of Pb and Cd in tap water, rain water, milk, and urine, *J. Assoc. Publ. Anal.,* 10, 110, 1972.
480. **Snodin, D. J.,** Pb and Cd in baby foods, *J. Assoc. Publ. Anal.,* 11, 112, 1973.
481. **Kaneko, Y.,** Trace elements in human teeth. I. Preliminary studies on the analytical method, *Koku Eisei Gakkai Zasshi,* 21, 227, 1971.
482. **Brovko, I. A., Nazarov, S. N., and Rish, M. A.,** AA determination of Zn, Cd, Co, Cu, and Ni after their extraction concentration in the diphenylcarbazon-pyridine-toluene system, *Zh. Anal. Khim.,* 29, 2387, 1974.
483. **Iguchi, T.,** Determination of Pb, Cd, and Zn in urine by AAS, *Gifu Daigaku Igakubu Kiyo,* 15, 840, 1968.
484. **Pulido, P., Fuwa, K., and Vallee, B. L.** Deterination of Cd in biological materials by AAS, *Anal. Biochem.,* 14, 393, 1966.
485. **Harnly, J. M. and O'Haver, T. C.,** Background correction for the analysis of high-solid samples by graphite furnace AA, *Anal. Chem.,* 49, 2187, 1977.
486. **Joselow, M. M. and Bogden, J. D.,**Multi-element microanalyses by Delves cup-AAS on chelate solvent extraction, *At. Absorpt. Newsl.,* 11, 127, 1972.
487. **Bogden, J. D. and Joselow, M. M.,** A micro-method for the collection and analysis of trace metals by paper-in-a-cup AAS, *Am. Ind. Hyg. Assoc. J.,* 35, 88, 1974.
488. **Carter, G. F. and Yeoman, W. B.,** Determination of Cd in blood after destruction of organic material by low-temperature ashing, *Analyst,* 105, 295, 1980.
489. **Delves, H. T.** Simple matrix modification procedure to allow the direct determination of Cd in blood by flame micro-sampling AAS, *Analyst,* 102, 403, 1977.
490. **Ediger, R. D. and Coleman, R. L.,** Determination of Cd in blood by a Delves cup technique, *At. Absorpt. Newsl.,* 12, 3, 1973.
491. **Lieberman, K. W.,** Determination of Cd in biological fluids by the Delves modification of AAS, *Clin. Chim. Acta,* 46, 217, 1973.
492. **Jackson, K. W. and Mitchell, D. G.,** Rapid determination of Cd in biological tissues by microsampling-cup AAS, *Anal. Chim. Acta,* 80, 39, 1975.
493. **Hauser, T. R., Hinners, T. A., and Kent, J. L.,** AA determination of Cd and Pb in whole blood by a reagent-free method, *Anal. Chem.,* 44, 1819, 1972.
494. **Julshamn, K. and Braekkan, O. R.,** Determination of trace elements in fish tissue by the standard addition method, *At. Absorpt. Newsl.,* 14, 49, 1975.
495. **Holak, W.,** Determination of traces of Pb and Cd in foods by AAS using the "Sampling Boat", *At. Absorpt. Newsl.,* 12, 63, 1973.
496. **Torres, F.,** AA Microsampling of Pb, Cd, In, and Tl in Urine, SC-RR-69-784, U.S. Atomic Energy Comm., Washington, D.C., 1969.
497. **Lundgren, G.,** Direct determination of Cd in blood with a temperature-controlled heated graphite-tube atomizer, *Talanta,* 23, 309, 1976.
498. **Cernik, A. A. and Sayers, M. H. P.,** Application of blood Cd determination to industry using a punched disc technique, *Br. J. Ind. Med.,* 32, 155, 1975.
499. **Herber, R. F. M. and de Boer, J. L. M.,** Simple background-monitoring device for AAS, *Anal. Chim. Acta,* 109, 177, 1979.
500. **De Castilho, P. and Herber, R. F. M.,** The rapid determination of Cd, Pb, Cu and Zn in whole blood by AAS with electrothermal atomization. Improvements in precision with a peak-monitoring device, *Anal. Chim. Acta,* 94, 269, 1977.
501. **Posma, F. D., Balke, J., Herber, R. F. M., and Stuik, E. J.,** Microdetermination of Cd and Pb in whole blood by flameless AAS using carbon-tube and carbon-cup as sample cell and comparison with flame studies, *Anal. Chem.,* 47, 834, 1975.
502. **Subramanian, K. S. and Méranger, J. C.,** A rapid electrothermal AAS method for Cd and Pb in human whole blood, *Clin. Chem.,* 27, 1866, 1981.

503. **Delves, H. T. and Woodward, J.,** Determination of low levels of Cd in blood by electrothermal atomization and AAS, *At. Spectrosc.*, 2, 65, 1981.
504. **Schumaker, E. and Umland, F.,** Improved fast destruction method for the determination of Cd in body fluids using a graphite-tube atomizer, *Fresenius Z. Anal. Chem.*, 270, 285, 1974.
505. **Langmyhr, F. J. and Aamodt, J.,** AAS determination of some trace metals in fish meal and bovine liver by the solid sampling technique, *Anal. Chim. Acta*, 87, 483, 1976.
506. **Parker, C. R., Rowe, J., and Sandoz, D. P.,** Methods of environmental Cd determination by flameless atomization, *Am. Lab.*, 5, 53, 1973.
507. **Ross, R. T. and Gonzalez, J. G.,** The direct determination of Cd in biological samples by selective volatization and graphite tube reservoir AAS, *Anal. Chim. Acta*, 70, 443, 1974.
508. **Ottaway, J. M. and Campbell, W. C.,** Determination of Cd in livers and kidneys of puffins by carbon furnace AAS, *Int. J. Environ. Anal. Chem.*, 4, 233, 1976.
509. **Chakrabarti, C. L., Wan, C. C., Hamed, H. A., and Bertels, P. C.,** Trace-element determination by capacitive-discharge AAS, *Nature (London)*, 288, 246, 1980.
510. **Carmack, G. D. and Evenson, M. A.,** Determination of Cd in urine by electrothermal AAS, *Anal. Chem.*, 51, 907, 1979.
511. **Buttgereit, G.,** Modified AAS method in analysis of trace metals, *Fresenius Z. Anal. Chem.*, 267, 81, 1973.
511a. **Bruhn, C. F. and Navarette, G. A.,** Matrix modification for the direct determination of Cd in urine by electrothermal AAS, *Anal. Chim. Acta*, 130, 209, 1981.
512. **Brodie, K. G. and Stevens, B. J.,** Measurements of whole blood Pb and Cd at low levels using an automated sample dispenser and furnace AA, *J. Anal. Toxicol.*, 1, 282, 1977.
513. **Poldoski, J. E.,** Determination of Pb and Cd in fish and clam tissue by AAS with a Mo and La treated pyrolytic graphite atomizer, *Anal. Chem.*, 52, 1147, 1980.
514. **Koops, J. and Westerbeek, D.,** Determination of Pb and Cd in pasteurized liquid milk by flameless AAS, *Neth. Milk Dairy, J.*, 32, 149, 1978.
515. **Evenson, M. A. and Anderson, C. T., Jr.,** Ultramicro analysis for Cu, Cd, and Zn in human liver tissue by use of AAS and the heated graphite tube atomizer, *Clin. Chem.*, 21, 537, 1975.
516. **Kawaraya, T., Kawasaki, M., Haruki, K., Tomita, K., Oka, M., and Horiguchi, S.,** Determination of trace metals in mammalian lungs by flameless AAS, *Bunseki Kagaku*, 25, 464, 1976.
517. **Slavin, S., Peterson, G. E., and Lindahl, P. C.,** Determination of heavy metals in meats by AAS, *At. Absorpt. Newsl.*, 14, 57, 1975.
518. **Schramel, P., Wolf, A., Seif, R., and Klose, B.-J.,** New device for ashing of biological material under pressure, *Fresenius Z. Anal. Chem.*, 302, 62, 1980.
519. **Robbins, W. B. and Caruso, J. A.,** Determination of Pb and Cd in normal human lung tissue by flameless AAS, *Spectrosc. Lett.*, 11, 333, 1978.
520. **Branuša, M. and Breški, D.,** Comparison of dry and wet ashing procedures for Cd and Fe determination in biological material by AAS, *Talanta*, 28, 681, 1981.
521. **Gross, S. B. and Parkinson, E. S.,** Analyses of metals in human tissues using base (TMAH) digests and graphite furnace AAS, *At. Absorpt. Newsl.*, 13, 107, 1974.
522. **Sperling, K.-R. and Bahr, B.,** Determination of extremely low concentrations of Cd in blood and urine by flameless AAS. I. Testing of a micromethod, *Fresenius Z. Anal. Chem.*, 301, 29, 1980.
523. **Sperling, K.-R. and Bahr, B.,** Determination of extremely low concentrations of Cd in blood and urine by flameless AAS. II. Extraction behavior of Cd, *Fresenius Z. Anal. Chem.*, 301, 31, 1980.
524. **Inoue, T. and Nagada, S.,** Determination of cadmium in blood by flameless atomic absorption method, *Kyoto-fu Eisei Kogai Kenkyusho Nempo*, 22, 53, 1978.
525. **Dornemann, A. and Kleist, H.,** Determination of Cd in whole blood, *Zentralbl. Arbeitsmed. Arbeitsschutz. Prophyl.*, 28, 165, 1978.
526. **Tulley, R. T. and Lehmann, H. P.,** Method for simultaneous determination of Cd and Zn in whole blood by AAS and measurement in normotensive and hypertensive humans, *Clin. Chim. Acta*, 122, 189, 1982.
527. **Sperling, K.-R.,** Determination of heavy metals in sea water and in marine organisms by flameless AAS. IX. Determination of Cd traces in biological materials by a simple extraction method, *Fresenius Z. Anal. Chem.*, 299, 103, 1979.
528. **Yasuda, K., Toda, S., Igarashi, C., and Tamura, S.,** Extraction system for solvent extraction — graphite furnace AAS, *Anal. Chem.*, 51, 161, 1979.
529. **Kubasik, N. P. and Volosin, M. T.,** Simplified determination of urinary Cd, Pb, and Tl with use of carbon rod atomization and AAS, *Clin. Chem.*, 19, 954, 1973.
530. **Boiteau, H. L. and Metayer, C.,** Mictodetermination of Pb, Cd, Zn, and Sn in biological materials by AAS after mineralization and extraction, *Analusis*, 6, 350, 1978.
531. **Stoeppler, M., Valenta, P., and Nürnberg, H. W.,** Application of independent methods and standard materials: an effective approach to reliable trace and ultratrace analysis of metals and metalloids in environmental and biological matrices, *Fresenius Z. Anal. Chem.*, 297, 22, 1979.

532. **Iyengar, G. V. and Kasperek, K.,** Application of the brittle fraction technique (BFT) to homogenise biological samples and some observations regarding the distribution behavior of the trace elements at different concentration levels in a biological matrix, *J. Radioanal. Chem.*, 39, 301, 1977.
533. **Nichols, J. A. and Hageman, L. R.,** Noncontaminating, representative sampling by shattering of cold, brittle, biological tissues, *Anal. Chem.*, 51, 1591, 1979.
534. **Koirtyohann, S. R. and Hopkins, C. A.,** Loss of trace metals during ashing of biological materials, *Analyst,* 101, 870, 1976.
535. **Koh, T.-S.,** Microwave drying of biological tissues for trace element determination, *Anal. Chem.*, 52, 1978, 1980.
536. **de Goeij, J. J. M., Volkers, K. J., and Tjioe, P. S.,** A search for losses of Cr and other trace elements during lyophilization of human liver tissue, *Anal. Chlm. Acta,* 109, 139, 1979.
537. **Feinberg, M. and Ducauze, C.,** High temperature dry ashing of foods for AAS determination of Pb, Cd, and Cu, *Anal. Chem.*, 52, 207, 1980.
538. **Menden, E. E., Brockman, D., Choudhury, H., and Petering, H. G.,** Dry-ashing of animal tissues for AAS determination of Zn, Cu, Cd, Pb, Fe, Mn, Mg, and Ca, *Anal. Chem.*, 49, 1644, 1977.
539. **Fernandez, F. J. and Kahn, H. L.,** Clinical methods for AAS, *Clin. Chem. Newsl.*, 3, 24, 1971.
540. **Scheubeck, E.,** A rapid enrichment method for the analysis of traces of metals in aqueous solutions, *Mikrochim. Acta,* II, 283, 1980.
541. **Frank, A.,** Automated wet ashing and multi-metal determination in biological materials by AAS, *Fresenius Z. Anal. Chem.*, 279, 101, 1976.
542. **Oehme, M.,** Inexpensive wet-ashing unit for routine trace analysis, *Anal. Chim. Acta,* 109, 195, 1979.
543. **Lekehal, N., Hanocq, M., and Helson-Cambier, M.,** Determination of traces of Cd by AA after extraction with dipivaloylmethane. Possible application to the study of Cd in urine, *J. Pharm. Belg.*, 32, 76, 1977.
544. **Kotz, L., Henze, G., Kaiser, G., Pahlke, S., Veber, M., and Tölg, G.,** Wet mineralisation of organic matrices in glassy-carbon vessels in pressure-bomb system for trace-element analysis, *Talanta,* 26, 681, 1979.
545. **Gaffin, S. L.,** Rapid solubilisation of human body tissues and tissue fluids for micro-determination of heavy metals, *Clin. Toxicol.*, 15, 293, 1979.
546. **Vens, M. D. and Lauwerys, R.,** Simultaneous determination of Pb and Cd in blood and urine by coupling ion exchange resin chromatography and AAS, *Arch. Mal. Prof. Med. Trav. Secur. Soc.*, 33, 97, 1972.
547. **Tsalev, D. L., Mzhel'skaya, T. I., and Larskii, E. G.,** Extraction AAA in biomedical studies (a review), *Med. Ref. Zh.*, 22(5), 41, 1976.
548. **Childs, E. A. and Gaffke, J. N.,** Organic solvent extraction of Pb and Cd from aqueous solutions for AAS measurements, *J. Assoc. Off. Anal. Chem.*, 57, 360, 1974.
549. **Sperling, K.-R.,** Determination of heavy metals in sea water and in marine organisms by flameless AAS. XIV. Comments on the usefulness of organohalides as solvents for the extraction of heavy metal (Cd) complexes, *Fresenius Z. Anal. Chem.*, 310, 254, 1982.
550. **Betz, M., Gücer, S., and Fuchs, F.,** Investigation on background absorption caused by organic solvents in flameless AAS, *Fresenius Z. Anal. Chem.*, 303, 4, 1980.
551. **Sturgeon, R. E., Chakrabarti, C. L., and Bertels, P. C.,** Atomization in graphite-furnace AAS. Peak height method vs. integration method of measuring absorbance: HGA 2100, *Anal. Chem.*, 47, 1250, 1975.
552. **Sturgeon, R. E., Chakrabarti, C. L., Maines, I. S., and Bertels, P. C.,** Atomization in graphite-furnace AAS. Peak height method vs. integration method of measuring absorbance: CRA 63, *Anal. Chem.*, 47, 1240, 1975.
553. **Slavin, W., Manning, D. C., and Carnick, G. R.,** The stabilized temperature platform furnace, *At. Spectrosc.*, 2, 137, 1981.
554. **Nomiyama, H., Yotoriyama, M., and Nomiyama, K.,** Normal Cr levels in urine and blood of Japanese subjects determined by direct flameless AAS and valency of Cr in urine after exposure to hexavalent Cr, *Am. Ind. Hyg. Assoc. J.*, 41, 98, 1980.
555. **Hambidge, K. M., Franklin, M. L., and Jacobs, M. A.,** Changes in hair Cr concentrations with increasing distances from the hair roots, *Am. J. Clin. Nutr.*, 25, 380, 1972.
556. **Sheard, E. A., Johnson, M. W., and Carter, R. J.,** The determination of Cr in hair and other biological materials, in *Hair, Trace Elements, and Human Illness,* Brown, A. C. and Crounse, R. G., Eds., Praeger Publishers, New York, 1980, 74.
557. **Pleban, P. A. and Matsushige, L.,** Determination of Cr, Cu, Mn, and Zn in Hair from Apparently Healthy Individuals and Diabetics, Paper 506, Jt. Meat. AACC and CSCC, Boston, July 20 to 25, 1980.
558. **Kumpulainen, J.,** Determination of Cr in human milk and urine by graphite-furnace AAS, *Anal. Chim. Acta,* 113, 355, 1980.
559. **Kumpulainen, J., Koivistoinen, P., Vuori, E., and Lehto, J.,** Direct Determination of Cr in Biological Fluids by Graphite Furnace AA, Paper 367, Euroanalysis-IV, Helsinki/Espoo, August 23, to 28, 1981.

560. **Kanabrocki, E. L., Kanabrocki, J. A., Greco, J., Kaplan, E., Oester, Y. T., Brar, S. S., Gustafson, P. S., Nelson, D. M., and Moore, C. E.,** Instrumental analysis of trace elements in thumbnails of human subjects, *Sci. Total Environ.*, 13, 131, 1979.
561. **Kayne, F. J., Komar, G., Laboda, H., and Vanderlinde, R. E.,** AAS of Cr in serum and urine with a modified Perkin-Elmer 603 AA spectrophotometer, *Clin. Chem.*, 24, 2151, 1978.
562. **Versieck, J., Hoste, J., Barbier, F., Steyaert, H., DeRudder, J., and Michels, H.,** Determination of Cr and Co in human serum by neutron activation analysis, *Clin. Chem.*, 24, 303, 1978.
563. **Vanderlinde, R. E., Kayne, F. J., Simmons, M. J., Tsou, J. Y., and Lavine, R. L.,** Measurement of Serum and Urinary Cr in Healthy Individuals, Paper 242, 31st Annu. Natl. Meet. Am. Assoc. Clin. Chem., New Orleans, July 15 to 20, 1979.
564. **Suddick, R. P., Hyde, R. J., and Feller, R. P.,** Salivary water and electrolytes and oral health, in *The Biological Basis of Dental Caries. An Oral Biology Textbook*, Menaker, L., Ed., Harper & Row, Hagerstown, Md., 1980, 132.
565. **Molokhia, A., Portnoi, B, and Dyer, A.,** Multi-element INAA of human skin, *Anal. Proc.*, 18, 160, 1981.
566. **Hambidge, K. M., Franklin, M. L., and Jacobs, M. A.,** Hair Cr concentration: effect of sample washing and external environment, *Am. J. Clin. Nutr.*, 25, 384, 1972.
567. **Kayne, F. J., Vanderlinde, R. S., Simmons, M., Milne, D., and Sandstead, H. H.,** The Measurement of Cr in Human Nutriture, Paper 501, Jt. Meet. AACC and CSCC, Boston, July 20 to 25, 1980.
568. **Veillon, C., Wolf, W. R., and Guthrie, B. E.,** Determination of Cr in biological materials by stable isotope dilution, *Anal. Chem.*, 51, 1022, 1979.
569. **Hayashi, Y., Hunakawa, K., Yoshida, N., Ishizawa, M., and Tsujino, R.,** Determination of the ppb levels of Cr in human urine by flameless dual channel AAS, *Bunseki Kagaku*, 25, 409, 1976.
570. **Shimomura, S., Morita, H., Kubo, M., Kondo, H., Kitamura, T., Kawano, M., Katata, M., Takata, I., and Kondo, M.,** Determination of Cr in urine samples by a flameless AA method, *Eisei Kagaku*, 21, 369, 1975.
571. **Routh, M. W.,** Analytical Parameters for determination of Cr in urine by electrothermal AAS, *Anal. Chem.*, 52, 182, 1980.
572. **Schaller, K. H., Essing, H.-G., Valentin, H., and Schäche, G.,** The quantitative determination of Cr in urine by flameless AAS, *Z. Klin. Chem. Klin. Biochem.*, 10, 434, 1972.
573. **Zdankiewicz, D. D. and Fasching, J. L.,** Analysis of whole blood by neutron activation: a search for a biochemical indicator of neoplasia, *Clin. Chem.*, 22, 1361, 1976.
574. **Behne D., Brätter, P., Gessner, H., Hübe, G., Mertz, W., and Rösick, V.,** Problems in the determination of Cr in biological materials. Comparison of flameless AAS and activation analysis, *Fresenius Z. Anal. Chem.*, 278, 269, 1976.
575. **Pierce, J. O., Lichte, F. E., Hastings Vogt, C. R., Abu-Samra, A., Ryan, T. R., Koirtyohann S. R., and Vogt, J. R.,** Comparison of Cr determinations in environmental and biological samples by NAA, AA, and GC, in *Measurement, Detection and Control of Environmental Pollutants*, IAEA, Vienna, 1976, 357.
576. **Agterdenbos, J., van Broekhoven, L., Jütte, B. A. H. G., and Schuring, J.,** Spectrophotometric determination of 0—50 ng of Cr in 1 mℓ of human serum, *Talanta*, 19, 341, 1972.
577. **Müller, H., Otto, M., and Werner, G.,** *Katalytischen Methoden in der Spurenanalyse*, Akad. Verlagsges. Geest & Portig, K.-G., Leipzig, 1980.
578. **Beyermann, K., Rose, H. J., Jr., andChristian, R. P.,** The determination of ng amounts of Cr in human urine by X-ray fluorescence spectroscopy, *Anal. Chim. Acta.*, 45, 51, 1969.
579. **Hambidge, K. M.,** Use of static argon atmosphere in emission spectrochemical determination of Cr in biological materials, *Anal. Chem.*, 43, 103, 1971.
580. **Black, M. S. and Browner, R. F.,** Volatile metal-chelate sample introduction for CIP-AES, *Anal. Chem.*, 53, 249, 1981.
581. **Guthrie, B. E., Wolf, W. R., and Veillon, C.,** Background correction and related problems in the determination of Cr in urine by graphite furnace AAS, *Anal. Chem.*, 50, 1900, 1978.
582. **Beaty, M., Barnett, W., and Grobenski, Z.,** Techniques for analysing difficult samples with the HGA graphite furnace, *At. Spectrosc.*, 1, 72, 1980.
583. **Chao, S. S. and Pickett, E. E.,** Trace Cr determination in furnace AAS following enrichment by extraction, *Anal. Chem.*, 52, 335, 1980.
584. **Pekarek, R. S., Hauer, E. C., Wannemacher, R. W., Jr., and Beisel, W. R.,** The direct determination of serum chromium by an AA spectrophotometer with a heated graphite atomizer, *Anal. Biochem.*, 59, 283, 1974.
585. **Veillon, C., Patterson, K. Y., and Bryden, N. A.,** Direct determination of Cr in human urine by electrothermal AAS, *Anal. Chim. Acta*, 136, 233, 1982.
586. **Matsusaki, K., Yoshino, T., and Yamamoto, Y.,** The removal of chloride interference in determination of chromate ion by AAS with electrothermal atomization, *Anal. Chim. Acta*, 113, 247, 1980.

587. **Devoto, G.,** Determination of Cr(VI) in biological fluids by AAS, *Boll. Soc. Ital. Biol. Sper.*, 44, 1251, 1968.
588. **Feldman, F. J., Knoblock, E. C., and Purdy, W. C.,** The determination of Cr in biological materials by AAS, *Anal. Chim. Acta*, 38, 489, 1967.
589. **Guněaga, J., Lentner, C., and Haas, H. G.,** Determination of Cr in feces by AAS, *Clin. Chim. Acta*, 57, 77, 1974.
590. **Jackson, F. J., Read, J. I., and Lucas, B. E.,** Determination of total Cr, Co and Ag in foodstuffs by flame AAS, *Analyst*, 105, 359, 1980.
591. **Méranger, J. C. and Smith, D. C.,** The heavy metal content of a typical Canadian diet, *Can. J. Public Health*, 63, 53, 1972.
592. **Castellari, G. and Fiorentini, P.,** Method for the determining Cr in urine by flame spectrophotometry in AA, *G. Clin. Med. (Bologna)*, 54, 116, 1973.
593. **Jeejeebhoy, K. N., Chu, R. C., Marliss, E. B., Greenberg, G. R., and Bruce-Robertson, A.,** Chromium deficiency, glucose intolerance, and neuropathy reversed by Cr supplementation, in patient receiving long-term total parenteral nutrition, *Am. J. Clin. Nutr.*, 30, 531, 1977.
594. **Davidson, I. W. F. and Secrest, W. L.,** Determination of Cr in biological materials by AAS using a graphite furnace atomizer, *Anal. Chem.*, 44, 1808, 1972.
595. **Langaard, S., Gundersen, N., Tsalev, D. L., and Gylseth, B.,** Whole blood Cr level and Cr excretion in the rat after zinc chromate inhalation, *Acta Pharmacol. Toxicol.*, 42, 142, 1978.
596. **Tessari, G. and Torsi, G.,** Determination of sub-ng amounts of Cr in different matrices by flameless AAS, *Talanta*, 19, 1059, 1972.
597. **Kumpulainen, J., Anderson, R. A., Polanski, M. M., and Wolf, W. R.,** Chromium content in biologically active extracts of SRMs, in *Chromium in Nutrition and Metabolism*, Shapcott, D. and Hubert, J., Eds., Elsevier/North-Holland, Amsterdam, 1979, 79.
598. **Seeling, W., Grünert, A., Kienle, K.-H., Opferkuch, R., and Swobodnik, M.,** Determination of Cr in human serum and plasma by flameless AAS, *Fresenius Z. Anal. Chem.*, 299, 368, 1979.
599. **Kumpulainen, J.,** Effects of volatility and adsorption during dry ashing on determination of Cr in biological materials, *Anal. Chim. Acta*, 91, 403, 1977.
600. **Kumpulainen, J. T., Wolf, W. R., Veillon, C., and Mertz, W.,** Determination of Cr in selected United States diets, *J. Agric. Food Chem.*, 27, 490, 1979.
601. **Shapcott, D.,** Preparation of samples for analysis, in *Chromium in Nutrition and Metabolism*, Shapcott, D. and Hubert, J., Eds., Elsevier/North-Holland, Amsterdam, 1979, 43.
602. **Rosson, J. W., Foster, K. J., Walton, R. J., Monro, P. P., Taylor, T. G., and Alberti, K. G. M. M.,** Hair Cr concentrations in adult insulin-treated diabetics, *Clin. Chim. Acta*, 93, 299, 1979.
603. **Gráf-Harsányi, E. and Langmyhr, F. J.,** AAS determination of the total content and distribution of Cr in blood serum, *Anal. Chim. Acta*, 116, 105, 1980.
604. **Gráf-Harsányi, E. and Langmyhr, F. J.,** Chromium determination in freese-dried serum and its protein fractions, *Magy. Kem. Foly*, 86, 412, 1980.
605. **Alcock, N. W.,** Precision of Absorbance Measurements from Serum Using an Automated Graphite Furnace System: Potential for Measuring Mn and Cr in Undiluted Serum, Paper 168, Jt. Meet. AACC and CSCC, Boston, July 20 to 25, 1980.
606. **Graglage, B., Buttgereit, G., and Kuebler, W.,** Use of flameless AA for the measurement of the trace elements in human serum, *Verh. Dtsch. Ges. Inn. Med.*, 80, 1710, 1974.
607. **Grafflage, B., Buttgereit, G., Kübler, W., and Mertens, H. M.,** Measurement of trace elements Cr and Mn in serum using flameless AA, *Z. Klin. Chem. Klin. Biochem.*, 12, 287,1974.
608. **Arnold, E. L., Hawthorne, S. B., and MacDonald, R. J.,** Chromium Transferrin Binding and Its Use in Biological Cr Analysis, Paper 241, 31st Annu. Natl. Meet. AACC, New Orleans, July 15 to 20, 1979.

608a. **Thompson, D. A.,** Flameless AAS of plasma Cr, *Ann. Clin. Biochem.*, 17, 144, 1980.

609. **Lidén, S. and Lundberg, E.,** Penetration of Cr in intact human skin in vivo, *J. Invest. Dermatol.*, 72, 42, 1979.
610. **Pfueler, U., Fuchs, V., Golbs, S., Ebert, S., and Pfeifer, D.,** Solubilization to prepare samples for determination of heavy metals in biological objects by flameless AAS, *Arch. Exp. Veterinaermed.*, 34, 367, 1980.
611. **Chao, S. S., Kanabrocki, E. L., Moore, C. E., Oester, Y. T., Greco, J., and von Smolinski, A.,** Determination of trace elements in human tissues. II. Cr in the pancreas and in raw and commercial sugar, *Appl. Spectrosc.*, 30, 155, 1976.
612. **Shimizu, T., Hayama, T., Shijo, Y., and Sakai, K.,** Determination of total Cr in human urine by graphite furnace AAS after coprecipitation treatment, *Bunseki Kagaku*, 29, 680, 1980.
613. **Laboda, H. M., Vanderlinde, R. E., and Kayne, F. J.,** Isolation of Cr Compounds from Human Urine, Paper 169, Jt. Meet. AACC and CSCC, Boston, July 20 to 25, 1980.

614. **Nise, G. and Vesterberg, O.,** Direct determination of Cr in urine by electrothermal AAS, *Scand. J. Work, Environ. Health,* 5, 404, 1979.
615. **Ross, R. T., Gonzalez, J. G., and Segar, D. A.,** Direct determination of Cr in urine by selective volatization and atom reservoir AA, *Anal. Chim. Acta,* 63, 205, 1973.
616. **Shimomura, S., Hayashi, Y., and Morita, H.,** Determination of trace amounts of Cr in human urine by flameless AAS, *Eisei Kagaku,* 21, 204, 1975.
617. **Nishijima, T., Dokiya, Y., Iizuka, H., and Toda, S.,** AAS determination of Cr in urine samples, *Bunseki Kagaku,* 26, 349, 1977.
618. **Shimizu, T., Kimoto, R., Shijo, Y., and Sakai, K.,** Determination of Cr(III) and Cr(VI) in human urine by graphite-furnace AAS after co-precipitation, *Bunseki Kagaku,* 30, 66, 1981.
618a. **Veillon, C., Guthrie, B. E., and Wolf, W. R.,** Retention of Cr by graphite furnace tubes, *Anal. Chem.,* 52, 457, 1980.
619. **Minoia, C., Colli, M., and Pozzoli, L.,** Determination of hexavalent Cr in urine by flameless AAS, *At. Spectrosc.,* 2, 163, 1981.
620. **Behne, D.,** Sources of error in sampling and sample preparation for trace element analysis in medicine, *J. Clin. Chem. Clin. Biochem.,* 19, 115, 1981.
621. **Lakomaa, E.-L.,** Use of NAA in the determination of elements in human cerebrospinal fluid, in *Trace Element Analytical Chemistry in Medicine and Biology,* Brätter, P. and Schramel, P., Eds., Walter de Gruyter & Co., New York, 1980, 97.
622. **Versieck, J., Speecke, A., Hoste, J., and Barbier, F.,** Trace element contamination in biopsies of the liver, *Clin. Chem.,* 19, 472, 1973.
623. **Masironi, R. and Parr, R. M.,** Collection and trace element analysis of post-mortem human samples: the WHO/IAEA research programme on trace elements in cardiovascular diseases, in *The Use of Biological Specimens for the Assessment of Human Exposure to Environmental Pollutants,* Berlin, A., Wolff, A. H., and Hasegawa, Y., Eds., Martinus Nijhoff Publishers, Boston, 1979, 275.
624. **Van Grieken, R., Van de Velde, R., and Robberecht, H.,** Sample contamination from a commercial grinding unit, *Anal. Chim. Acta,* 118, 137, 1980.
625. **Chittleborough, G.,** A chemist's view on the analysis of human hair for trace elements, *Sci. Total Environ.,* 14, 53, 1980.
625a. **Kumpulainen, J., Salmela, S., Vuori, E., and Lehto, J.,** Effects of various washing procedures on the Cr content of human hair, *Anal. Chim. Acta,* 138, 361, 1982.
626. **Lutz, G. J., Stemple, J. S., and Rook, H. L.,** Evaluation by activation analysis of elemental retention in biological samples after low-temperature ashing, *J. Radioanal. Chem.,* 39, 277, 1977.
627. **Kumpulainen, J.,** personal communication, 1980.
628. **Fernandez, F. J., Bohler, W., Beaty, M. M., and Barnett, W. B.,** Correction for high background levels using the Zeeman effect, *At. Spectrosc.,* 2, 73, 1981.
629. **Tsalev, D. L. and Simovska, M.,** Spectral lines for successive background correction in AAS, *God. Sofii. Univ., Khim. Fak.,* 71, 67, 1976/77.
630. **Lidums, V. V.,** Determination of Co in blood and urine by electrothermal AAS, *At. Absorpt. Newsl.,* 18, 71, 1979.
631. **Krivan, V., Geiger, H., and Franz, H. E.,** Determination of Fe, Co, Cu, Zn, Se, Rb and Cs in NBS bovine liver, blood plasma, and erythrocytes by INAA and AAS, *Fresenius Z. Anal. Chem.,* 305, 399, 1981.
632. **Nodiya, P. I.,** Study of body Co and Ni balance in students of a technical trade school, *Gig. Sanit.,* 37, 108, 1972.
633. **Parr, R. M. and Taylor, D. M.,** The concentrations of Co, Cu, Fe, and Zn in some normal human tissues as determined by NAA,, *Biochem. J.,* 91, 424, 1964.
634. **Hubbard, D. M., Greech, F. M., and Cholak, J.,** Determination of Co in air and biological material, *Arch. Environ. Health,* 13, 190, 1966.
635. **Mansurov, K., Khatamov, Sh., Kist, A. A., and Zhuk, L. I.,** Preparation of fur and hair for NAA, in *Nuclear Techniques in Environmental Control,* Gidrometeoizdat, Leningrad, 1980, 238.
636. **Irons, R. D., Schrenk, E. A., and Giauque, R. D.,** EDXRF spectroscopy and ICP-ES evaluated for multielement analysis in complex biological matrices, *Clin. Chem.,* 22, 2018, 1976.
637. **Wang, P.-Y.,** Polarographic catalytic-wave method for determination of trace Co in biological tissues, *Fen Hsi, Hua Hsuen,* 8, 249, 1980.
638. **Riekkola, M.-L. and Mäkitie, O.,** Capillary GC determination of Co as diethyldithiocarbamate chelate in human tissue material, *Finn. Chem. Lett.,* 2, 56, 1980.
639. **Julshamn, K. and Braekkan, O. R.,** The determination of Co in fish tissue by AAS, *At. Absorpt. Newsl.,* 12, 139, 1973.
640. **Tsalev, D. L., Mzhel'skaya, T. I., and Larskii, E. G.,** Extraction of microelements from blood serum and their subsequent AA determination, *Lab. Delo,* 7, 390, 1973.

641. **Wilson, P. E.,** A Venturi sampler for aspirating small sample volumes in AAS, *At. Absorpt. Newsl.*, 18, 115, 1979.
642. **Jago, J., Wilson, P. E., and Lee, B. M.,** Determination of sub-μg amounts of Co in plants and animal tissues by extraction and AAS, *Analyst,* 96, 349, 1971.
643. **Aihara, M. and Kuboku, M.,** AAS of Co and Ni by using solvent extraction with K ethylxanthate-MIBK, *Bunseki Kagaku,* 24, 501, 1975.
644. **Suzuki, M., Hayashi, K., and Wacker, W. E.,** Determination of Co in biological materials by AAS, *Anal. Chim. Acta,* 104, 389, 1979.
645. **Bartfoot, R. A. and Pritchard, J. G.,** Determination of Co in blood, *Analyst,* 105, 551, 1980.
646. **Ishizaki, M., Oyamada, N., Fujiki, M., and Yamaguchi, S.,** Determination of Co in blood by flameless AAS using a carbon-tube atomizer, *Sangyo Igaku,* 20, 174, 1978.
647. **Oyamada, N. and Ishizaki, M.,** Determination of Co in biological samples by trioctylamine extraction-AAS using a carbon-tube atomizer, *Bunseki Kagaku,* 28, 289, 1979.
648. **Oyamada, N., Ishizaki, M., Ueno, S., Kataoka, F., Murakami, R., Kubota, K., and Katsumura, K,.** Determination of Co in biological samples by flameless AAS using carbon tube atomizer, *Ibaraki-ken Eisei Kenkyusho Nempo,* 15, 59, 1977.
649. **Alder, J. F., Pankhurst, C. A., Samuel, A. J., and West, T. S.,** The use of silk and animal hairs as standards for hair analysis, *Anal. Chim. Acta,* 91, 407, 1977.
650. **Alder, J. F., Samuel, A. J., and West, T. S.,** The simultaneous multi-element analysis of hair: a non-parametric method for evaluating the ability of the data to distinguish between individuals, *Anal. Chim. Acta,* 94, 187, 1977.
651. **Lagathu, J. and Desirant, J.,** Determination of metalic elements in milk by AA with a flameless atomization method, *Rev. Fr. Corps Gras,* 19, 169, 1972.
652. **Muzzarelli, R. A. A. and Rocchetti, R.,** AA determination of Mn, Co and Cu in whole blood and serum with a graphite atomizer, *Talenta,* 22, 683, 1975.
653. **Maessen, F. J. M. J., Posma, F. D., and Balke, J.,** Direct determination of Au, Co and Li in blood plasma using mini-Massmann carbon rod atomizer, *Anal. Chem.*, 46, 1445, 1974.
654. **Alt, F. and Massmann, H.,** Determination of trace elements (Mn, Co) in sera by AAS, *Fresenius Z. Anal. Chem.*, 279, 100, 1976.
655. **Lundgren, G. and Johansson, G.,** A temperature-controlled graphite tube furnace for the determination of trace metals in solid biological tissue, *Talanta,* 21, 257, 1974.
656. **Kasperek, K., Kiem, J., Iyengar, G. V., and Feinendegen, L. E.,** Concentration differences between serum and plasma of the elements Co, Fe, Hg, Rb, Se, and Zn determined by NAA, *Sci. Total Environ.*, 17, 133, 1981.
657. **Tsalev, D. L.,** AAS Determination of Trace Elements after Preconcentration by Extraction, Ph.D. thesis, Moscow State University, Moscow, 1972.
658. **Willems, M., Borggaard, O. K., Christensen, H. E. M., and Nielsen, T. K.,** Determination of Co in Plant Material by Solvent Extraction Graphite Furnace AA — Interference from Fe, Mn, Zn, and Cu, Paper 413p, Euroanalysis-IV, Helsinki/Espoo, 1981.
659. **Busev, A. I., Byrko, V. M., Terestchenko, A. P., Novikova, N. N., Naidina, V. P., and Terent'ev, P. B.,** AA and spectrographic determination of traces of heavy metals after their concentration by extraction with hexamethyleneammonium hexamethylenedithiocarbamate, *Zh. Anal. Khim.*, 25, 665, 1970.
660. **Cartwright, G. E. and Wintrobe, M. M.,** Copper metabolism in normal subjects, *Am. J. Clin. Nutr.*, 14, 224, 1964.
661. **Morita, K., and Shimizu, M.,** Evaluation of health on the basis of heavy metal concentration in blood, *Okayama-ken Kankyo Hoken Senta Nempo,* 3, 338, 1979.
662. **Backer, E. T.** Chloric acid digestion in the determination of trace metals (Fe, Zn, Cu) in brain and hair by AAS, *Clin. Chim. Acta,* 24, 233, 1969.
663. **Smeyers-Verbeke, J., Defrise-Gussenhoven, E., Ebinger, G., Löwenthal, A., and Massart, D. L.,** Distribution of Cu and Zn in human brain tissue, *Clin. Chim. Acta,* 51, 309, 1974.
664. **Meret, S. and Henkin, R. I.,** Simultaneous direct estimation by AAS of Cu and Zn in serum, urine and CSF, *Clin. Chem.*, 17, 369, 1971.
665. **Versieck, J., Speecke, A., Hoste, J., and Barbier, F.,** Determination of Mn, Cu and Zn in serum and packed blood cells by NAA, *Z. Klin. Chem. Klin. Biochem.*, 11, 193, 1973.
666. **Mishima, M., Kawamura, E., Hitosugi, N., Sugawara, K., and Suzuki, T.,** The metal concentration in Nepalese hair, *Koshu Eiseiin Kenkyu Hokoku,* 23, 227, 1974.
667. **Nayman, R., Thomson, M. E., Scriver, C. R., and Clow, C. L.,** Observation on the composition of milk-substitute products for treatment of inborn errors of amino acid metabolism. Comparisons with human milk. A proposal to rationalize nutrient content of treatment products, *Am. J. Clin. Nutr.*, 32, 1279, 1979.
668. **Picciano, M. F. and Guthrie, H. A.,** Copper, Fe and Zn contents of mature human milk, *Am. J. Clin. Nutr.*, 29, 242, 1976.

669. **Barnett, W. B. and Kahn, H. L.,** Determination of Cu in fingernails by AA with the graphite furnace, *Clin. Chem.*, 18, 923, 1972.
670. **Mahler, D. J., Scott, A. F., Walsh, J. R., and Haynie, G.,** A study of trace metals in fingernails and hair using NAA, *J. Nucl. Med.*, 11, 739, 1970.
671. **Arwill, T., Myrberg, N., and Söremark, R.,** The concentration of Cl, Na, Br, Cu, Sr, and Mn in human mixed saliva, *Odontologisk Revy (Malmö)*, 18, 1, 1967.
672. **Kanabrocki, E. L., Case, L. F., Fields, T., Graham, L., Miller, E. B., Oester, Y. T., and Kaplan, E.,** Nondialysable Mn, Cu and Au levels in saliva of normal adult subjects, *J. Nucl. Med.* 6, 489, 1965.
673. **Faulkner, W. R., King, J. W., and Damm, H. C., Eds.,** *Handbook of Clinical Laboratory Data*, 2nd. ed., Chemical Rubber Co., Cleveland, 1968.
674. **Sunderman, F. W., Jr. and Roszel, N. O.,** Measurement of Cu in biological materials by AAS, *Am. J. Clin. Pathol.*, 48, 286, 1967.
675. **Hohnadel, D. G., Sunderman, F. W., Jr., Nechay, M. W., and McNeely, M. D.,** AAS of Ni, Cu, Zn and Pb in sweat collected from healthy subjects during sauna bathing; *Clin. Chem.*, 19, 1288, 1973.
676. **Söremark, R. and Samsahl, K.,** Gamma-ray spectrometric analysis of elements in normal human dentin, *J. Dent. Res.*, 41, 603, 1962.
677. **Spector, H., Glusman, S., Jatlow, P., and Seligson, D.,** Direct determination of Cu in urine by AAS, *Clin. Chim. Acta*, 31, 5, 1971.
678. **Rice, E. W.,** Copper in serum, *Stand. Methods Clin. Chem*,. 4, 57, 1963.
679. **Landers, W. J. and Zak, B.,** Determination of serum Cu and Fe in a single small sample, *Am. J. Clin. Pathol.*, 29, 590, 1958.
680. **Smeyers-Verbeke, J., Massart, D. L., Versieck, J., and Speecke, A.,** The determination of Cu and Zn in biological materials. A comparison of AA with spectrophotometry and neutron activation, *Clin. Chim. Acta*, 44, 243, 1973.
681. **Roschnik, M. R.,** Determination of traces of Cu in liquid milk and milk powder by AAS, *Mitt. Geb. Lebensmittelunters. Hyg.*, 63, 206, 1972.
682. **Dittel, F.,** Quantitative selective Bestimmung kleinster Mengen von Mangan und Kupfer in der Asche biologischer Substanzen unter Verwendung von Katalytischen Reaktionen, *Z. Anal. Chem.*, 229, 193, 1967.
683. **Batley, G. E. and Farrar, Y. J.,** Irradiation techniques for the release of bound heavy metals in natural waters and blood, *Anal. Chim. Acta*, 99, 283, 1978.
684. **Lund, W. and Eriksen, R.,** The determination of Cd, Pb and Cu in urine by DPASV, *Anal. Chim. Acta*, 107, 37, 1979.
685. **Kocsis, E. and Kovács, M.,** Radiochemical Determination of Several Trace Metals in Blood with Using Selective Adsorbent, 8th Int. Microchim. Symp., Graz, August 25 to 30, 1980, 152.
686. **Ward, N. I., Stephens, R., and Ryan, D. E.,** Comparision of three analytical methods for the determination of trace elements in whole blood, *Anal. Chim. Acta*, 110, 9, 1979.
687. **Kanabrocki, E. L., Case, L. F., Fields, T., Graham, L., Oester, Y. T., and Kaplan, E.,** Manganese and Cu determinations in body fluids, in *Developments in Applied Spectroscopy*, Vol. 5, Pearson, L. A. and Grove, E. L., Eds., Plenum Press, New York, 1965, 471.
688. **Chen, R., Lambert, C., Wilcox, A., Knall, J., and Dahlquist, R.,** Simultaneous Determination of Na, K, Li, Ca, Mg, Fe, Cu, Zn, and Au in 500 $\mu\ell$ of Serum by ICP-AES, Paper 249, 31st Annu. Natl. Meet. AACC, New Orleans, July 15 to 20, 1979.
689. **Uchida, H., Nojiri, Y., Haraguchi, H., and Fuwa, K.,** Simultaneous multi-element analysis by ICP-ES utilizing microsampling techniques with internal standard, *Anal. Chim. Acta*, 123, 57, 1981.
690. **Jackson, C. J., Porter, D. G., Dennis, A. L., and Stockwell, P. B.,** Automated digestion and extraction apparatus for use in the determination of trace metals in foodstuffs, *Analyst*, 103, 317, 1978.
691. **Paradelis, T.,** Determination of trace elements in whole blood by photon-induced X-ray fluorescence, *Eur. J. Nucl. Med.*, 2, 277, 1977.
692. **Bank, H. L., Robson, J., Bigelow, J. B., Morrison, J., Spell, L., H., and Kantor, R.,** Preparation of fingernails for trace element analysis, *Clin. Chim. Acta*, 116, 179, 1981.
693. **Flint, R. W., Lawson, C. D., and Standil, S.,** Application of trace element analysis by XRF to human blood serum, *J. Lab. Clin. Med.*, 85, 155, 1975.
694. **Knoth, J., Schwenke, H., Marten, R., and Glauer, J.,** Determination of Cu and Fe in human blood serum by energy dispersive X-ray analysis, *J. Clin. Chem. Clin. Biochem.*, 15, 557, 1977.
695. **Vos, L., Robberecht, H., Van Dyck, P., and Van Grieken, R.,** Multielement analysis of urine by energy-dispersive X-ray fluorescence spectrometry, *Anal. Chim. Acta*, 130, 167, 1981.
696. **Bearse, R. C., Close, D. A., Malanifi, J. J., and Umbarger, C. J.,** Elemental analysis of whole blood using proton-induced X-ray emission, *Anal. Chem.*, 46, 499, 1974.
697. **Khan, A. H., Khaliquzzmann, M., Zaman, M. B., Husain, M., Abdullah, M., and Akhter, A.,** Trace element composition of blood in adult population in Bangladesh, *J. Radioanal. Chem.*, 57, 157, 1980.

698. **Grant, G. C. and Buckle, D. C.,** Simultaneous Determination of S, Cl, K, Ca, Fe, Cu, Zn, Se, Br, Rb, and Pb in Blood Plasma and Erythrocytes by PIXE, Paper 509, Jt. Meet. AACC and CSCC, Boston, July 20 to 25, 1980.
699. **Manning, D. C.,** Aspirating small volume samples in flame AAS, *At. Absorpt. Newsl.*, 14, 99, 1975.
700. **Berndt, H. and Jackwerth, E.,** Automated micromethod for the determination of 8 elements in human sera by flame AAS, *Fresenius Z. Anal. Chem.*, 290, 105, 1978.
701. **Berndt, H. and Jackwerth, E.,** Determination of Fe, Cu and Zn by a mechanized micromethod of flame photometry. II. Mechanized micromethod (''injection method'') of flame photometry (AA-AE) for the determination of serum electrolytes and trace elements (Fe, Cu, Zn), *J. Clin. Chem. Clin. Biochem.*, 17, 489, 1979.
702. **Makino, T. and Takahara, K,.** Determination of trace metals in blood by ''one-drop'' method of flame AAS. II. Direct determination of Cu or Zn in 10 μt of serum, *Rinsho Byori*, 28, 483, 1980.
703. **Makino, T. and Takahara, K.,** Direct determination of plasma-Cu and Zn in infants by AA with discrete nebulization, *Clin. Chem.*, 27, 1445, 1981.
704. **Voth, L. M.,** *Determination of Li, Zn, and Cu in Blood Serum by Flame Microsampling*, Varian Instruments at Work, No. AA-16, Varian Techtron, Springvale, Victoria, 1981.
705. *AS-3 Automatic Micro Sampling System*, Perkin-Elmer Corp., Norwalk, Conn., 1978.
706. **Smeyers-Verbeke, J., Michotte, Y., Van den Winkel, P., and Massart, D. L.,** Matrix effects in the determination of Cu and Mn in biological materials using carbon furnace AAS, *Anal. Chem.*, 48, 125, 1976.
707. **Smeyers-Verbeke, J., Michotte, Y., and Massart, D. L.,** Influence of some matrix effects on the determination of Cu and Mn by furnace AAS, *Anal. Chem.*, 50, 10, 1978.
708. **Churella, D. J. and Copeland, T. R.,** Interference of salt matrices in the determination of Cu by AAS with electrothermal atomization, *Anal. Chem.*, 50, 309, 1978.
709. **Bogden, J. D., Troiano, R. A., and Joselow, M. M.,** Copper, Zn, Mg, and Ca in plasma and CSF of patients with neurological diseases, *Clin. Chem.*, 23, 485, 1977.
710. **Arpadjan, S., and Stojanova, D.,** Application of detergents to the direct determination of Fe, Zn and Cu in milk by means of flame AAS, *Fresenius Z. Anal. Chem.*, 302, 206, 1980.
711. **Kahn, H. L. and Kerber, J. D.,** Sampling improvements in AAS, *J. Am. Oil Chem. Soc.*, 48, 434, 1971.
712. **Arpadjan, S. and Nakova, D.,** Direct determination of Fe, Zn, Cu and Mn in evaporated milk by AA, *Nahrung*, 25, 359, 1981.
713. **Rosenthal, R. W. and Blackburn, A.** Higher Cu concentrations in serum than in plasma, *Clin. Chem.*, 20, 1233, 1974.
714. **Peaston, R. T.,** Determination of Cu and Zn in plasma and urine by AAS, *Med. Lab. Technol.*, 30, 249, 1973.
715. **Dawson, J. B., Ellis, D. J., and Newton-John, H.,** Direct estimation of Cu in serum and urine by AAS, *Clin. Chim. Acta*, 21, 33, 1968.
716. **Ichida, T. and Nobuoka, M.,** Determination of serum Cu with AAS, *Clin. Chim. Acta*, 24, 299, 1969.
717. **Lawrence, C. B. and Phillippo, M.,** Rapid semi-automatic AAS determination of Cu in bovine serum, *Anal. Chim. Acta*, 118, 153, 1980.
718. **Parker, M. M., Humoller, F. L., and Mahler, D. J.,** Determination of Cu and Zn in biological material, *Clin. Chem.*, 13, 40, 1967.
719. **Tsalev, D. L. and Mzhel'skaya, T. I.,** Comparison of three AA methods for the determination of Fe, Cu, and Zn in blood serum, *Khig. Zdraveop.*, 18, 326, 1975.
720. **Sprague, S. and Slavin, W.,** Determination of Fe, Cu and Zn in blood serum by an AA method requiring only dilution, *At. Absorpt. Newsl.*, 4, 228, 1965.
721. **Salmela, S. and Vuori, E.,** Direct Determination of Cu and Zn in a Single Serum Dilution by AAS, Paper 168, Euroanalysis-IV, Helsinki/Espoo, August 23 to 28, 1981.
722. **Taylor, A. and Bryant, T. N.,** Comparision of procedures for the determination of Cu and Zn in serum by AAS, *Clin. Chim. Acta*, 110, 83, 1981.
723. **Mzhel'skaya, T. I.,** Determination of Cu, Fe, and Zn content in blood serum using the Spectr-1 AA spectrophotometer, *Lab. Delo*, 229, 1976.
724. **Healy, P. J., Turvey, W. S., and Willats, H. G.,** Interference in estimation of serum copper concentration resulting from use of silicone-coated tubes for collection of blood, *Clin. Chim. Acta*, 88, 573, 1978.
725. **Williams, D. M.,** Trace metal determinations in blood obtained in evacuated collection tubes, *Clin. Chim. Acta*, 99, 23, 1979.
726. **Roos, J. T. H.,** Vapor-phase dissolution of blood samples prior to AAA, in *The Analysis of Biological Materials*, Butler, L. R. P., Ed., Pergamon Press, Oxford, 1979, 91.
727. **Sakla, A. B., Badran, A. H., and Shalaby, A. M.,** Determination of elements by AAS after destruction of blood in the oxygen flask, *Mikrochim. Acta*, I, 483, 1982.

728. **Szpunar, C. B., Lambert, J. B., and Buikstra, J. E.,** Analysis of excavated bone by AA, *Am. J. Phys. Anthropol.*, 48, 199, 1978.
729. **Maurer, J.,** Method of extraction for simultaneous determination of Na, K, Ca, Mg, Fe, Cu, Zn, and Mn in organic material by AAS, *Z. Lebensm.-Unters. Forsch.*, 165, 1, 1977.
730. **Evans, W. H., Dellar, D., Lucas, B. E., Jackson, F. J., and Read, J. T.,** Observations on the determination of total Cu, Fe, Mn, and Zn in foodstuffs by flame AAS, *Analyst*, 105, 529, 1980.
731. **Soman, S. D., Panday, V. K., Joseph, K. T., and Raut, S. J.,** Daily intake of some major and trace elements, *Health Phys.*, 17, 35, 1969.
732. **Harrison, W. W., Yurachek, J. P., and Benson, C.,** The determination of trace elements in human hair by AAS, *Clin. Chim. Acta*, 23, 83, 1969.
733. **Salmela, S., Vuori, E., and Kilpiö, J. O.,** The effect of washing procedures on trace element content of human hair, *Anal. Chim. Acta*, 125, 131, 1981.
734. **Hilderbrand, D. C. and White, D. H.,** Trace-element analysis in hair: an evaluation, *Clin. Chem.*, 20, 148, 1974.
735. **Mattera, V. D., Jr., Arbige, V. A., Jr., Tomellini, S. A., Erbe, D. A., Doxtader, M. M., and Forcé, R. K.,** Evaluation of wash solutions as a preliminary step for Cu and Zn determinations in hair, *Anal. Chim. Acta*, 124, 409, 1981.
736. **Murthy, G. K. and Rhea, U. S.,** Cadmium, Cu, Fe, Pb, Mn, and Zn in evaporated milk, infant products, and human milk, *J. Dairy Sci.*, 54, 1001, 1971.
737. **Vellar, O. D.,** Composition of human nail substance, *Am. J. Clin. Nutr.*, 23, 1272, 1970.
738. **Harrison, W. W. and Tyree, B. B.,** The determination of trace elements in human fingernails by AAS, *Clin. Chim. Acta*, 31, 63, 1971.
739. **Soman, S. D., Joseph, K. T., Raut, S. J., Mulay, G. D., Parameshwaran, M., and Panday, V. K.,** Studies on major and trace element content in human tissues, *Health Phys.*, 19, 641, 1970.
740. **Harrison, W. W., Netsky, M. G., and Brown, M. D.,** Trace elements in human brain: Cu, Zn, Fe, and Mg, *Clin. Chim. Acta*, 21, 55, 1968.
741. **Jackson, A. J., Michael, L. M., and Schumacher, H. J.,** Improved tissue solubilization for AA, *Anal. Chem.*, 44, 1064, 1972.
742. **Devoto, G.,** Determining Cu in biological fluids by AAS, *Boll. Soc. Ital. Biol. Sper.*, 44, 1249, 1968.
743. **Bromfield, J. and MacMahon, R. A.,** Microdetermination of plasma and erythrocyte Cu by AAS, *J. Clin. Pathol.*, 22, 136, 1969.
744. **Schuller, P. L. and Coles, L. E.,** The determination of Cu in foodstuffs, *Pure Appl. Chem.*, 51, 385, 1979.
745. **Antila, P. and Antila, V.,** Trace elements in Finnish cow's milk, *Suomen Kemistilehti*, B44, 161, 1971.
746. **Helsby, C. A.,** Determination of Cu and Mo in hard dental tissues of rats by AAS, *Talanta*, 20, 779, 1973.
747. **Berge, D. G. and Pflaum, R. T.,** Determination of Cu in urine by AAS, *Am. J. Med. Technol.*, 34, 725, 1968.
748. **Pirke, K. M. and Stamm, D.,** Measurement of urinary Cu excretion by AAS, *Z. Klin. Chem. Klin. Biochem.*, 8, 449, 1970.
749. **Fry, R. C. and Denton, M. B.,** High solids sample introduction for flame AAS, *Anal. Chem.*, 49, 1413, 1977.
750. **Roussos, G. G. and Morrow, B. H.,** Direct method for determination of microquantities of Mo, Fe and Cu in milk xanthine oxidase fractions by AAS, *Appl. Spectrosc.*, 22, 769, 1968.
751. **Dawson, J. B., Bahreyni-Toosi, M. H., Ellis, D. J., and Hodgkinson, A.,** Separation of protein-bound Cu and Zn in human plasma by means of gel filtration — ion-exchange chromatography, *Analyst*, 106, 153, 1981.
752. **Hartley, T. F. and Ellis, D. J.,** Combined electrolysis and AA for the determination of metals in biological materials, *Proc. Soc. Anal. Chem.*, 9, 281, 1972.
753. **Imanari, T., Okubo, N., Hayakawa, K., and Miyazaki, M.,** Studies on the electrodialytic extraction of metals from biological materials, *Bunseki Kagaku*, 28, 285, 1979.
754. **Van Loon, J. C., Radziuk, B., Kahn, N., Lichwa, J., Fernandez, F. J., and Kerber, J. D.,** Metal speciation using AAS, *At. Absorpt. Newsl.*, 16, 79, 1977.
755. **Popken, J. L., Brooks, R. A., and Foy, R. B.,** AAA of selected metals associated with serum protein fractions, *Am. J. Med. Technol.*, 40, 260, 1974.
756. **Weinstock, N.,** *Routine Determination of Cu and Zn from Untreated and Undiluted Serum by Flame AAS*, Jt. Congr. Scand. Germ. Soc. Clin. Chem., Hamburg, Oct. 8 to 13, 1980.
757. **Berndt, H. and Messerschmidt, J.,** Loop method of AAS/AES for the determination of elements in biological materials. A new possibility for trace and micro-analysis, *Fresenius Z. Anal. Chem.*, 311, 365, 1982.

758. **Furuno, K.,** Simultaneous determination of Cu and Zn in synovial fluid by AAS, *Okayama Daigaku Onseu Kenkyusho Hokoku,* 13, 1979.
759. **Uchida, T., Kojima, I., and Iida, C.,** Application of an automated triggered digital integrator to flame AAS of Cu using a discrete nebulization technique, *Analyst,* 106, 206, 1981.
760. **Uchida, T., Kojima, I., and Iida, C.,** "One-drop" method in flame AAS, *Bunseki Kagaku,* 27, T44, 1978.
761. **Uchida, T., Kojima, I., and Iida, C.,** Determination of metals in small samples by AA and emission spectrometry with discrete nebulization, *Anal. Chim. Acta,* 116, 205, 1980.
762. **Matoušek, J. P. and Stevens, B. J.,** Biological applications of the Carbon Rod Atomizer in AAS. I. Preliminary studies on Mg, Fe, Cu, Pb, and Zn in blood and plasma, *Clin. Chem.,* 17, 363, 1971.
763. **Mzhel'skaya, T. I. and Borisova, T. V.,** Flameless AAA for microelements in biological fluids, *Lab. Delo,* 10, 596, 1977.
764. **Murakami, K., Ito, Y., Taguchi, K., Ogata, K., and Imanari, T.,** Determination of Cu in rabbit plasma and red cells by flameless AAS, *Bunseki Kagaku,* 30, 200, 1981.
765. **Kamel, H., Teape, J., Brown, D. H., Ottaway, J. M., and Smith, W. E.,** Determination of Cu in plasma ultrafiltrate by AAS using carbon furnace atomization, *Analyst,* 103, 921, 1978.
766. **Baily, P., Kilroe-Smith, T. A., and Rollin, H. B.,** Effect of sample preparation and matrix on Cu determinations in biological specimens by flameless AAS, *Lab. Pract.,* 29, 141, 1980.
767. **Evenson, M. A. and Warren, B. L.,** Determination of serum Cu by AA with use of the graphite cuvette, *Clin. Chem.,* 21, 619, 1975.
768. **Glen, M., Savory, J., Hart, L., Glenn, T., and Winefordner, J.,** Determination of Cu in serum with graphite rod atomizer for AAS, *Anal. Chim. Acta,* 57, 263, 1971.
769. **Welz, B. and Wiedeking, E.,** Determination of trace elements in serum and urine with flameless atomization, *Z. Anal. Chem.,* 252, 111, 1970.
770. **Wawschinek, O.,** Determination of Cu and Au in serum by flameless absorptiometry, *Mikrochim. Acta,* II, 111, 1979.
771. **Halls, D. J., Fell, G. S., and Dunbar, P. M.,** Determination of Cu in urine by graphite furnace AAS, *Clin. Chim. Acta,* 114, 21, 1981.
772. **Robbins, W. B., De Koven, B. M., and Caruso, J. A.,** Copper in erythrocytes by flameless AAS, *Biochem. Med.,* 14, 184, 1975.
773. **Tatro, M. E., Raynolds, W. L., and Costa, F. M.,** Determination of Cu and Fe in biological specimens by flameless AA, *At. Absorpt. Newsl.,* 16, 143, 1977.
774. **Smeyers-Verbeke, J., Segabarth, G., and Massart, D. L.,** The determination of Cu and Mn in small biological samples with graphite furnace AAS, *At. Absorpt. Newsl.,* 14, 153, 1975.
775. **Jönsson, H.,** Determination of Cu, Fe, and Mn in milk with flameless AAS and a survey of the contents of these metals in Swedish market milk, *Milchwissenschaft,* 31, 210, 1976.
776. **Goldberg, W. J. and Allen, N.,** Determination of Cu, Mn, Fe, and Ca in six regions of normal human brain by AAS, *Clin. Chem.,* 27, 562, 1981.
777. **Nakamura, K., Fujimori, M., Tsuchiya, H., and Orii, H.,** Determination of Ga in biological materials by electrothermal AAS, *Anal. Chim. Acta,* 138, 129, 1982.
778. **Odinochkina, T. F.,** Studies of small samples of hairs and their discrimination by AAA, *Sud. Med. Expertiza,* 14, 22, 1971.
779. **Renshaw, G. D., Pounds, C. A., and Pearson, E. E.,** Determination of Pb and Cu in hair by non-flame AAS, *J. Forens. Sci.,* 18, 143, 1973.
780. **Van Steklenburg, G. J., Van de Laar, A. J. B., and Van der Laag, J.,** Copper analysis of nail clippings: an attempt to differentiate between normal children and patients suffering from cystic fibrosis, *Clin. Chim. Acta,* 59, 233, 1975.
781. **Pickford, C. J. and Rossi, G.,** Determination of some trace elements in NBS (SRM 1577) Bovine Liver using flameless AA and solid sampling, *At. Absorpt. Newsl.,* 14, 78, 1975.
782. **Lagesson, H. V.,** Analysis by means of AAS using a Ta boat, *Mikrochim. Acta,* 527, 1974.
783. **Delves, H. T.,** The microdetermination of Cu in plasma protein fractions, *Clin. Chim. Acta,* 71, 495, 1976.
784. **Teape, J., Kamel, H., Brown, D. H., Ottaway, J. M., and Smith, W. E.,** An evaluation of the use of electrophoresis and carbon furnace AAS to determine the Cu levels in separated protein fractions, *Clin. Chim. Acta,* 94, 1, 1979.
785. **Bahreni-Toosi, M. H., Dawson, J. B., and Ellis, D. J.,** Techniques for reducing the cycle time in AAS with electrothermal atomization, *Analyst,* 107, 124, 1982.
786. **Gardiner, P. E., Ottaway, J. M., Fell, G. S., and Burns, R. R.,** The application of gel filtration and electrothermal AAS to the speciation of protein-bound Zn and Cu in human blood serum, *Anal. Chim. Acta,* 124, 281, 1981.
787. **Pinta, M., Baron, D., Riandey, C., and Ghidalia, W.,** Non-flame AAA of trace metal elements fixation by seric proteins of crustacean decapods, *Spectrochim. Acta,* 33B, 489, 1978.

788. **Jonsen, A. C., Wibetoe, G., Langmyhr, F. J., and Aaseth, J.,** AAS determination of the total content and distribution of Cu and Au in synovial fluid from patients with rheumatoid arthritis, *Anal. Chim. Acta,* 135, 243, 1982.
789. **Iyengar, G. V.,** Post-mortem changes of the elemental composition of autopsy specimens: variations of K, Na, Mg, Ca, Cl, Fe, Zn, Cu, Mn, and Rb in rat liver, *Sci. Total Environ.,* 15, 217, 1980.
790. **Omang, S. H. and Vellar, O. D.,** Concentration gradients in biological samples during storage, freezing and thawing, *Fresenius Z. Anal. Chem.,* 269, 177, 1974.
791. **Wuyts, L., Smeyers-Verbeke, J., and Massart, D. L.,** AAS of Cu and Zn in human brain tissue. A critical investigation of two digestion techniques, *Clin. Chim. Acta,* 72, 405, 1976.
792. **Adrian, W. J.,** A comparison of a wet pressure digestion method with other commonly used wet and dry ashing methods, *Analyst,* 98, 213, 1973.
793. **Smeyers-Verbeke, J., May, C., Drochmans, P., and Massart, D. L.,** The determination of Cu, Zn, and Mn in subcellular rat liver fractions, *Anal. Biochem.,* 83, 746, 1977.
794. **Subramanian, K. S. and Méranger, J. C.,** Stability of tetramethylene-dithiocarbamate chelates in the 4-methylpentan-2-one phase after extraction from an aqueous phase, *Analyst,* 105, 620, 1980.
795. **Kundu, M. K. and Prévot, A.,** Oxygen-rich atmosphere for direct determination of Cu in oils by nonflame AAS, *Anal. Chem.,* 46, 1591, 1974.
796. **Lange, H. H.,** Natural concentration of Ga in human tissues, *Nucl. Med.,* 12, 178, 1973.
797. **Stulzaft, O., Mazière, B., and Ly, S.,** Gallium determination in biological samples, *J. Radioanal. Chem.,* 55, 291, 1980.
798. **Zweidinger, R. A., Barnett, L., and Pitt, C. G.,** Fluorimetric quantitation of Ga in biological materials at ng levels, *Anal. Chem.,* 45, 1563, 1973.
799. **Caroli, S., Alimonti, A., and Violante, N.,** Determination of Ga in biological samples by means of a hollow-cathode discharge, *Spectrosc. Lett.,* 13, 313, 1980.
800. **Caroli, S., Alimonti, A., Delle Femmine, P., and Shukla, S. K.,** Determination of Ga in tumor-affected tissues by means of spectroscopic techniques. A comparative study, *Anal. Chim. Acta,* 136, 225, 1982.
801. **Newman, R. A.,** Flameless AAS determination of Ga in biological materials, *Clin. Chim. Acta,* 86, 195, 1978.
802. **Regan, J. G. T. and Warren, J.,** A novel approach to the elimination of matrix interferences in flameless AAS using a graphite furnace, *Analyst,* 101, 220, 1976.
803. **Popova, S. A., Bezur, L., Pólos, L., and Pungor, E.,** Determination of Ga by emission and AA measurement, *Fresenius Z. Anal. Chem.,* 270, 180, 1974.
804. **Pelosi, C. and Attolini, G.,** Determination of Ga by AAS with a graphite furnace atomizer, *Anal. Chim. Acta,* 84, 179, 1976.
805. **Schroeder, H. A. and Balassa, J. J.,** Abnormal trace elements in man: germanium, *J. Chron, Dis.,* 20, 211, 1967.
806. **Castillo, J. R., Lanaja, J., and Aznárez, J.,** Determination of Ge in coal ashes by hydride generation and flame AAS, *Analyst,* 107, 89, 1982.
807. **Castillo, J. R., Lanaja, J., Belarra, M. A., and Aznarez, J.,** Flame AA determination of Ge in lignite ash with extraction of $GeCl_4$ into n-hexane, *At. Spectrosc.,* 2, 159, 1981.
808. **Mino, Y., Ota, N., Sakao, S., and Shimomura, S.,** Determination of Ge in medicinal plants by AAS with electrothermal atomization, *Chem. Pharm. Bull.,* 28, 2687, 1980.
809. **Studnicki, M.,** Determination of Ge, V, and Ti by carbon furnace AAS, *Anal. Chem.,* 52, 1762, 1980.
810. **Thorburn Burns, D. and Dadgar, D.,** Investigations on the determination of Ge in organogermanium compounds using carbon furnace atomization, *Analyst,* 107, 452, 1982.
811. **Pollock, E. N. and West, S. J.,** The generation and determination of covalent hydrides by AA, *At. Absorpt. Newsl.,* 12, 6, 1973.
812. **Fernandez, F. J.,** AA determination of gaseous hydrides utilizing $NaBH_4$ reduction, *At. Absorpt. Newsl.,* 12, 93, 1973.
813. **Andreae, M. O. and Froelich, P. N., Jr.,** Determination of Ge in natural waters by graphite furnace AAS with hydride generation, *Anal. Chem.,* 53, 287, 1981.
814. **Verghese, G. C., Kishore, R., and Guinn, V. P.,** Differences in trace element concentrations in hair between males and females, *J. Radioanal. Chem.,* 15, 329, 1973.
815. **Akashi, J., Fukushima, I., and Imahori, A.,** Elution behavior of elements from the hair, *Radioisotopes,* 30, 150, 1981.
816. **Ward, R. J., Danpure, C. J., and Fyfe, D. A.,** Determination of Au in plasma and plasma fractions by AAS, *Clin. Chim. Acta,* 81, 87, 1977.
817. **Schmid, G. M. and Bolger, G. W.,** Determination of Au in drugs and serum by use of ASV, *Clin. Chem.,* 19, 1002, 1973.
818. **Walton, R. J., Thibert, R. J., Bozic, J., and Holland, W. J.,** Spectrophotometric determination of Au in biological materials, *Can. J. Biochem.,* 48, 823, 1970.

819. **Rowston, W. B. and Ottaway, J. M.,** Determination of noble metals by carbon furnace AAS. I. Atom formation processes, *Analyst,* 104, 645, 1979.
820. **Rodgers, A. I. A., Brown, D. H., Smith, W. E., Lewis, D., and Capell, H. A.,** Distribution of Au in blood following administration of Myocrisin and Auranofin, *Anal. Proc.,* 19, 87, 1982.
821. **Schattenkirchner, M. and Grobenski, Z.,** The measurement of Au in blood and urine by AA in the treatment of rheumatoid arthritis, *At. Absorpt. Newsl.,* 16, 84, 1977.
822. **Melethil, S., Poklis, A., and Sagar, V. A.,** Binding of Au to bovine serum albumin using flameless AA, *J. Pharm. Sci.,* 69, 585, 1980.
823. **Kamel, H., Brown, D. H., Ottaway, J. M., and Smith, W. E.,** Determination of Au in tissue by carbon-furnace AAS, *Talanta,* 24, 309, 1977.
824. **Harth, M., Haines, D. S. M., and Boudy, D. C.,** Simple method for the determination of Au in serum, blood and urine by AAS, *Am. J. Clin. Pathol.,* 59, 423, 1973.
825. **Lorber, A., Cohen, R. L., Chang, C. C., and Anderson, H. E.,** Gold determination in biological fluids by AAS; application to chrysotherapy in rheumatoid arthritis patients, *Arthritis Rheum.,* 11, 170, 1968.
826. **Dunckley, J. V.,** Estimation of Au in serum by AAS, *Clin. Chem.,* 17, 992, 1971.
827. **Balázs, N. D. H., Pole, D. J., and Masarei, J. R.,** Determination of Au in body fluids by AAS, *Clin. Chim. Acta,* 40, 213, 1972.
828. **Dunckley, J. V., Grennan, D. M., and Palmer, D. G.,** Estimation of serum and urinary Au by AAS in rheumatoid patients receiving Au therapy, *J. Anal. Toxicol.,* 3, 242, 1979.
829. **Dunckley, J. V.,** Estimation of Au in urine by AAS, *Clin. Chem.,* 19, 1081, 1973.
830. **Kamel, H., Brown, D. H., Ottaway, J. M., and Smith, W. E.,** Determination of Au in blood fractions by AAS using carbon rod and carbon furnace atomization, *Analyst,* 101, 790, 1976.
831. **Barrett, M.J., DeFries, R., and Henderson, W. M.,** Rapid determination of Au in whole blood of arthritis patients using flameless AAS, *J. Pharm. Sci.,* 67, 1332, 1978.
832. **Turkall, R. M. and Bianchine, J. R.,** Determination of Au in tissue and feces by AAS using carbon rod atomization, *Analyst,* 106, 1096, 1981.
833. **Aggett, J.,** The determination of Au in serum by AAS with a carbon filament atom reservoir, *Anal. Chim. Acta,* 63, 473, 1973.
834. **Wawschinek, O. and Rainer, F.,** Determination of Au in serum and urine by AA after extraction with dimorpholinethiuramdisulphide in MIBK, *At. Absorpt. Newsl.,* 18, 50, 1979.
835. **Kamel, H., Brown, D. H., Ottaway, J. M., and Smith, W. E.,** Determination of Au in separate protein fractions of blood serum by carbon furnace AAS, *Analyst,* 102, 645, 1977.
836. **Dunckley, J. V. and Staynes, F. A.,** The estimation of Au in urine by flameless AAS, *Ann. Clin. Biochem.,* 14, 53, 1977.
837. **Torres, F.,** Determination of Pb, Cd, and In in Urine in μg Amounts by AAS, Rep. SC-TM-68-4, U.S. At. Energy Comm., Washington, D.C., 1968.
838. **Diem, K. and Lentner, C., Eds.,** Documenta Geigy. Scientific Tables, *7th ed., J. R. Geigy, S. A., Basel,* 1970.
839. **Tsvetanova, E. Ts.,** *Liquorologia,* 2nd ed., Medizina i Fizkultura, Sofia, 1980.
840. **Versieck, J., Hoste, J., Barbier, F., Michels, H., and De Rudder, J.,** Simultaneous determination of Fe, Zn, Se, Rb, and Cs in serum and packed blood cells by NAA, *Clin. Chem.,* 23, 1301, 1977.
841. **De Antonio, S. M., Katz, S. A., Scheiner, D. M., and Wood, J. D.,** Anatomical variations of trace metal levels in hair, *Anal. Proc.,* 18, 162, 1981.
842. **Fransson, G.-B. and Lönnerdal, B.,** Iron in human milk, *J. Pediatr.,* 96, 380, 1980.
843. **Vellar, O. D.,** Studies on sweat losses of nutrients. I. Fe content of whole body sweat and its association with other sweat constituents, serum Fe levels, hematological indices, body surface area, and sweat rate, *Scand. J. Clin. Lab. Invest.,* 21, 157, 1968.
844. **Hopps, H. C.,** The biological basis for using hair and nail for analyses of trace elements, *Trace Subst. Environ. Health,* 8, 59, 1974.
845. **Ben-Aryeh, H. an Gutman, D.,** Teeth for biological monitoring, in *The Use of Biological Specimens for the Assessment of Human Exposure to Environmental Pollutants,* Berlin, A., Wolff, A. H., and Hasegawa, Y., Eds., Martinus Nijhoff Publishers, Boston, 1979, 71.
846. **Zak, B., Eugene, S., and Epstein, E.,** Modern Fe ligands useful for measurement of serum Fe, *Ann. Clin. Lab. Sci.,* 10, 276, 1980.
847. **Lauber, K.,** Determination of serum Fe; a comparison of two methods: AA and bathophenanthroline without deproteinization, *J. Clin. Chem. Clin. Biochem.,* 16, 315, 1978.
848. **Bouda, J.,** Determination of Fe with bathophenanthroline without deproteinization, *Clin. Chim. Acta,* 21, 159, 1968.
849. **Giovanniello, T. J. and Pecci, J.,** Measurement of serum Fe and TIBC: manual and automated techniques, *Stand. Methods Clin. Chem.,* 7, 127, 1972.
850. **Giovanniello, T. J. and Peters, T., Jr.,** Serum Fe and serum iron-binding capacity, *Stand. Methods Clin. Chem.,* 4, 139, 1963.

851. **O'Connor, B. H., Kerrigan, G. C., Taylor, K. R., Morris, P. D., and Wright, C. R.,** Levels and temporal trends of trace element concentrations in vertebral bone, *Arch. Environ. Health,* 35, 21, 1980.
852. **Lindh, U. and Tveit, A. B.,** Proton microprobe determination of F depth distribution and surface multielemental characterization in dental enamel, *J. Radioanal. Chem.,* 59, 167, 1980.
853. **Taguchi, M., Takagi, H., Iwashima, K., and Yamagata, N.,** Metal content of shark muscle powder biological reference material, *J. Assoc. Off. Anal. Chem.,* 64, 260, 1981.
854. **Seamonds, B. and Anderson, K. M.,** Serum Fe and TIBC as measured radiometrically, *Clin. Chem.,* 27, 1946, 1981.
855. **Guest, L.,** Quantitative determination of exogenous and endogenous storage Fe content of haematite workers' lungs, *Analyst,* 106, 663, 1981.
856. **Kaliomäki, P.-L., Sutinen, S., Kelhä, V., Lakomaa, E., Sortti, V., and Sutinen, S.,** Amount and distribution of fume contaminants in the lungs of an arc welder post-mortem, *Br. J. Ind. Med.,* 36, 224, 1979.
857. **Thompson, K. C. and Wagstaff, K.,** Some observations on the determination of Fe by AAS using air-acetylene flames, *Analyst,* 105, 641, 1980.
858. **Olsen, E. D., Jatlow, P. I., Fernandez, F. J., and Kahn, H. L.,** Ultramicro method for determination of Fe in serum with the graphite furnace, *Clin. Chem.,* 19, 326, 1973.
859. **Mitsui, T. and Fujimura, Y.,** Determination of Fe in blood by AAS, *Bunseki Kagaku,* 21, 37, 1972.
860. **Loetterle, J.,** Measurement of hemoglobin content of fresh blood by AAS Fe determination, *Z. Rechtsmed.,* 85, 283, 1980.
861. **Zettner, A. and Mensch, A. H.,** The use of AAS in hemoglobinometry. I. The determination of Fe in hemoglobin, *Am. J. Clin. Pathol.,* 48, 225, 1967.
862. **Van Assendelft, O. W., Zijlstra, W. G., Buursma, A., Van Kampen, E. J., and Hoek, W.,** The use of AAS for the measurement of haemoglobin-Fe, with special reference to the determination of ϵ^{540}_{HiCN}, *Clin. Chim. Acta,* 22, 281, 1968.
863. **Zaino, E. C.,** Plasma Fe and iron-binding capacity determinations by AAS, *At. Absorpt. Newsl.,* 6, 93, 1967.
864. **Zettner, A. and Mansbach, L.,** Application of AAS in the determination of Fe in urine, *Am. J. Clin. Pathol.,* 44, 517, 1965.
865. **Dawczynski, H., Paul, J., Preu, E., and Yersin, A.,** Proposals to the 2nd ed. of the Pharmacopea of the DDR (diagnostic laboratory methods). Determination of Fe in blood serum and urine (AAS), with commentary, *Zentralbl. Pharm. Pharmakother. Laboratoriumsdiagn.,* 119, 1270, 1980.
866. **Nakamura, K., Watanabe, H., and Orii, H.,** Direct measurement of Fe in serum by electrothermal AAS, *Anal. Chim. Acta,* 120, 155, 1980.
867. **Olson, A. D. and Hamlin, W. B.,** A new method for serum Fe and TIBC by AAS, *Clin. Chem.,* 15, 438, 1969.
868. **Tavenier, P. and Hellendoorn, H. B. A.,** Comparison of serum iron values determined by AA and by some spectrophotometric methods, *Clin. Chim. Acta,* 23, 47, 1969.
869. **Pronk, C., Oldenziel, H., and Lequin, H. C.,** A method for determination of serum Fe, TIBC, and Fe in urine by AAS with Mn as internal standard, *Clin. Chim. Acta,* 50, 35, 1974.
870. **Zettner, A., Silvia, L. C., and Capacho-Delgado, L.,** Determination of serum Fe and IBC by AAS, *Am. J. Clin. Pathol.,* 45, 533, 1966.
871. **Devoto, G.,** Determination of Fe in urine by AAS, *Rass. Med. Sarda,* 71, 357, 1968.
872. **Cantle, J. E.,** Peak area measurement of transient signals in AAS, *Proc. Anal. Div. Chem. Soc.,* 13, 276, 1976.
873. **Makino, T. and Takahara, K.,** The effect of TCA-ascorbic acid deproteinization method on the determination of serum Fe by flame AAS, *Rinsho Byori,* 28, 239, 1980.
874. **Makino, T. and Takahara, K.,** Reference values: accurate determination of plasma-Fe in infants by discrete nebulization in AA, *Clin. Chem.,* 27, 2073, 1981.
875. **Baily, P., Rollin, H. B., and Kilroe-Smith, T. A.,** Comparison of a graphite-tube micro method for determination of serum Fe and TIBC with spectrophotometric techniques, *Microchem. J.,* 26, 250, 1981.
876. **Tanaka, T., Hayashi, Y., Funakawa, K., and Ishizawa, M.,** Simultaneous determination of Fe and Mn in human hair by graphite-furnace two channel AAS, *Nippon Kagaku Kaishi,* 1, 169, 1981.
877. **Selden, C. and Peters, T. J.,** Separation and assay of Fe proteins in needle biopsy specimens of human liver, *Clin. Chim. Acta,* 98, 47, 1979.
878. **Fielding, J. and Ryall, R. G.,** Some characteristics of TCA-precipitated proteins and their effects on biochemical assay, *Clin. Chim. Acta,* 33, 235, 1971.
879. **Siertsema, L. H.,** A continuous UV spectrum of NaCl enhanced by TCA and HCl in AAS, *Clin. Chim. Acta,* 69, 533, 1976.
880. **Karwowska, R., Bulska, E., and Hulanicki, A.,** Effect of chlorinated solvents on Fe atomization in graphite-furnace AAS, *Talanta,* 27, 397, 1980.

881. **Manton, W. I. and Cook, J. D.,** Lead content of CSF and other tissues in amyotrophic lateral sclerosis, *Neurology*, 29, 611, 1979.
882. **Barry, P. S. I.,** Concentrations of Pb in the tissues of children, *Br. J. Ind. Med.*, 38, 61, 1981.
883. **Evenson, M. A. and Pendergast, D. D.,** Rapid ultramicro direct determination of erythrocyte Pb concentration by AAS with use of a graphite-tube furnace, *Clin. Chem.*, 20, 163, 1974.
884. **Grandjean, P.,** Lead content of scalp hairs as an indicator of occupational Pb exposure, in *Toxicology and Occupational Medicine*, Deichmann, W. B. (Organizer), Elsevier/North Holland, New York, 1979, 311.
885. **Fergusson, J. E., Hibbard, K. A., and Lau, Hie Ting, R.,** Lead in human hair: general survey — battery factory employees and their families, *Environ. Pollut. (Ser. B)*, 2, 235, 1981.
886. **Graef, V.,** AAS determination of Pb in beard hair, *J. Clin. Chem. Clin. Biochem.*, 14, 181, 1976.
887. **Stringer, C. A., Jr., Zingaro, R. A., Creech, B., and Kolar, F. L.,** Lead concentrations in human lung samples, *Arch. Environ. Health*, 29, 268, 1974.
888. **Everson, J. and Patterson, C. C.,** ''Ultra-clean'' IDMS analyses for Pb in human blood plasma indicate that most reported values are artificially high, *Clin. Chem.*, 26, 1603, 1980.
889. **deSilva, P. E.,** Determination of Pb in plasma and studies on its relationship to Pb in erythrocytes, *Br. J. Ind. Med.*, 38, 209, 1981.
890. **Di Gregorio, G. J., Ferko, A. P., Sample, R. G., Bobyock, E., McMichael, R., and Chernick, W. S.,** Lead and δ-aminolevulinic acid concentrations in human parotid saliva, *Toxicol. Appl. Pharmacol.*, 27, 491, 1974.
891. **Rytömaa, I. and Tuompo, H.,** Lead levels in deciduous teeth, *Naturwissenschaft*, 61, 363, 1974.
892. **Curzon, M. E. J., Spector, P. C., Losee, F. L., and Crocker, D. C.,** Dental caries related to Cd and Pb in whole human dental enamel, *Trace Subst. Environ. Health*, 11, 23, 1977.
893. **Needleman, H. L., Tuncay, O. C., and Shapiro, I. M.,** Lead levels in deciduous teeth of urban and suburban American children, *Nature (London)*, 235, 111, 1972.
894. **Steenhout, A. and Pourtois, M.,** Lead accumulation in teeth as a function of age with different exposures, *Br. J. Ind. Med.*, 38, 297, 1981.
895. **Lazarev, N. V. and Gadaskina, I. D., Eds.,** Noxious Substances in Industry, *Vol. 3, 7th ed.*, Khimia, Leningrad, 1977.
896. **Morita, K., Shimizu, M., and Kamashiro, K.,** Heavy metal (Zn, Cu, Mn, and Pb) concentrations in normal human urine, *Okayama-ken Kankyo Hoken Senta Nempo*, 3, 336, 1979.
897. **Morrell, G. and Giridhar, G.,** Rapid micromethod for blood Pb analysis by ASV, *Clin. Chem.*, 22, 221, 1976.
898. **Fiorino, J. A., Moffitt, R. A., Woodson, A. L., Gajan, R. J., Huskey, G. E., and Scholz, R. G.,** Determination of Pb in evaporated milk by AAS and ASV. Collaborative study, *J. Assoc. Off. Anal. Chem.*, 56, 1246, 1973.
899. **Zeng, B.-W.,** Determination of tetraalkyllead in environmental samples, *Huang Ching K'o Hsueh*, 1, 55, 1980.
900. **Rice, E. W., Fletcher, D. C., and Stumpff, A.,** Lead in blood and urine, *Stand. Methods Clin. Chem.*, 5, 121, 1965.
901. **Eller, P. M. and Haartz, J. C.,** A study of methods for the determination of Pb and Cd, *Am. Ind. Hyg. Assoc. J.*, 38, 116, 1977.
902. **Hislop, J. S., Parker, A., Spicer, G. S., and Webb, M. S. W.,** Determination of Pb in Human Rib Bone, Rep. AERE-R7321, U.K. Atomic Energy Res. Estab., London, 1973.
903. **Pernis, B.,** The presence and significance of Pb in CSF, *Med. Lavoro*, 43, 251, 1952.
904. **Teraoka, H.,** Distribution of 24 elements in the internal organs of normal males and the metalic workers in Japan, *Arch. Environ. Health*, 36, 155, 1981.
905. **Rabinovitz, M., Weatherill, G., and Kopple, J.,** Delayed appearance of tracer Pb in facial hair, *Arch. Environ. Health*, 31, 220, 1976.
906. **Bowen, H. J. M.,** Problems in the elementary analysis of standard biological materials, *J. Radioanal. Chem.*, 19, 215, 1974.
907. **Human, H. G. C. and Norval, E.,** The determination of Pb in blood by atomic fluorescence flame spectrometry, *Anal. Chim. Acta*, 73, 73, 1974.
908. **Camara Rica, C. and Kirkbright, G. F.,** Determination of trace concentrations of Pb and Ni in human milk by electrothermal atomization AAS and ICP-ES, *Sci. Total Environ.*, 22, 193, 1982.
909. **Delves, H. T.,** Analytical techniques for blood-lead measurements, *J. Anal. Toxicol.*, 1, 261, 1977.
910. **Pierce, J. O., Koirtyohann, S. R., Clevenger, T. E., and Lichte, F. E.,** *The Determination of Pb in Blood. A Review and Critique of the State of the Art, 1975*, Int. Lead Zinc Res. Org., 1976.
911. **Stevens, B. J., Sanders, J. B., and Stux, R.,** *Lead Determination in Blood and Urine by AAS*, Varian Techtron, Springvale, Victoria, 1972.
912. **Delves, H. T.,** A micro-sampling method for the rapid determination of Pb in blood by AAS, *Analyst*, 95, 431, 1970.

913. **Kahn, H. L., Peterson, G. E., and Schallis, J. E.,** AA microsampling with the sampling boat technique, *At. Absorpt. Newsl.*, 7, 35, 1968.
914. **Berndt, H., Gücer, S., and Messerschmidt, J.,** Direct determination of Pb in urine with a new micromethod of flame AAS (loop-AAS), *J. Clin. Chem. Clin. Biochem.*, 20, 85, 1982.
915. **Frech, W. and Cedergren, A.,** Investigation of the reactions involved in flameless AA procedures. III. A study of factors influencing the determination of Pb in strong NaCl solutions, *Anal. Chim. Acta*, 88, 57, 1977.
916. **Manning, D. C. and Slavin, W.,** Determination of Pb in a chloride matrix, *At. Absorpt. Newsl.*, 17, 43, 1978.
917. **Pleban, P. A. and Pearson, K. H.,** Determination of Pb in whole blood and urine using Zeeman effect flameless AAS, *Anal. Lett.*, 12, 935, 1979.
918. **Cruz, R. B. and Van Loon, J. C.,** A critical study of the application of graphite-furnace non-flame AAS to the determination of trace base metals in complex heavy-matrix sample solutions, *Anal. Chim. Acta*, 72, 231, 1974.
919. **Slavin, W. and Manning, D. C.,** Reduction of matrix interferences for Pb determination with the L'vov platform and graphite furnace, *Anal. Chem.*, 51, 261, 1979.
920. **Hauck, G.,** Experiences with flameless AAS in the investigation of traces of heavy metals in biological materials, *Fresenius Z. Anal. Chem.*, 267, 337, 1973.
921. **Shamberger, R. J. and Willis, C. E.,** Matrix Effects on Blood and Urine Lead Analysis by Zeeman Effect Flameless AA, Paper 428, 31st Annu. Natl. Meet. Am. Assoc. Clin. Chem., New Orleans, 1979.
922. **Hodges, D. J.,** Observations on the direct determination of Pb in complex matrices by carbon furnace AAS, *Analyst*, 102, 67, 1977.
923. **Hodges, D. J. and Skelding, D.,** Determination of Pb in urine by AAS with electrothermal atomization, *Analyst*, 106, 299, 1981.
924. **Ebert, J. and Jungmann, H.,** Rapid and selective determination of Pb in urine by flameless AAS, *Fresenius Z. Anal. Chem.*, 272, 287, 1974.
925. **Nise, G. and Vesterberg, O.,** Blood lead determination by flameless AAS, *Clin. Chim. Acta*, 84, 129, 1978.
926. **Frigieri, P. and Trucco, R.,** The use of the Zeeman effect for the background correction in AAA of biological fluids, *Spectrochim. Acta*, 35B, 113, 1980.
927. **Fernandez, F. J., Myers, S. A., and Slavin, W.,** Background correction in AA utilizing the Zeeman effect, *Anal. Chem.*, 52, 741, 1980.
928. **Hadeishi, T. and McLaughlin, R. D.,** Zeeman AA determination of Pb with a dual chamber furnace, *Anal. Chem.*, 48, 1009, 1976.
929. **Knutti, R.,** Matrix effects and matrix modifications in graphite furnace-AAS — direct determination of Pb in blood, *Mitt. Gebiete Lebensm. Hyg.*, 72, 183, 1981.
930. **Baily, P., Norval, E., Kilroe-Smith, T. A., Skikne, M. I., and Roellin, H. B.,** The application of metal-coated graphite tubes to the determination of trace metals in biological materials. I. The determination of Pb in blood using a W-coated graphite tube, *Microchem. J.*, 24, 107, 1979.
931. **Robinson, J. W., Wolcott, D. K., and Rhodes, L.,** Direct analysis of blood, urine, sea water, filter paper, and polyethylene by AAS with the "Hollow-T" atomizer, *Anal. Chim. Acta*, 78, 285, 1975.
932. **Vickrey, T. M., Howell, H. E., Harrison, G. V., and Ramelow, G. J.,** Post column digestion methods for liquid chromatography — graphite furnace AA speciation of organolead and organotin compounds, *Anal. Chem.*, 52, 1743, 1980.
933. **Brodie, K. G. and Rowland, J. J.,** Trace analysis of As, Pb and Sn, Application of vapor generation AA, *Eur. Spectrosc. News*, 41, 1981.
934. **Ikeda, M., Nishibe, J., Hamada, S., and Tujino, R.,** Determination of Pb at ng/mℓ level by reduction to plumbane and measurement by ICP-ES, *Anal. Chim. Acta*, 125, 109, 1981.
935. **Kopito, L., Shwachman, H., and Williams, L. A.,** Measurement of Pb in blood, urine, and scalp hair by AAS, *Stand. Methods Clin. Chem.*, 7, 151, 1972.
936. **Renshaw, G. D., Pounds, C. A., and Pearson, E. F.,** Variation in Pb concentration along single hairs as measured by non-flame AAS, *Nature (London)*, 238, 162, 1972.
937. **Lyons, H. and Quinn, F. E.,** Measurement of Pb in biological materials by combined axion-exchange chromatography and AAS, *Clin. Chem.*, 17, 152, 1971.
938. **Auermann, E., Heidel, G., Gumbrowski, J., Jacobi, J., and Meckel, U.,** Critical observations on blood lead estimations, *Dtsch. Gesundheitswes.*, 33, 1769, 1978.
939. **Bratzel, M. P., Jr. and Reed, A. J.,** Microsampling for blood-Pb analysis, *Clin. Chem.*, 20, 217, 1974.
940. **Joselow, M. M. and Singh, N. P.,** Microanalysis for Pb in blood with built-in contamination control, *Am. Ind. Hyg. Assoc. J.*, 35, 793, 1974.
941. **Mitchell, D. G., Aldous, K. M., and Ryan, F. J.,** Mass screening for Pb poisoning. Capillary blood sampling and automated Delves-Cup AAA, *N.Y. State J. Med.*, 74, 1599, 1974.

942. **Tillery, J. B. and Johnson, D. E.,** Determination of Pt, Pd, and Pb in biological samples by AAS, *Environ. Health Perspect.,* 12, 19, 1975.
943. **Clarke, A. N. and Wilson, D. J.,** Preparation of hair for Pb analysis, *Arch. Environ. Health,* 28, 292, 1974.
944. **De Boer, J. L. M. and Maessen, F. J. M. J.,** Optimum experimental conditions of the brittle fracture technique for homogenization of biological materials, *Anal. Chim. Acta,* 117, 371, 1980.
945. **Unger, B. C. and Green, V. A.,** Blood lead analyses — Pb losses to storage containers, *Clin. Toxicol.,* 11, 237, 1977.
946. **Moore, M. R. and Meredith, P. A.,** The storage of samples for blood and water lead analysis, *Clin. Chim. Acta,* 75, 167, 1977.
947. **de Haas, E. J. M. and de Wolff, F. A.,** Microassay of Pb in blood, with an improved procedure for silanization of reagent tubes, *Clin. Chem.,* 27, 205, 1981.
948. **Jackson, K. W., Fuller, T. D., Mitchell, D. G., and Aldous, K. M.,** A rapid microsampling-cup AA procedure for the determination of Pb in urine, *At. Absorpt. Newsl.,* 14, 121, 1975.
949. **Mitchell, D. G., Ryan, F. J., and Aldous, K. M.,** The precise determination of Pb in whole blood by solvent extraction-AAS, *At. Absorpt. Newsl.,* 11, 120, 1972.
950. **Fernandez, F. J.,** Automated micro determination of Pb in blood, *At. Absorpt. Newsl.,* 17, 115, 1978.
951. **Matsumoto, T., Abe, H., and Tsukamoto, H.,** The effect of EDTA on the stability of Pb content in the blood samples stored for a long term, *Igaku to Seibutsugaku,* 98, 315, 1979.
952. **Thompson, K. C. and Godden, R. G.,** A simple method for monitoring excessive levels of Pb in whole blood using AAS and a rapid, direct nebulisation technique, *Analyst,* 101, 174, 1976.
953. **Sabet, S., Ottaway, J. M., and Fell, G. S.,** Comparison of the Delves cup and carbon furnace atomization used in AAS for determination of Pb in blood, *Proc. Anal. Div. Chem. Soc.,* 14, 300, 1977.
954. **Filkova, L.,** Reproducibility, accuracy and detection limit in determination of Pb in blood and hair by flameless AAS, *Chem. Listy,* 74, 533, 1980.
955. **Tsuchiya, K.,** Biological monitoring of chemical pollutants in the general population in Japan with special reference to Cd, Pb, and Hg, in *The Use of Biological Specimens for the Assessment of Human Exposure to Environmental Pollutants,* Berlin, A., Wolff, A. H., and Hasegawa, Y., Eds., Martinus Nijhoff Publishers, Boston, 1979, 333.
956. **Cernik, A. A.,** Some observations on the filter paper punched disc method for the determination of Pb in capillary blood, *At. Absorpt. Newsl.,* 12, 42, 1973.
957. **Cernik, A. A.,** Determination of blood lead using a 4.0 mm paper punched disc carbon sampling cup technique, *Br. J. Ind. Med.,* 31, 239, 1974.
958. **Cernik, A. A. and Sayers, M. H. P.,** Determination of Pb in capillary blood using a paper punched disc AA technique: application to supervision of lead workers, *Br. J. Ind. Med.,* 28, 392, 1971.
959. **Riner, J. C., Wright, F. C., and McBeth, C. A.,** A technique for determining lead in feces of cattle by flameless AAS, *At. Absorpt. Newsl.,* 13, 129, 1974.
960. **Berman, E.,** The determination of Pb in blood and urine by AAS, *At. Absorpt. Newsl.,* 3, 111, 1964.
961. **Berman, E., Valavanis, V., and Dulin, A.,** A micro method for determination of Pb in blood, *Clin. Chem.,* 14, 239, 1968.
962. **Blankman, L.,** AA method for Pb in blood and urine, *Clin. Lab. Diagn.,* 7, 461, 1968.
963. **Cernik, A. A.,** Determination of Pb chelated with EDTA in blood after precipitation of protein with perchloric acid, *Br. J. Ind. Med.,* 27, 40, 1970.
964. **Devoto, G.,** Determination of Pb in urine and blood by AAS, *Boll. Soc. Ital. Biol. Sper.,* 44, 421, 1968.
965. **Farelly, R. O. and Pybus, J.,** Measurement of Pb in blood and urine by AAS, *Clin. Chem.,* 15, 566, 1969.
966. **Hessel, D. W.,** Quantitative determination of Pb in blood, *At. Absorpt. Newsl.,* 7, 55, 1968.
967. **Lehnert, G., Shaller, K. H., and Sazdkowski, D.,** Determination of Pb in small quantities of blood, *Z. Klin. Chem. Klin. Biochem.,* 7, 310, 1969.
968. **Mishima, M., Hoshiai, T., and Suzuki, T.,** Determination of Pb in blood by AAS, *Koshu Eiseiin Kenkyu Hokoku,* 20, 187, 1971.
969. **Selander, S., Cramer, K., Borjesson, B., and Mandorf, G.,** Determination of Pb in blood by AAS, *Br. J. Ind. Med.,* 25, 209, 1968.
970. **Watanabe, T., Iwahama, T., and Ikeda, M.,** Comparative study on determination of Pb in blood by flame and flameless AAS with and without wet digestion, *Int. Arch. Occup. Environ. Health,* 39, 121, 1977.
971. **Yeager, D. W., Cholak, J., and Henderson, E. W.,** Determination of Pb in biological and related materials by AAS, *Environ. Sci. Technol.,* 5, 1020, 1971.

971a. **de Ruig, W. G.,** Experiences with various methods of trace analysis in organic matrices, with special reference to lead blank values, *Mikrochim. Acta,* II, 199, 1981.

972. **Zinterhofer, L. J. M., Jatlow, P. I., and Fappiano, A.,** AA determination of Pb in blood and urine in the presence of EDTA, *J. Lab. Clin. Med.,* 78, 664, 1971.

973. **Kubasik, N. P., Volosin, M. T., and Murray, M. H.,** Carbon rod atomizer applied to measurement of Pb in whole blood by AAS, *Clin. Chem.*, 18, 410, 1972.
974. **Ikeda, M., Kaneko, I., Watanabe, T., Ishihara, N., and Miura, T.,** An automated system for the determination of Pb in blood, Mn in urine and Ni in waste water, *Am. Ind. Hyg. Assoc. J.*, 39, 226, 1978.
975. **Pierce, J. O. and Cholak, J.,** Lead, Cr, and Mo by AA, *Arch. Environ. Health,* 13, 208, 1966.
976. **Strehlow, C. D. and Kneip, T. J.,** Distribution of Pb and Zn in human skeleton, *Am. Ind. Hyg. Assoc. J.*, 30, 372, 1969.
977. **Murthy, G. K., Rhea, U., and Peeler, J. T.,** Rubidium and Pb content of market milk, *J. Dairy Sci.*, 50, 651, 1967.
978. **Kuboku, M. and Aihara, M.,** AAS determination of microamounts of Pb by using solvent extraction with K ethylxanthate — MIBK, *Bunseki Kagaku,* 22, 1581, 1973.
979. **Roosels, D. and Vanderkeel, J. V.,** An AA determination of Pb in urine after extraction with dithizone, *At. Absorpt. Newsl.*, 7, 9, 1968.
980. **Selander, S., Cramer, K., Borjesson, B., and Mandorf, G.,** Determination of Pb in urine by AAS, *Br. J. Ind. Med.*, 25, 139, 1968.
981. **Willis, J. B.,** Determination of Pb in urine by AAS, *Nature (London),* 191, 381, 1961.
982. **Einarsson, O. and Lindstedt, G.,** A non-extraction AA method for the determination of Pb in blood, *Scand. J. Clin. Lab. Invest.*, 23, 367, 1969.
983. **Gabrielli, L. F., Marletta, G. P., and Favretto, L.,** Determination of Pb in mussels by AAS and solid microsampling, *At. Spectrosc.*, 1, 35, 1980.
984. **Hoover, W. L., Reagor, J. C., and Garner, J. C.,** Extraction and AAA of Pb in plant and animal products, *J. Assoc. Off. Anal. Chem.*, 52, 708, 1969.
985. **Hoover, W. L.,** Collaborative study of a method for determining Pb in plant and animal products, *J. Assoc. Off. Anal. Chem.*, 55, 737, 1972.
986. **Kopito, L. and Shwachman, H.,** Determination of Pb in urine by AAS using coprecipitation with Bi, *J. Lab. Clin. Med.*, 70, 326, 1967.
987. **Lorimier, D. F. and Fernandez-Garcia, J. G.,** Determination of Pb in urine by AAS, *Helv. Chim. Acta,* 53, 1990, 1970.
988. **Zurlo, N., Griffini, A. M., and Colombo, G.,** Determination of Pb in urine by AAS after coprecipitation with thorium, *Anal. Chim. Acta,* 47, 203, 1969.
989. **Singh, N. P., Bogden, J. D., and Joselow, M. M.,** Distribution of Tl and Pb in children's blood, *Arch. Environ. Health,* 30, 557, 1975.
990. **Aldous, K. M., Mitchell, D. G., and Ryan, F. J.,** Computer-controlled AA spectrometer for measurement of transient populations, *Anal. Chem.*, 45, 1990, 1973.
991. **Barthel, W. F., Smrek, A. L., Angel, G. P., Liddle, J. A., Landrigan, P. J., Gehlbach, S. H., and Chisolm, J. J.,** Modified Delves cup AA determination of Pb in blood, *J. Assoc. Off. Anal. Chem.*, 56, 1252, 1973.
992. **Ediger, R. D. and Coleman, R. L.,** A modified Delves cup AA procedure for the determination of Pb in blood, *At. Absorpt. Newsl.*, 11, 33, 1972.
993. **Elfbaum, S. G., Juliano, R., MacFarland, R. E., and Pfeil, D. L.,** Blood Pb determinations by AAS with use of the Delves sampling-cup technique: effect of various anticoagulants, *Clin. Chem.*, 18, 316, 1972.
994. **Fernandez, F. J.,** Some observation on the determination of Pb in blood with the Delves cup method, *At. Absorpt. Newsl.*, 12, 70, 1973.
995. **Fernandez, F. and Kahn, H. L.,** The determination of Pb in whole blood by AAS with the "Delves sampling cup" technique, *At. Absorpt. Newsl.*, 10, 1, 1971.
996. **Heinemann, G.,** Determination of Pb in blood and urine using the Delves sampling system, *Z. Klin. Chem. Klin. Biochem.*, 11, 197, 1973.
997. **Hicks, J. M., Gutierrez, A. N., and Worthy, B. E.,** Evaluation of the Delves micro system for blood Pb analysis, *Clin. Chem.*, 19, 322, 1973.
998. **Joselow, M. M. and Bogden, J. D.,** A simplified micromethod for collection and determination of Pb in whole blood using a paper disk-in-Delves cup technique, *At. Absorpt. Newsl.*, 11, 99, 1972.
999. **Joselow, M. M. and Singh, N. P.,** Loss of "sensitivity" of the Delves cup with use, *At. Absorpt. Newsl.*, 12, 128, 1973.
1000. **Olsen, E. D. and Jatlow, P. I.,** An improved Delves cup AA procedure for determination of Pb in blood and urine, *Clin. Chem.*, 18, 1312, 1972.
1001. **Rose, G. A. and Willden, E. G.,** An improved method for the determination of whole blood lead by using an AA technique, *Analyst,* 98, 243, 1973.
1002. **Haelen, P., Cooper, G., and Pampel, C.,** The determination of Pb in evaporated milk by Delves cup AAS, *At. Absorpt. Newsl.*, 13, 1, 1974.
1003. **Manning, D. C.,** AAA by flameless sampling: Pb in milk, *Am. Lab.*, 37, 1973.

1004. **Lamm, S., Cole, B., Glynn, K., and Ullmann, W.,** Lead content of milk fed to infants — 1971 to 1972, *New Engl. J. Med.*, 289, 574, 1973.
1005. **Di Gregorio, G. J., Ferko, A. P., Sample, R. G., Bobyock, E., McMichael, R., and Chernick, W. S.,** Lead determination in human parotid saliva, *J. Dent. Res.*, 52, 1152, 1973.
1006. **Jackson, K. W., Marczak, E., and Mitchell, D. G.,** Rapid determination of Pb in biological tissues by microsampling-cup AAS, *Anal. Chim. Acta,* 97, 37, 1978.
1007. **Anderson, M. P. and Mesman, B. B.,** Determination of Pb in untreated urine by the Delves cup system, *Am. Ind. Hyg. Assoc. J.*, 34, 310, 1973.
1008. **Hilderbrand, D. C., Koirtyohann, S. R., and Pickett, E. E.,** The Sampling-Boat technique for the determination of Pb in blood and urine by AA, *Biochem. Med.*, 3, 437, 1970.
1009. **Kahn, H. L. and Sebestyen, J. S.,** Determination of Pb in blood and urine by AAS with the Sampling Boat system, *At. Absorpt. Newsl.*, 9, 33, 1970.
1010. **Marletta, G. P., Gabrielli, L. F., and Favretto, L.,** Pollution of mussels by particulate lead from sea water, *Z. Lebensm.-Unters. Forsch.*, 168, 181, 1979.
1011. **Amos, M. D., Bennett, P. A., Brodie, K. G., Lung, P. W., and Matoušek, J. P.,** Carbon Rod Atomizer in AA and fluorescence spectrometry and its clinical application, *Anal. Chem.*, 43, 211, 1971.
1012. **Therrel, B. L., Jr., Drosche, J. M., and Dziuk, T. W.,** Analysis for Pb in undiluted whole blood by Ta ribbon AAS, *Clin. Chem.*, 24, 1182, 1978.
1013. **Hwang, J. Y., Ullucci, P. A., and Mokeler, C. J.,** Direct flameless AA determination of Pb in blood, *Anal. Chem.*, 45, 795, 1973.
1014. **Volosin, M. T., Kubasik, N. P., and Sine, H. E.,** Use of the Carbon Rod Atomizer for analysis of Pb in blood. Three methods compared, *Clin. Chem.*, 21, 1986, 1975.
1015. **Paschal, D. C. and Bell, C. J.,** Improved accuracy in the determination of blood lead by electrothermal atomization, *At. Spectrosc.*, 2, 146, 1981.
1016. **Herber, R. F. M. and van Deyck, W.,** On the optimization of blood-lead standards in electrothermal atomization-AAS, *Clin. Chim. Acta,* 120, 313, 1982.
1017. **Aungst, B. J., Dolce, J., and Fung, H.-L.,** Solubilization of rat whole blood and erythrocytes for automated determination of Pb using AAS, *Anal. Lett.*, 13, 347, 1980.
1018. **Alt, F. and Massmann, H.,** Determination of Pb in blood by AAS, *Spectrochim. Acta,* 33B, 337, 1978.
1019. **Alt, F.,** A simple, fast and reliable determination of Pb in blood by AAS, *Fresenius Z. Anal. Chem.*, 290, 108, 1978.
1020. **Döllefeld, E.,** Micro-method for determination of the Pb content of blood and urine by AAS, *Ärztl. Lab.*, 17, 369, 1971.
1021. **Ealy, J. A., Bolton, N. E., McElheny, R. J., and Morrow, R. W.,** Determination of Pb in whole blood by graphite furnace AAS, *Am. Ind. Hyg. Assoc. J.*, 35, 566, 1974.
1022. **Fernandez, F. J.,** Micromethod for Pb determination in whole blood by AA, with use of the graphite furnace, *Clin. Chem.*, 21, 558, 1975.
1023. **Garnys, V. P. and Matoušek, J. P.,** Correction for spectral interference with determination of Pb in blood by non-flame AAS, *Clin. Chem.*, 21, 891, 1975.
1024. **Minoia, C., Catenacci, G., Baruffini, A., and Prestinoni, A.,** Determination of Pb in capillary blood using an automatic micro sampler and a graphite furnace, *Ann. 1st. Super. Sanita,* 14, 753, 1978.
1025. **Kubasik, N. P. and Volosin, M. T.,** Use of the CRA for direct analysis of Pb in blood, *Clin. Chem.*, 20, 300, 1974.
1026. **Kilroe-Smith, T. A.,** Linearization of calibration curves with the HGA-72 flameless cuvette for the determination of Pb in blood, *Anal. Chim. Acta,* 82, 421, 1976.
1027. **Norval, E. and Butler, L. R. P.,** The determination of Pb in blood by AA with the high-temperature graphite tube, *Anal. Chim. Acta,* 58, 47, 1972.
1028. **Rosen, J. F. and Trinidad, E. E.,** The microdetermination of blood lead in children by flameless AA. The Carbon Rod Atomizer, *J. Lab. Clin. Med.*, 80, 567, 1972.
1029. **Wittmers, L. E., Jr., Alich, A., and Aufderheide, A. C.,** Lead in bone. I. Direct analysis for Pb in mg quantities of bone by graphite furnace AAS, *Am. J. Clin. Pathol.*, 75, 80, 1981.
1030. **Bertholf, R. L. and Renoe, B. W.,** The determination of bismuth in serum and urine by electrothermal AAS, *Anal. Chim. Acta,* 139, 287, 1982.
1031. **Garnys, V. P. and Smythe, L. E.,** Fundamental studies on improvement of precision and accuracy in flameless AAS using the graphite-tube atomizer. Lead in blood, *Talanta,* 22, 881, 1975.
1032. **Huffman, H. L., Jr. and Caruso, J. A.,** Analysis of Pb in evaporated milk by flameless AAS, *J. Agric. Food Chem.*, 22, 824, 1974.
1033. **Velghe, G., Verloo, M., and Cottenie, A.,** Determination of Pb in milk and butter by flameless AAS, *Z. Lebensm. Unters.-Forsch.*, 156, 77, 1974.
1034. **Baczyk, S., Mielcarz, G., and Stawinski, K.,** Method for determination of Pb in tartar using flameless AAS, *Czas. Stomatol.*, 32, 409, 1979.

1035. **Garnys, V. P. and Smythe, L. E.,** Filament in furnace atomization AAS, *Anal. Chem.*, 51, 62, 1979.

1036. **Stoeppler, M.,** Automated blood lead determinations by electrothermal AAS, *J. Clin. Chem. Clin. Biochem.*, 16, 58, 1978.

1037. **Stoeppler, M., Brandt, K., and Rains, T. C.,** Contributions to automated trace analysis. II. Rapid method for the determination of Pb in whole blood by electrothermal AAS, *Analyst,* 103, 714, 1978.

1038. **Barlow, P. J. and Khera, A. K.,** Sample preparation using tissue solubilization by Soluene-350® for Pb determinations by graphite furnace AAS, *At. Absorpt. Newsl.*, 14, 149, 1975.

1039. **Bohm, B. and Kovats, A.,** Blood and urine lead determination by AAS method with oven atomization, *Rev. Ig., Bacteriol., Virusol., Parazitol., Epidemiol., Pneumoftiziol., Ig.*, 24, 235, 1975.

1040. **Carelli, G., Rimatori, V., and Sperduto, B.,** Determination of Pb in blood by wet-ashing, solvent extraction and flameless AA, *Med. Lav.*, 70, 313, 1979.

1041. **Hwang, J. Y., Ullucci, P. A., Smith, S. B., Jr., and Malenfant, A. L.,** Microdetermination of Pb in blood by flameless AAS, *Anal. Chem.*, 43, 1319, 1971.

1042. **Schmidt, C.,** Lead determination in blood by AAS, *Am. Ind. Hyg. Assoc. J.*, 40, 1085, 1979.

1043. **Cavalleri, A., Minoia, C., Pozzoli, L., and Baruffini, A.,** Determination of plasma lead levels in normal subjects and Pb-exposed workers, *Br. J. Ind. Med.*, 35, 21, 1978.

1044. **Conley, M. K. and Sotera, J. J.,** *An Automated Method for the Determination of Pb in Blood,* Rep. No. 10, Instrumentation Laboratory, Wilmington, Mass., July 1979.

1045. **Matoušek, J. P.,** Aerosol deposition in furnace atomization, *Talanta,* 24, 315, 1977.

1046. **Raghavan, S. R. V., Dwight Culver, B., and Gonick, H. C.,** Erythrocyte Pb-binding protein after occupational exposure. I. Relationship to Pb toxicity, *Environ. Res.*, 22, 264, 1980.

1047. **Chau, Y. K., Wong, P. T. S., Bengert, G. A., and Kramar, D.,** Determination of tetraalkyllead compounds in water, sediment, and fish samples, *Anal. Chem.*, 51, 186, 1979.

1048. **Sirota, G. R. and Uthe, J. F.,** Determination of tetraalkyllead compounds in biological materials, *Anal. Chem.*, 49, 823, 1977.

1049. **Cruz, R. B., Lorouso, C., George, S., Thomassen, Y., Kinrade, J. D., Butler, L. R., Lye, J., and Van Loon, J. C.,** Determination of total, organic solvent extractable, volatile, and tetraalkyl lead in fish, vegetation, sediment, and water samples, *Spectrochim. Acta,* 35B, 775, 1980.

1050. **Brudevold, F., Reda, A., Aasenden, R., and Bakhos, Y.,** Determination of trace elements in surface enamel of human teeth by a new biopsy procedure, *Arch. Oral Biol.*, 20, 667, 1975.

1051. **Dinischiotu, G. T., Nestorescu, B., Radulescu, I. C., Ionescu, N., Preda, N., and Ientza, G.,** Studies on the chemical forms of urinary lead, *Br. J. Ind. Med.*, 17, 141, 1960.

1052. **Baily, P. and Kilroe-Smith, T. A.,** Effect of sample preparation of blood lead values, *Anal. Chim. Acta,* 77, 29, 1975.

1053. **Kopito, L. E., Davis, M. A., and Shwachman, H.,** Sources of error in determining Pb in blood by AAS, *Clin. Chem.*, 20, 205, 1974.

1054. **Behne, D., Brätter, P., and Wolters, W.,** Determination of Pb in biological material by means of flameless AAS, *Fresenius Z. Anal. Chem.*, 277, 355, 1975.

1055. **Anderson, W. W., Broughton, P. M. G., Dawson, J. B., and Fisher, G. W.,** An evaluation of some AA systems for the determination of Pb in blood, *Clin. Chim. Acta,* 50, 129, 1974.

1056. **Segal, R. J.,** Non-specificity of urinary Pb measurements by AAS. A spectrophotometric method for correction, *Clin. Chem.*, 15, 1124, 1969.

1057. **Grime, J. K. and Vickers, T. J.,** Determination of Li in microliter samples of blood serum using flame AES with a Ta filament vaporizer, *Anal. Chem.*, 47, 432, 1973.

1058. **Lang, W. and Herrmann, R.,** Determination of Li in serum by flame AAS, *Z. Ges. Exp. Med.*, 139, 200, 1965.

1059. **Christian, G. D. and Feldman, F. J.,** *Atomic Absorption Spectroscopy: Applications in Agriculture, Biology and Medicine,* Wiley-Interscience, New York, 1970.

1060. **Matusiewicz, H.,** Determination of natural levels of Li and Sr in human blood serum by discrete injection and AES with a N_2O-C_2H_2 flame, *Anal. Chim. Acta,* 136, 215, 1982.

1061. **Pickett, E. E. and Hawkins, J. L.,** Determination of Li in small animal tissue at physiological levels by flame emission photometry, *Anal. Biochem.*, 112, 213, 1981.

1062. **Velapoldi, R. A., Paule, R. C., Schaffer, R., Mandel, J., Machlan, L. A., Garner, E. L., and Rains, T. C.,** A Reference Method for the Determination of Li in Serum, NBS Spec. Publ. No. 260-69, National Bureau of Standards, U.S. Department of Commerce, Washington, D.C., 1980.

1063. **Babaskin, P. M.,** Determination of the Li content in blood by flame photometry, *Lab. Delo,* 10, 622, 1976.

1064. **Blijenberg, B. G. and Leijnse, B.,** The determination of Li in serum by AAS and flame emission spectroscopy, *Clin. Chim. Acta,* 19, 97, 1968.

1065. **Eiseneg, R. and Lantz, R.,** Erythrocyte lithium analysis, *Clin. Chem.*, 23, 900, 1977.

1066. **Levy, A. L. and Katz, E. M.,** Comparison of serum lithium determinations by flame photometry and AAS, *Clin. Chem.*, 16, 840, 1970.

1067. **Robertson, R., Fritze, K., and Grof, P.,** On the determination of Li in blood and urine, *Clin. Chim. Acta,* 45, 25, 1973.
1068. **Pybus, J. and Bowers, G. N., Jr.,** Measurement of serum-lithium by AAS, *Clin. Chem.,* 16, 139, 1970.
1069. **Stafford, D. T. and Saharovici, F.,** Serum lithium determinations using flameless AAS, *Spectrochim. Acta,* 29B, 277, 1974.
1070. **Cooper, T. B., Simpson, G. M., and Allen, D.,** Rapid direct micro method for determination of plasma Li, *At. Absorpt. Newsl.,* 13, 119, 1974.
1071. **Hasayasu, G. H., Cohen, J. L., and Nelson, R. W.,** Determination of plasma and erythrocyte Li concentrations by AAS, *Clin. Chem.,* 23, 41, 1977.
1072. **Little, B. R., Platman, S. R., and Fieve, R. R.,** The measurement of Li in biological samples by AAS, *Clin. Chem.,* 14, 1211, 1968.
1073. **Lazarus, J. H., Fell, G. S., Robertson, J. W. K., and Millar W. T.,** Secretion of Li in human parotid saliva in manic depressive patients treated with Li_2CO_3, *Arch. Oral. Biol.,* 18, 329, 1973.
1074. **Groth, U., Prellwitz, W., and Jähnchen, E.,** Estimation of pharmacokinetic parameters of Li from saliva and urine, *Clin. Pharmacol. Ther.,* 16, 490, 1974.
1075. **Matsushita, K., Kohno, K., Koshiro, J., Kodama, Y., and Matsumoto, K.,** Direct determination of serum lithium using an AA spectrophotometer, *Rinsho Kensa,* 19, 636, 1975.
1076. **Arroyo, M., De Paz, J. F., and Coca, M. C.,** Analytical determination of Li in blood serum by AAS, *Rev. Clin. Espan.,* 123, 433, 1971.
1077. **Bowman, J. A.,** The application of resonance monochromators to the determination of Li in blood serum by AAS, *Anal. Chim. Acta,* 37, 465, 1967.
1078. **Brost, D. F., Brackett, J. M., and Busch, K. W.,** Determination of Li by optically monitored stable-isotope dilution, *Anal. Chem.,* 51, 1512, 1979.
1079. **Hansen, J. L.,** Measurement of serum and urine lithium by AAS, *Am. J. Med. Technol.,* 34, 625, 1968.
1080. **Lehmann, V.,** Direct determination of Li in serum by AAS, *Clin. Chim. Acta,* 20, 523, 1968.
1081. **Woods, A. E., Crowder, R. D., Coates, J. T., and Wittrig, J. J.,** Determination of microquantities of Li in biological fluids. I. Recovery studies in serum, *At. Absorpt. Newsl.,* 7, 85, 1968.
1082. **Zettner, A., Rafferty, K., and Jarecki, H. J.,** Determination of Li in serum and urine by AAS, *At. Absorpt. Newsl.,* 7, 32, 1968.
1083. **Pybus, J. and Bowers, G. N.,** Serum Li determination by AAS, *Stand. Methods Clin. Chem.,* 6, 189, 1970.
1084. **Rocks, B. F., Sherwood, R. A., and Riley, C.,** Direct determination of therapeutic concentrations of Li in serum by flow-injection analysis with AAS detection, *Clin. Chem.,* 28, 440, 1982.
1085. **McGovern, A. J., Makanjuola, R., Arbuthnott, G. W., Loudon, J. B., and Glen, A. I. M.,** Lithium neurotoxicity. I. The concentration of Li in dopaminergic systems of rat brain determined by flameless AAS, *Acta Pharmacol. Toxicol.,* 42, 259, 1978.
1086. **Weissman, N. and Babich-Armstrong, M.,** Determination of Mn in Human Whole Blood, Paper 334, Jt. Meet. AACC and CSCC, Boston, July 20 to 25, 1980.
1087. **Cotzias, G. C. and Papavasiliou, P. S.,** State of binding of natural Mn in human CSF, blood and plasma, *Nature (London),* 195, 823, 1962.
1088. **Pleban, P. A. and Pearson, K. H.,** Determination of Mn in whole blood and serum, *Clin. Chem.,* 25, 1915, 1979.
1089. **Tsalev, D. L., Langmyhr, F. J., and Gundersen, N.,** Direct AAS determination of Mn in whole blood of unexposed individuals and exposed workers in a Norwegian Mn alloy plant, *Bull. Environ. Contam. Toxicol.,* 17, 660, 1977.
1090. **Buchet, J. P., Lauwerys, R., Roels, H., and de Vos, C.,** Determination of Mn in blood and urine by flameless AAS, *Clin. Chim. Acta,* 73, 481, 1976.
1091. **Smeyers-Verbeke, J., Bell, P., Lowenthal, A., and Massart, D. L.,** Distribution of Mn in human brain tissue, *Clin. Chim. Acta,* 68, 343, 1976.
1092. **D'Amico, D. J. and Klawans, H. L.,** Direct microdetermination of Mn in normal serum and CSF by flameless AAS, *Anal. Chem.,* 48, 1469, 1976.
1093. **Ross, R. T. and Gonzalez, J. G.,** Direct determination of trace quantities of Mn in blood and serum samples using selective volatization and graphite tube reservoir AAS, *Bull. Environ. Contam. Toxicol.,* 12, 470, 1974.
1094. **McLeod, B. E. and Robinson, M. F.,** Metabolic balance of Mn in young women, *Br. J. Nutr.,* 27, 221, 1972.
1095. **Grund, W., Schneider, W. D., and Wiesner, W.,** Der Mangangehalt des Haares, ein Kriterium für die Bewertung des Expositions Riskos der ElectroschweiBer, *J. Radioanal. Chem.,* 58, 319, 1980.
1096. **McLeod, B. E. and Robinson, M. F.,** Dietary intake of Mn by New Zealand infants during the first six months of life, *Br. J. Nutr.,* 27, 229, 1972.
1097. **Halls, D. J. and Fell, G. S.,** Determination of Mn in serum and urine by electrothermal AAS, *Anal. Chim. Acta,* 129, 205, 1981.

1098. **Fernandez, A. A., Sobel, C., and Jacobs, S. L.**, Sensitive method for determination of sub-μg quantities of Mn and its application to human serum, *Anal. Chem.*, 35, 1721, 1963.
1099. **Langmyhr, F. J., Lind, T., and Jonsen, J.**, AAS determination of Mn, Ag and Zn in dental material by atomization directly from the solid state, *Anal. Chim. Acta*, 80, 297, 1975.
1100. **Watanabe, T., Tokunaga, R., Iwahana, T., Tati, M., and Ikeda, M.**, Determination of urinary Mn by the direct chelation-extraction method and flameless AAS, *Br. J. Ind. Med.*, 35, 73, 1978.
1101. **Ajemian, R. S. and Whitman, N. E.**, Determination of Mn in urine by AAS, *Am. Ind. Hyg. Assoc. J.*, 30, 52, 1969.
1102. **Papavasiliou, P. S. and Cotzias, G. C.**, NAA: determination of Mn, *J. Biol. Chem.*, 236, 2365, 1961.
1103. **Kocsis, E. and Kovats, M.**, Radiochemical Determination of Several Trace Metals in Blood Using Selective Adsorption, 8th Int. Microchem. Symp., Graz, August 25 to 30, 1980, 152.
1104. **Camara Rica, C., Kirkbright, G. F., and Snook, R. D.**, Determination of Mn and Ni in whole blood by optical emission spectrometry with an ICP source and sample introduction by electrothermal atomization, *At. Spectrosc.*, 2, 172, 1981.
1105. **Cholak, J. and Hubbard, D. M.**, Determination of Mn in air and biological materials, *Am. Ind. Hyg. Assoc. J.*, 21, 356, 1960.
1106. **Carnrick, G. R., Slavin, W., and Manning, D. C.**, Direct determination of Mn in sea water with the L'vov platform and Zeeman background correction in the graphite furnace, *Anal. Chem.*, 53, 1866, 1981.
1107. **Paynter, D. I.**, Microdetermination of Mn in animal tissues by flameless AAS, *Anal. Chem.*, 51, 2086, 1979.
1108. **Grobenski, Z., Lehman, R., Tamm, R., and Welz, B.**, Improvements in graphite-furnace AA microanalysis with solid sampling, *Mikrochim. Acta*, I, 115, 1982.
1109. **Iida, C., Uchida, T., and Kojima, I.**, Decomposition of "Bovine Liver" in a sealed Teflon® vessel for determination of metals by AAS, *Anal. Chim. Acta*, 113, 365, 1980.
1110. **Shearer, D. A., Cloutier, R. O., and Hidroglou, M.**, Chelate extraction and flame AAS determination of ng amounts of Mn in blood and animal tissue, *J. Assoc. Off. Anal. Chem.*, 60, 155, 1977.
1111. **Aihara, M. and Kiboku, M.**, AAS of Mn using solvent extraction with K benzylxanthate — MIBK, *Bunseki Kagaku*, 26, 111, 1977.
1112. **Kambayashi, K.**, Determination of Mn in urine by AAS, *Igaku to Seibutsagaku*, 82, 181, 1971.
1113. **Lekehal, N. and Hanocq, M.**, Determination de traces de Mn urinaire après extraction à l'aide de 2,2,6,6,-tetramethyl-heptane-3,5-dione et dosage par spectrometrie d'absorption atomique, *Anal. Chim. Acta*, 83, 93, 1976.
1114. **Van Ormer, D. G. and Purdy, W. C.**, The determination of Mn in urine by AAS, *Anal. Chim. Acta*, 64, 93, 1973.
1115. **Shamberger, R. J., Zydowicz, S., and Willis, C. E.**, Manganese Analyses by Flameless AA, Paper 050, 29th Natl. Meet. Am. Assoc. Clin. Chem., Chicago, July 17 to 22, 1977.
1116. **Bek, F., Janouskova, J., and Moldan, B.**, Determination of Mn and Sr in blood serum using the Perkin-Elmer HGA-70 graphite furnace, *At. Absorpt. Newsl.*, 13, 47, 1974.
1117. **Ishizaki, M., Ueno, S., Murakami, R., Kataoka, F., Oyamada, N., Kubota, K., and Katsumura, K.**, Manganese determination in blood by flameless AAS, *Ibaraki-ken Eisei Kenkyusho Nempo*, 13, 61, 1975.
1118. **Joselow, M. M., Tobias, E., Koehler, R., Coleman, S., Bogden, J., and Gause, D.**, Manganese pollution in city environment and its relationship to traffic density, *Am. J. Publ. Health*, 68, 557, 1978.
1119. **Bonilla, E.**, Flameless AAS determination of Mn in rat brain and other tissues, *Clin. Chem.*, 24, 471, 1978.
1120. **Belling, G. B. and Jones, G. B.**, The determination of Mn in small samples of biological tissue by flameless AAS, *Anal. Chim. Acta*, 80, 279, 1975.
1121. **Suzuki, M. and Wacker, W. E. C.**, Determination of Mn in biological materials by AAS, *Anal. Biochem.*, 57, 605, 1974.
1122. **Shimada, K.**, Manganese concentration in the teeth of Japanese, *Shikoku Igaku Zasshi*, 27, 516, 1971.
1123. **Koh, T. S., Benson, T. H., and Judson, G. J.**, Trace element analysis of bovine liver: interlaboratory survey in Australia and New Zealand, *J. Assoc. Off. Anal. Chem.*, 63, 809, 1980.
1124. **Jenne, E. A. and Ball, J. W.**, Time stability of aqueous APDC and its Mn and Ni complexes in MIBK, *At. Absorpt. Newsl.*, 11, 90, 1972.
1125. **Tsalev, D. L.**, AAS investigation of hexamethyleneammonium hexamethylenedithiocarbamate-*n*-butylacetate extraction system, *Dokl. Bolg. Akad. Nauk*, 32, 779, 1979.
1126. **Theroux, R. E., Correia, B. C., and Daley, H. O., Jr.**, Removal of dispersed water from MIBK for AAA, *At. Absorpt. Newsl.*, 15, 144, 1976.
1127. **Sharma, D. C. and Davis, P. S.**, Direct determination of Hg in blood by $NaBH_4$ reduction and AAS, *Clin. Chem.*, 25, 769, 1979.

1128. **Cigna Rossi, L., Clemente, G. F., and Santaroni, G.,** Mercury and Se distribution in a defined area and its population, *Arch. Environ. Health*, 24, 160, 1976.
1129. **Dogan, S. and Haerdi, W.,** Rapid separation on Cu powder of total Hg in blood and determination of Hg by flameless AAS, *Int. J. Environ. Anal. Chem.*, 6, 327, 1979.
1130. **Yamamura, Y., Yamamura, S., and Yoshida, M.,** Mercury concentration in blood and urine of workers in small plant manufacturing Hg thermometers, *Sangyo Igaku*, 14, 455, 1972.
1131. **Brune, D.,** *Low Temperature Irradiation Applied to NAA of Hg in Human Whole Blood*, Rep. AE-213, Aktiebolaget Atomenergi, Stockholm, 1966.
1132. **Pallotti, G., Bencivenga, B., and Simonetti, T.,** Total Hg levels in whole blood, hair and fingernails for the population group from Rome and its surroundings, *Sci. Total Environ.*, 11, 69, 1979.
1133. **Cross, J. D., Dale, I. M., Smith, H., and Smith, L. B.,** Dietary Hg in the Glasgow area, *J. Radioanal. Chem.*, 48, 159, 1979.
1134. **Dubinskaya, N. A., Pelekis, L. L., and Kostenko, I. V.,** Application of INAA to the determination of Hg in human hair in conditions of possible occupational contamination, in *Nuclear Physical Analytical Methods in Environmental Control*, Gidrometeoizdat, Leningrad, 1980, 180.
1135. **Gutenmann, W. H., Silvin, J. J., and Lisk, D. J.,** Elevated concentrations of Hg in dentists' hair, *Bull. Environ. Contam. Toxicol.*, 9, 318, 1973.
1136. **Shimomura, S., Kimura, A., Nakagawa, H., and Takao, M.,** Mercury levels in human hair and sex factors, *Environ. Res.*, 22, 22, 1980.
1137. **Kitamura, S., Sumino, K., Hayakawa, K., and Shibata, T.,** Mercury content in human tissues from Japan, in *Effects and Dose-Response Relationships of Toxic Metals*, Nordberg, G. F., Ed., Elsevier, Amsterdam, 1976, 290.
1138. **Nord, P. J., Kabada, M. P., and Sorenson, J. R. J.,** Mercury in human hair. Study of the residents of Los Alamos, NM, and Pasadena, CA, by cold vapor AAS, *Arch. Environ. Health*, 27, 40, 1973.
1139. **Stein, P. C., Campbell, E. E., Moss, W. D., and Trujillo, P.,** Mercury in man, *Arch. Environ. Health*, 29, 25, 1974.
1140. **Mottet, N. K. and Body, R. L.,** Mercury burden of human autopsy organs and tissues, *Arch. Environ. Health*, 29, 18, 1974.
1141. **Kaiser, G., Götz, D., Tölg, G., Knapp, G., Maichin, B., and Spitzy, H.,** Study of systematic errors in the determination of total Hg levels in the range $<10^{-5}\%$ in inorganic and organic matrices with two reliable spectrometrical determination procedures, *Fresenius Z. Anal. Chem.*, 291, 278, 1978.
1142. **Malissa, H., Maly, K., and Till, T.,** On the AAS determination of Hg in roots of teeth and in jawbones, *Fresenius Z. Anal. Chem.*, 293, 141, 1978.
1143. **Campe, A., Velghe, N., and Claeys, A.,** Determination of inorganic, phenyl and total Hg in urine, *At. Absorpt. Newsl.*, 17, 100, 1978.
1144. **Schaller, K-H., Strasser, P., Woitowitz, R., and Szadkowski, D.,** Quantitative determination of traces of Hg in urine after electrolytic concentration, *Fresenius Z. Anal. Chem.*, 256, 123, 1971.
1145. **Least, C. J., Jr., Rejent, T. A., and Lees, H.,** Modification of a cold vapor technique for the determination of Hg in urine, *At. Absorpt. Newsl.*, 13, 4, 1974.
1146. **Kubasik, N. P., Sine, H. E., and Volosin, M. T.,** Rapid analysis for total Hg in urine and plasma by flameless AAA, *Clin. Chem.*, 18, 1326, 1972.
1147. **Kosta, L. and Byrne, A. R.,** Activation analysis for Hg in biological samples at ng level, *Talanta*, 16, 1297, 1969.
1148. **Al-Shahristani, H. and Al-Haddad, I. K.,** Mercury content of hair from normal and poisoned persons, *J. Radioanal. Chem.*, 15, 59, 1973.
1149. **Johansson, A. and Uhrnell, H.,** Determination of Hg in urine, *Acta Chem. Scand.*, 9, 583, 1955.
1150. **Lindström, O.,** Rapid microdetermination of Hg by spectrophotometric flame combustion, *Anal. Chem.*, 31, 461, 1959.
1151. **Nobel, S.,** Mercury in urine, *Stand. Methods Clin. Chem.*, 3, 176, 1961.
1152. **Nishikata, A., Tanzawa, K., Takeda, Y., Osawa, T., and Ukita, T.,** Identification and determination of Hg compounds in the hair by extraction with acids, *Eisei Kagaku*, 14, 211, 1968.
1153. **Tölg, G.,** On the analysis of trace elements in biological material, *Fresenius Z. Anal. Chem.*, 283, 257, 1977.
1154. **Alexandrov, S.,** Study of conditions for spectrochemical determination of Hg in urine after preconcentration, *Dokl. Bolg. Akad. Nauk*, 31, 691, 1978.
1155. **Greenwood, M. R., Dhahir, P., Clarkson, T. W., Farant, J. P., Chartrand, A., and Khayat, A.,** Epidemiological experience with the Magos' reagents in the determination of different forms of Hg in biological samples by flameless AA, *J. Anal. Toxicol.*, 1, 265, 1977.
1156. **Sakashita, H., Oda, S., and Kamada, H.,** Determination of trace Hg in human hair by non-dispersive atomic-fluorescence spectrometry with a cold-vapor technique, *Bunko Kenkyu*, 28, 140, 1979.

1157. **Ahmed, R., Valenta, P., and Nürnberg, H. W.,** Voltametric determination of Hg levels in tuna fish, *Mikrochim. Acta,* I, 171, 1981.
1158. **Rohm, T. J. and Purdy, W. C.,** Kinetic-coulometric determination of Hg in biological samples, *Anal. Chim. Acta,* 72, 177, 1974.
1159. **Rava, R., Brunetti, A., and Spinosi, G.,** Proposed enzymic method for determining Hg(II), *Boll. Soc. Ital. Biol. Sper.,* 57, 1702, 1981.
1160. **Harms, U.,** Trace analysis for Hg at the ng/g level, *Z. Lebensm.-Unters. Forsch.,* 172, 118, 1981.
1161. **Rains, T. C. and Menis, O.,** Determination of submicrogram amounts of Hg in SRMs by flameless AAS, *J. Assoc. Off. Anal. Chem.,* 55, 1339, 1972.
1162. **Cappon, C. J. and Smith, J. C.,** Gas-chromatographic determination of inorganic Hg and organomercurials in biological materials, *Anal. Chem.,* 49, 365, 1977.
1163. **Goolvard, L. and Smith, H.,** Determination of methylmercury in human blood, *Analyst,* 105, 726, 1980.
1164. **Castello, G. and Kanitz, S.,** Determination of organomercury compounds in food and man using gas chromatography and flameless AAS, *Boll. Chim. Unione Ital. Lab. Prov.,* 4, 57, 1978.
1165. **Westöö, G.,** Determination of methylmercury compounds in foodstuffs. I. Methylmercury compounds in fish, identification and determination, *Acta Chim. Scand.,* 20, 2131, 1966.
1166. Analytical Methods Committee, Determination of Hg and methylmercury in fish, *Analyst,* 102, 769, 1977.
1167. **Huckabee, J. W., Feldman, C., and Talmi, Y.,** Mercury concentrations in fish from the Great Smoky Mountains National Park, *Anal. Chim. Acta,* 70, 41, 1974.
1168. **Giovanoli-Jakubczak, T., Greenwood, M. R., Smith, J. C., and Clarkson, T. W.,** Determination of total and inorganic Hg in hair by flameless AA, and of methylmercury by gas chromatography, *Clin. Chem.,* 20, 222, 1974.
1169. **Cappon, C. J. and Smith, J. C.,** Breakdown of methylmercury in sodium hydroxide solution, *Anal. Chem.,* 52, 1527, 1980.
1170. **Margler, L. W. and Mah, R. A.,** Thin layer chromatographic and AAS determination of methylmercury, *J. Assoc. Off. Anal. Chem.,* 64, 1017, 1981.
1171. **Baluja, G. and Gonzalez, M. J.,** Contribution to the simultaneous residual analysis of inorganic and organic mercury compounds by chromatographic-spectroscopic techniques, *Rev. Agroquim. Technol. Aliment.,* 19, 270, 1979.
1172. **Collett, D. L., Fleming, D. E., and Taylor, G. A.,** Determination of alkylmercury in fish by steam-distillation and cold-vapor AAS, *Analyst,* 105, 897, 1980.
1173. **Mitani, K.,** Environmental hygienic studies on Hg compounds. I. Selective determination of methylmercury and inorganic Hg by steam distillation and flameless AAS, *Eisei Kagaku,* 22, 65, 1976.
1174. **Chilov, S.,** Determination of small amounts of Hg, *Talanta,* 22, 205, 1975.
1175. **Ure, A. M.,** The determination of Hg by non-flame AA and fluorescence spectrometry. A review, *Anal. Chim. Acta,* 76, 1, 1975.
1176. **Hoffman, E., Lüdke, C., and Tilch, J.,** Determination of Hg by flameless AAS using the resonance line 184.9 nm, *Spectrochim. Acta,* 34B, 301, 1979.
1177. **Clark, D., Dagnall, R. M., and West, T. S.,** The determination of Hg by AAS with the "Delves sampling cup" technique, *Anal. Chim. Acta,* 58, 339, 1972.
1178. **Rattonetti, A.,** *Determination of Hg in Water and Urine by Furnace Atomizer AAS,* Rep. No. 12, Instrumentation Laboratory, Wilmington, Mass., 1980.
1179. **Fujiwara, K., Sato, K., and Fuwa, K.,** AAS of Hg using a graphite furnace atomizer, *Bunseki Kagaku,* 26, 772, 1977.
1180. **Kirkbright, G. F., Shan Hsiao-Chuan, and Snook, R. D.,** An evaluation of some matrix modification procedures for use in the determination of Hg and Se by AAS with a graphite tube electrothermal atomizer, *At. Spectrosc.,* 1, 85, 1980.
1181. **Alder, J. F. and Hickman, D. A.,** Determination of Hg by AAS with graphite tube atomization, *Anal. Chem.,* 49, 336, 1977.
1182. **Kunert, I., Komarek, J., and Sommer, L.,** Determination of Hg by AAS with cold vapor and electrothermal techniques, *Anal. Chim. Acta,* 106, 285, 1979.
1183. **Shum, G. T. C., Freeman, H. C., and Uthe, J. F.,** Determination of organic (methyl) mercury in fish by graphite furnace AAS, *Anal. Chem.,* 51, 414, 1979.
1184. **Magyar, B. and Vonmont, H.,** Anwendbarkeit der Zeeman-Effect/Atomabsorption in der Komplexchemie und Spurenanalyse, *Spectrochim. Acta,* 35B, 177, 1980.
1185. **Halasz, A. and Polyak, K.,** Determination of Hg by Flameless AAS, 8th Int. Microchem. Symp., Graz, August 25 to 30, 1980, 182.
1186. **Slovák, Z. and Dočekalová, H.,** Electrothermal atomization of Hg in the presence of thiols, *Anal. Chim. Acta,* 115, 111, 1980.
1187. **Yamamoto, Y., Kumamaru, T., and Shiraki, A.,** Comparative study of $NaBH_4$ tablet and $SnCl_2$ reducing systems in the determination of Hg by AAS, *Fresenius Z. Anal. Chem.,* 292, 273, 1978.

1188. **Poluektov, N. S., Vitkun, R. A., and Zelyukova, Yu. V.,** Determination of milligamma amounts of Hg by AA in the gaseous phase, *Zh. Anal. Khim.*, 19, 937, 1964.
1189. **Oda, C. E. and Ingle, J. D., Jr.,** Speciation of Hg by cold vapor AAS with selective reduction, *Anal. Chem.*, 53, 2305, 1981.
1190. **Tong, S. L., Chu, C. K., and Goh, S. H.,** A versatile AAS microattachment for cold-vapor atomic and gas-phase UV absorptiometric measurements, *Mikrochim. Acta*, I, 99, 1981.
1191. **Bourcier, D. R. and Sharma, R. P.,** Stationary cold vapor technique for determination of sub-μg amounts of Hg in biological tissues by flameless AAS, *J. Anal. Toxicol.*, 5, 65, 1981.
1192. **Tuncel, G. and Ataman, O. Y.,** Design and evaluation of a new absorption cell for cold vapor Hg determination by AAS, *At. Spectrosc.*, 1, 126, 1980.
1193. **Helsby, C. A.,** Determination of Hg in fingernails and body hair, *Anal. Chim. Acta*, 82, 427, 1976.
1194. **Coles, L. E. and Guthenberg, H.,** Determination of Hg in foodstuffs, *Pure Appl. Chem.*, 51, 2527, 1979.
1195. **Hatch, W. R. and Ott, W. L.,** Determination of sub-μg quantities of Hg by AAS, *Anal. Chem.*, 40, 2085, 1968.
1196. **Richardson, R. A.,** Automated method for determination of Hg in urine, *Clin. Chem.*, 22, 1604, 1976.
1197. **Kothandaraman, P. and Dallmeyer, J. F.,** Improved desiccator for Hg cold vapor technique, *At. Absorpt. Newsl.*, 15, 120, 1976.
1198. **Farant, J.-P., Brissette, D., Moncion, L., Bigras, L., and Chartrand, A.,** Improved cold-vapor AA technique for micro-determination of total and inorganic Hg in biological samples, *J. Anal. Toxicol.*, 5, 47, 1981.
1199. **Magos, L.,** Selective determination of inorganic Hg and methylmercury in undigested biological samples, *Analyst*, 96, 847, 1971.
1200. **Stuart, D. C.,** Radiotracer investigation of the cold-vapor AA method of analysis for trace Hg, *Anal. Chim. Acta*, 101, 429, 1978.
1201. **Gardner, D.,** The use of $Mg(ClO_4)_2$ as desiccant in the syringe injection technique for determination of Hg by cold-vapor AAS, *Anal. Chim. Acta*, 119, 167, 1980.
1202. **Oda, C. E. and Ingle, J. D., Jr.,** Continuous flow cold vapor AA determination of Hg, *Anal. Chem.*, 53, 2030, 1981.
1203. **Melcher, M. and Welz, B.,** Vergleich von Störmöglichkeiten bei der Bestimmung von Hg mit $SnCl_2$ oder $NaBH_4$ als Reduktionsmittel, AA Lab. Notes No. 20/D, Bodenseewerk Perkin-Elmer & Co. GmbH, Überlingen, 1978.
1204. **Oster, O.,** The direct determination of Hg in urine by cold vapor AAS, *J. Clin. Chem. Clin. Biochem.*, 19, 471, 1981.
1205. **Stuart, D. C.,** Factors affecting peak shape in cold-vapor AAS for Hg, *Anal. Chim. Acta*, 106, 411, 1979.
1206. **Gardner, D. and Dal Pont, G.,** Rapid, simple method for determination of total Hg in fish and hair by cold-vapor AAS, *Anal. Chim. Acta*, 108, 13, 1979.
1207. **Poluektov, N. S. and Zeljukova, Yu. V.,** Flameless AA method for determination of Hg and its application in environmental control, in *Chemical Pollution of Marine Environment*, Simonov, A. I. and Prokofiev, A. K., Eds., Gidrometeoizdat, Leningrad, 1979, 112.
1208. **Magos, L. and Cernik, A. A.,** A rapid method for estimating Hg in undigested biological samples, *Br. J. Ind. Med.*, 26, 144, 1969.
1209. **Gage, J. C. and Warren, J. M.,** The determination of Hg and organic mercurials in biological samples, *Ann. Occup. Hyg.*, 13, 115, 1970.
1210. **Ebbestad, U., Gundersen, N., and Torgrimsen, T.,** A simple method for the determination of inorganic Hg and methylmercury in biological samples by flameless AAS, *At. Absorpt. Newsl.*, 14, 142, 1975.
1211. **Yamamoto, R., Satoh, H., Sudzuki, T., Naganuma, A., and Imura, N.,** The applicable conditions of Magos' method for Hg measurement under coexistence of Se, *Anal. Biochem.*, 101, 254, 1980.
1212. **Hwang, J. Y., Ullucci, P. A., and Melenfant, A. L.,** Determination of Hg by flameless AAS technique, *Can. Spectrosc.*, 16, 100, 1971.
1213. **Velghe, N., Campe, A., and Claeys, A.,** Determination of Cu in undiluted serum and whole blood by AAS with graphite furnace, *At. Spectrosc.*, 3, 48, 1982.
1214. **Manthey, G. and Berge, H.,** Differential determination of inorganic and organic bound Hg in foods, water and biological materials, *Nahrung*, 24, 413, 1980.
1215. **Littlejohn, D. L., Fell, G. S., and Ottaway, J. M.,** Modified determination of total and inorganic Hg in urine by cold vapor AAS, *Clin. Chem.*, 22, 1719, 1976.
1216. **Rooney, R. C.,** Use of $NaBH_4$ for cold-vapor AA determination of trace amounts of inorganic Hg, *Analyst*, 101, 678, 1976.
1217. **Mizunuma, H., Morita, H., Sakarai, H., and Shimomura, S.,** Selective AA determination of inorganic and organic Hg by combined use of Fe(III) and $NaBH_4$, *Bunseki Kagaku*, 28, 695, 1979.
1218. **Toffaletti, J. and Savory, J.,** Use of $NaBH_4$ for determination of total Hg in urine by AAS, *Anal. Chem.*, 47, 2091, 1975.

1219. **Melcher, M.,** Interference Possibilities in the Determination of As, Se, and Hg with the MHS-Techniques, *AA Lab. Notes,* No. 17/E, Bodenseewerk Perkin-Elmer & Co. GmbH, Überlingen, 1978.
1220. **Melcher, M.,** Influence of Different Acids on the Determination of As, Se, and Hg with the MHS-Techniques, *AA Lab. Notes,* No. 16/E, Bodenseewerk Perkin-Elmer & Co. GmbH, Überlingen, 1978.
1221. **Vitkun, R. A., Poluektov, N. S., and Zelyukova, Yu. V.,** Ascorbic acid as a reductant in flameless AA determination of Hg, *Zh. Anal. Khim.,* 29, 691, 1974.
1222. **Vitkun, R. A., Didorenko, T. O., Zelyukova, Yu. A., and Poluektov, N. S.,** Determination of Hg by non-flame AAS using dihydroxymaleic acid, *Zh. Anal. Khim.,* 37, 833, 1982.
1223. **Wigfield, D. C., Croteau, S. M., and Perkins, S. L.,** Elimination of matrix effects in cold-vapor AAA of Hg in human hair samples, *J. Anal. Toxicol.,* 5, 52, 1981.
1224. **Lindstedt, G.,** A rapid method for the determination of Hg in urine, *Analyst,* 95, 264, 1970.
1225. **Naganuma, A., Satoh, H., Yamamoto, R., Suzuki, T., and Imura, N.,** Effect of Se on determination of Hg in animal tissues, *Anal. Biochem.,* 98, 287, 1979.
1226. **Omang, S. H.,** Trace determination of Hg in biological materials by flameless AAS, *Anal. Chim. Acta,* 63, 247, 1973.
1227. **Taguchi, M., Yasuda, K., Hashimoto, M., and Toda, S.,** Some improvements for Hg determination in marine organisms by AAS, *Bunseki Kagaku,* 28, T33, 1979.
1228. **Chapman, J. F. and Dale, L. S.,** The use of alkaline permanganate in the preparation of biological materials for the determination of Hg by AAS, *Anal. Chim. Acta,* 134, 379, 1982.
1229. **Stuart, D. C.,** Trace Hg analysis in biological material: use of ^{203}Hg labeled methylmercury chloride for in vivo labeling of fish to study the efficacy of various wet ashing procedures, *Anal. Chim. Acta,* 96, 83, 1978.
1230. **Fawkes, J., Folsom, M., and Oswald, E. O.,** The determination of total Hg in biological tissues by a modified $KMnO_4$ procedure, *Anal. Chim. Acta,* 82, 55, 1976.
1231. **Rathie, A. O.,** A rapid UV absorption method for the determination of Hg in urine, *Am. Ind. Hyg. Assoc. J.,* 30, 126, 1969.
1232. **Melcher, M.,** Bestimmung von Hg in Urin mit dem MHS-1, *AA Lab. Notes,* No. 23/D, Bodenseewerk Perkin-Elmer & Co. GmbH, Überlingen, 1978.
1233. **La Fleur, P. D.,** Retention of Hg when freeze-drying biological materials, *Anal. Chem.,* 45, 1534, 1973.
1234. **Dumarey, M., Heindryckx, R., and Dams, R.,** Determination of Hg in environmental SRMs by pyrolysis, *Anal. Chim. Acta,* 118, 381, 1980.
1235. **Mesman, B. B., Smith, B. S., and Pierce, J. O., II,** Determination of Hg in urine by AA, *Am. Ind. Hyg. Assoc. J.,* 31, 701, 1970.
1236. **Mesman, B. B. and Smith, B. S.,** Determination of Hg in urine by AA, utilizing the APDC-MIBK extraction system and boat technique, *At. Absorpt. Newsl.,* 9, 81, 1970.
1237. **Moffitt, A. E., Jr. and Kupel, R. E.,** A rapid method employing impregnated charcoal and AAS for the determination of Hg in atmospheric, biological and aquatic samples, *At. Absorpt. Newsl.,* 9, 113, 1970.
1238. **Jacobs, M. B., Yamaguchi, S., Goldwater, L. J., and Gilbert, H.,** Determination of Hg in blood, *Am. Ind. Hyg. Assoc. J.,* 21, 475, 1960.
1239. **Joselow, M. M., Ruiz, R., and Goldwater, L. J.,** The use of salivary (parotid) fluid for biochemical monitoring, *Am. Ind. Hyg. Assoc. J.,* 30, 77, 1969.
1240. **Joselow, M. M., Ruiz, R., and Goldwater, L. J.,** Adsorption and excretion of Hg in man. XIV. Salivary excretion of Hg and its relation to blood and urine Hg, *Arch. Environ. Health,* 17, 35, 1968.
1241. **Magos, L. and Clarkson, T. W.,** AA determination of total, inorganic and organic Hg in blood, *J. Assoc. Off. Anal. Chem.,* 55, 966, 1972.
1242. **Roschig, M., Bach, W., Craesser, K., and Rackow, S.,** Proposals for the 2nd ed. of the pharmacopea of the DDR (diagnostic lab. methods) Determination of Hg in biological materials (flameless AAS), *Zentralbl. Pharm. Pharmakother. Laboratoriumsdiagn.,* 119, 1375, 1980.
1243. **Barlow, P. J., Crump, D. R., Khera, A. K., and Wibberley, D. G.,** Some analytical problems in the determination of Hg in biological materials by cold vapor technique, *Proc. Anal. Div. Chem. Soc.,* 16, 15, 1979.
1244. **Ambe, M. and Niikura, N.,** Determination of Hg in biological and environmental samples by cold vapor AAS: comparison of sample pretreatments, *Bunseki Kagaku,* 29, 5T, 1980.
1245. **Bouchard, A.,** Determination of Hg after room temperature digestion by flameless AAS, *At. Absorpt. Newsl.,* 12, 115, 1973.
1246. **Knechtel, J. R. and Fraser, J. L.,** Wet digestion method for the determination of Hg in biological and environmental samples, *Anal. Chem.,* 51, 315, 1979.
1247. **Armstrong, F. A. J. and Uthe, J. F.,** Semi-automated determination of Hg in animal tissue, *At. Absorpt. Newsl.,* 10, 101, 1971.
1248. **Malaiyandi, M. and Barrette, J. P.,** Determination of Hg in soils and biological matrices by the V_2O_5 digestion procedure, *Arch. Environ. Contam. Toxicol.,* 2, 315, 1974.

1249. **Kruse, R.,** Determination of total Hg in fish by means of a digestion method with $HNO_3/HClO_3/HClO_4$ which does not lead to low results, *Z. Lebensm.-Unters. Forsch.*, 169, 259, 1979.
1250. **Munns, R. K. and Holland, D. C.,** Rapid digestion and flameless AAS of Hg in fish: collaborative study, *J. Assoc. Off. Anal. Chem.*, 60, 833, 1977.
1251. **Barrett, P., Davidowski, L. J., Jr., Penaro, K. W., and Copeland, T. R.,** Microwave oven-based wet digestion technique, *Anal. Chem.*, 50, 1021, 1978.
1252. **Reamer, D. C. and Veillon, C.,** Preparation of biological materials for determination of Se by hydride generation AAS, *Anal. Chem.*, 53, 1192, 1981.
1253. **Suzuki, T. and Yamamoto, R.,** Organic Hg levels in human hair with and without storage for 11 years, *Bull. Environ. Contam. Toxicol.*, 28, 186, 1982.
1254. Mercury Analysis Working Party of BITC, Standardization of methods for the determination of traces of Hg. II. Determination of total Hg in materials containing organic matter, *Anal. Chim. Acta*, 84, 231, 1976.
1255. **Cappon, C. J. and Smith, J. C.,** Hg and Se content and chemical form in human and animal tissue, *J. Anal. Toxicol.*, 5, 90, 1981.
1256. **Lindstedt, G. and Skare, I.,** Microdetermination of Hg in biological samples. II. An apparatus for rapid automatic determination of Hg in digested samples, *Analyst*, 96, 223, 1971.
1257. **Ramelow, G. and Hornung, H.,** An investigation into possible Hg losses during lyophilization of marine biological samples, *At. Absorpt. Newsl.*, 17, 59, 1978.
1258. **Korunová, V. and Dědina, J.,** Determination of trace concentrations of Hg in biological materials after digestion under pressure in HNO_3 catalysed by V_2O_5, *Analyst*, 105, 48, 1980.
1259. **Matthes, W., Flucht, R., and Stoeppler, M.,** Contributions to automated trace analysis. III. A sensitive automated method for the determination of Hg in biological and environmental samples, *Fresenius Z. Anal. Chem.*, 291, 20, 1978.
1260. **Gaffin, S. L. and Hornung, H.,** Rapid determination of Hg in urine by flameless AAS, *Clin. Toxicol.*, 10, 345, 1977.
1261. **Malaiyandi, M. and Barrette, J. P.,** Determination of submicro quantities of Hg in biological materials, *Anal. Lett.*, 3, 579, 1970.
1262. **Tamura, Y., Maki, T., Yamada, H., Shimamura, Y., Nishigaki, S., and Kimura, Y.,** Rapid method for the determination of Hg in biological samples, *Tokyo Toritsu Eisei Kenkyusho Kenkyu Nempo*, 25, 227, 1974.
1263. **Ozawa, T.,** Determination of trace amounts of Hg in organic materials by the combustion method, *Tokyo-Toritsu Kogyo Gijutsu Senta Kenkyu Hokoku*, 6, 103, 1976.
1264. **Watling, R. J.,** A simple oxidation procedure for biological material prior to analysis for Hg, *Anal. Chim. Acta*, 99, 357, 1978.
1265. **Narasaki, H.,** Determination of trace Hg in milk products and plastics by combustion in an oxygen bomb and cold-vapor AAS, *Anal. Chim. Acta*, 125, 187, 1981.
1266. **Okuno, I., Wilson, R. A., and White, R. E.,** Determination of Hg in biological samples by flameless AA after combustion and Hg-Ag amalgamation, *J. Assoc. Off. Anal. Chem.*, 55, 96, 1972.
1267. **Sadin, Y. and Deldime, P.,** Determination of Hg by flameless AAS. Study of solution preparation, *Anal. Lett.*, 12B, 563, 1979.
1268. **Thillez, G.,** Rapid and accurate determination of traces of Hg in the air and in biological media by AA, *Chim. Anal.*, 50, 226, 1968.
1269. **Lidums, V. and Ulfvarson, U.,** Mercury analysis in biological material by direct combustion in oxygen and photometric determination of Hg vapor, *Acta Chem. Scand.*, 22, 2150, 1968.
1270. **Imaeda, K. and Ohsawa, K.,** Determination of Hg using porous gold as collector, *Bunseki Kagaku*, 28, 239, 1979.
1271. **Toribara, T. Y. and Shields, C. P.,** The analysis of sub-μg amounts of Hg in tissues, *Am. Ind. Hyg. Assoc. J.*, 29, 87, 1968.
1272. **Zelyukova, Yu. V., Vitkun, R. A., Didorenko, T. O., and Poluektov, N. S.,** Nonflame AA determination of Hg in environmental samples, *Zh. Anal. Khim.*, 36, 454, 1981.
1273. **Hadeishi, T.,** Isotope-shift Zeeman effect for trace-element detection: an application of atomic physics to environmental problems, *Appl. Phys. Lett.*, 21, 438, 1972.
1274. **Church, D. A., Hadeishi, T., Leong, L., McLaughlin, R. D., and Zak, B. D.,** Two-chamber furnace for flameless AAS, *Anal. Chem.*, 46, 1352, 1974.
1275. **Coyle, P. and Hartley, T.,** Automated determination of Hg in urine and blood by the Magos reagent and cold vapor AAS, *Anal. Chem.*, 53, 354, 1981.
1276. **Skare, I.,** Microdetermination of Hg in biological samples. III. Automated determination of Hg in urine, fish and blood samples, *Analyst*, 97, 148, 1972.
1277. **Suzuki, T., Miyama, T., and Katsunuma, H.,** Mercury contents in red cells, plasma, urine, and hair from workers exposed to Hg vapor, *Ind. Health*, 8, 39, 1970.
1278. **Clarkson, T. W. and Greenwood, M. R.,** Selective determination of inorganic Hg in the presence of organomercurial compounds in biological material, *Anal. Biochem.*, 37, 236, 1970.

1279. **Chvojka, R. and Kacprzak, J. L.,** Use of background correction to improve the accuracy of the selective-reduction method for the determining methylmercury in fish, *J. Assoc. Off. Anal. Chem.*, 62, 1179, 1979.
1280. **Clarkson, T. W. and Greenwood, M. R.,** Mercury, biological specimen collections, in *The Use of Biological Specimens for the Assessment of Human Exposure to Environmental Pollutants,* Berlin, A., Wolff, A. H., and Hasegawa, Y., Eds., Martinus Nijhoff Publishers, Boston, 1979, 109.
1281. **Suzuki, T. and Shishido, S.,** Desquamation of renal epithelian cells as a route of Hg excretion in man: a preliminary study, *Tohoku J. Exp. Med.*, 117, 397, 1975.
1282. **Wall, H. and Rhodes, C.,** The effect of bacterial contamination and aging on the volatility of Hg in urine specimens, *Clin. Chem.*, 12, 837, 1966.
1283. **Magos, L., Tuffery, A. A., and Clarkson, T. W.,** Volatization of Hg by bacteria, *Br. J. Ind. Med.*, 21, 294, 1964.
1284. **Robinson, J. W. and Skelly, E. M.,** Direct determination of Hg in hair by AAS at the 184.9 nm resonance line, *Spectrosc. Lett.*, 14, 519, 1981.
1285. **Watling, R. J.,** The analysis of volatile metals in biological materials, in *The Analysis of Biological Materials,* Butler, L. R. P., Ed., Pergamon Press, Oxford, 1979, 81.
1286. **Duve, R. N., Chandra, J. P., and Singh, S. B.,** Open digestion modification of AOAC method for determination of Hg in fish, *J. Assoc. Off. Anal. Chem.*, 64, 1027, 1981.
1287. **Egaas, E. J. and Julshamn, K.,** A method for the determination of Se and Hg in fish products using the same digestion procedure, *At. Absorpt. Newsl.*, 17, 135, 1978.
1288. **Stoeppler, M. and Backhaus, F.,** Pretreatment studies with biological and environmental materials. I. Systems for pressurized multisample decomposition, *Fresenius Z. Anal. Chem.*, 291, 116, 1978.
1289. **Stoeppler, M., Müller, K. P., and Backhaus, F.,** Pretreatment studies with biological and environmental materials. III. Pressure evaluation and carbon balance in pressurized decomposition with nitric acid, *Fresenius Z. Anal. Chem.*, 297, 107, 1979.
1290. **Kotz, L., Kaiser, G., Tschöpel, P., and Tölg, G.,** Decomposition of biological materials for the determination of extremely low contents of trace elements in limited amounts with HNO_3 under pressure in a Teflon® tube, *Fresenius Z. Anal. Chem.*, 260, 207, 1972.
1291. **Brandenberger, H. and Bader, H.,** The determination of ng quantities of Hg in solutions by means of flameless AA method, *Helv. Chim. Acta,* 50, 1409, 1967.
1292. **Lo, J. M. and Wal, C. M.,** Mercury loss from water during storage: mechanisms and prevention, *Anal. Chem.*, 47, 1869, 1975.
1293. **Morgan, A. and Holmes, A.,** Determination of Mo in blood and biological materials by NAA, *Radiochim. Radioanal. Lett.*, 9, 329, 1972.
1294. **Healy, W. B. and Bate, L. C.,** Determination of Mo in hair and wool by NAA, *Anal. Chim. Acta,* 33, 443, 1965.
1295. **Versieck, J., Hoste, J., Barbier, F., Vanballenberghe, L., and de Rudder, J.,** Determination of Mo in human serum by NAA, *Clin. Chim. Acta,* 87, 135, 1978.
1296. **Byrne, A. R.,** Simultaneous RNAA of V, Mo and As in biological samples, *Radiochem. Radioanal. Lett.*, 52, 99, 1982.
1297. **Christian, G. D. and Patriarche, G. J.,** Catalytic determination of Mo in blood and urine, *Anal. Lett.*, 12, 11, 1979.
1298. **Tsongas, T. A., Meglen, R. R., Walravens, P. A., and Chappell, W. R.,** Molybdenum in the diet: an estimate of average daily intake in the U.S., *Am. J. Clin. Nutr.*, 33, 1103, 1980.
1299. **Allaway, W. H., Kubota, J., Losee, F., and Roth, M.,** Selenium, Mo, and V in human blood, *Arch. Environ. Health,* 16, 342, 1968.
1300. **Christian, G. D. and Patriarche, G. J.,** Cathode-ray polarographic determination of Mo in serum, plasma and urine, *Analyst,* 104, 680, 1979.
1301. **Barbooti, M. M. and Jasim, F.,** Electrothermal AAS determination of Mo, *Talanta,* 28, 359, 1981.
1302. **Manning, D. C. and Ediger, R. D.,** Pyrolysis graphite surface treatment for HGA-2100 sample tubes, *At. Absorpt. Newsl.*, 15, 42, 1976.
1303. **Bentley, G. E., Markowitz, L., and Meglen, R. R.,** Analysis of Mo in biological materials, *Adv. Chem. Ser.*, 172, 33, 1979.
1304. **Studnicki, M.,** Matrix effects in the determination of Mo in plants by carbon furnace AAS, *Anal. Chem.*, 51, 1336, 1979.
1305. **Khan, S. U., Cloutier, R. O., and Hidiroglou, M.,** AAS determination of Mo in plant tissue and blood plasma, *J. Assoc. Off. Anal. Chem.*, 62, 1062, 1979.
1306. **Abbasi, S. A.,** Toxicity of Mo and its trace analysis in animal tissues and plants, *Int. J. Environ. Anal. Chem.*, in press.
1307. **Manning, D. C.,** *Determination of Mo in Plant and Animal Tissue Using HGA-2000,* AA Application Study 518, Perkin-Elmer, Norwalk, Conn., 1973.

1308. **Nomoto, S. and Sunderman, F. W., Jr.,** AAS of Ni in serum, urine, and other biological materials, *Clin. Chem.*, 16, 477, 1970.
1309. **Zachariasen, H., Andersen, I., Kostol, C., and Barton, R.,** Technique for determining Ni in blood by flameless AAS, *Clin. Chem.*, 21, 562, 1975.
1310. **Horak, E. and Sunderman, F. W., Jr.,** Fecal nickel excretion by healthy adults, *Clin. Chem.*, 19, 429, 1973.
1311. **Nechay, M. W. and Sunderman, F. W., Jr.,** Measurements of Ni in hair by AAS, *Ann. Clin. Lab. Sci.*, 3, 30, 1973.
1312. **Chen, J. R., Francisco, R. B., and Miller, T. E.,** Legionnaires' disease: Ni levels, *Science*, 196, 906, 1977.
1313. **Andersen, I., Torjussen, W., and Zachariasen, H.,** Analysis for Ni in plasma and urine by electrothermal AAS, with sample preparation by protein precipitation, *Clin. Chem.*, 24, 1198, 1978.
1314. **Sunderman, F. W., Jr. and Nechay, M. W.,** Measurements of serum Ni by flameless AAS, *J. Clin. Chem. Clin. Biochem.*, 12, 220, 1974.
1315. **Mikac-Dević, D., Sunderman, F. W., Jr., and Nomoto, S.,** Furildioxime method for Ni analysis in serum and urine by electrothermal AAS, *Clin. Chem.*, 23, 948, 1977.
1316. **Catalanatto, F. A., Sunderman, F. W., Jr., and Macintosh, T. R.,** Nickel concentrations in human parotid saliva, *Ann. Clin. Lab. Sci.*, 7, 146, 1977.
1317. **Ader, D. and Stoeppler, M.,** Radiochemical and methodological studies on the recovery and analysis of Ni in urine, *J. Anal. Toxicol.*, 1, 252, 1977.
1318. **Brown, S. S. and Sunderman, F. W., Jr.,** *Nickel Toxicology*, Proc. 2nd Int. Conf. Ni Toxicol., 3-5 Sept. 1980, Swansea, Wales, Academic Press, London, 1980.
1319. **Pihlar, B., Valenta, P., and Nürnberg, H. W.,** New high-performance analytical procedure for voltametric determination of Ni in routine analysis of water, biological materials and food, *Fresenius Z. Anal. Chem.*, 307, 337, 1981.
1320. IUPAC, Commission on Toxicology, Sub-committee on Environmental and Occupational Toxicology of Ni, Analytical biochemistry of Ni, *Pure Appl. Chem.*, 52, 527, 1980.
1321. **Uden, P. C., Henderson, D. E., and Kamalizad, A.,** Electron capture gas chromatographic determination of Cu and Ni beta-ketoamine chelates, *J. Chromatogr. Sci.*, 12, 591, 1974.
1322. **Kingston, H. and Pella, P. A.,** Preconcentration of trace metals in environmental and biological samples by cation exchange resin filters for X-ray spectrometry, *Anal. Chem.*, 53, 223, 1981.
1323. **Völlkopf, U., Grobenski, Z., and Welz, B.,** Determination of Ni in serum using graphite furnace AA, *At. Spectrosc.*, 2, 68, 1981.
1324. **Sutter, E. M. and Leroy, M. J. F.,** Nature of interferences of HNO_3 in the determination of Ni and V by AAS with electrothermal atomization, *Anal. Chim. Acta*, 96, 243, 1978.
1325. **Hagedorn-Götz, H., Küppers, G., and Stoeppler, M.,** On Ni contents in urine and hair in a case of exposure to nickel carbonyl, *Arch. Toxicol.*, 38, 275, 1977.
1326. **Myron, D. R., Zimmerman, T. J., Shuler, T. R., Klevay, L. M., Lee, D. E., and Nielsen, F. H.,** Intake of Ni and V by humans. A survey of selected diets, *Am. J. Clin. Nutr.*, 31, 527, 1978.
1327. **Ikebe, K. and Tanaka, R.,** Determination of V and Ni in marine samples by flameless and flame AAS, *Bull. Environ. Contam. Toxicol.*, 21, 526, 1979.
1328. **Sprague, S. and Slavin, W.,** The determination of Ni in urine by AAS. Preliminary study, *At. Absorpt. Newsl.*, 3, 160, 1964.
1329. **Adams, D. B., Brown, S. S., Sunderman, F. W., Jr., and Zachariasen, H.,** Interlaboratory comparisons of Ni analyses in urine by AAS, *Clin. Chem.*, 24, 862, 1978.
1330. IUPAC, Commission on Toxicology, Sub-committee on Environmental and Occupational Toxicology of Ni, IUPAC reference method for analysis of Ni in serum and urine by electrothermal AAS, *Pure Appl. Chem.*, 53, 773, 1981.
1331. **Elakhovskaya, N. P., Ershova, K. P., and Itskova, A. I.,** Determination of Ni in biological materials by AAS, *Gig. Sanit.*, 64, 1978.
1332. **Torjussen, W., Andersen, I., and Zachariasen, H.,** Nickel content of human palatine tonsils: analysis of small tissue samples by flameless AAS, *Clin. Chem.*, 23, 1018, 1977.
1333. **Sunderman, F. W., Jr.,** Measurements of Ni in biological materials by AAS, *Am. J. Clin. Pathol.*, 44, 182, 1965.
1334. **Dornemann, A. and Kleist, H.,** Determination of nanotraces of Ni in biological samples, *Fresenius Z. Anal. Chem.*, 300, 197, 1980.
1335. **Schaller, K. H., Kuchner, A., and Lehnert, G.,** Nickel as a trace element in human blood, *Blut*, 17, 155, 1968.
1336. Analytical Methods Committee, Determination of small amounts of Ni in organic matter by AAS, *Analyst*, 104, 1070, 1979.

1337. **Zachariasen, H., Andersen, I., Kostol, C., and Barton, R. T.,** Technique for determining Ni in blood by flameless AAS, *Arztl. Lab.*, 22, 172, 1976.
1338. **Johnson, D. E., Tillery, J. B., and Prevost, R. J.,** Levels of Pt, Pd, and Pb in populations of Southern California, *Environ. Health Perspect.*, 12, 27, 1975.
1339. **Jones, A. H.,** Determination of Pt and Pd in blood and urine by flameless AAS, *Anal. Chem.*, 48, 1472, 1976.
1340. **Smith, I. C., Carson, B. L., and Ferguson, T. L.,** *Trace Metals in the Environment,* Vol. 4. Ann Arbor Science Publishers, Ann Arbor, Mich., 1978.
1341. **Miller, R. G. and Doerger, J. U.,** Determination of Pt and Pd in biological samples, *At. Absorpt. Newsl.*, 14, 66, 1973.
1342. **Byrne, A. R.,** Determination of Pd in biological samples by NAA, *Mikrochim. Acta,* I, 323, 1981.
1343. **Fazakas, J.,** Critical study of the determination of Pd by graphite furnace-AAS, *Anal. Lett.*, 14, 535, 1981.
1344. **Fazakas, J.,** Determination of Pd by graphite furnace AAS with vaporization from a platform, *Anal. Lett.*, 15, (A3), 1982.
1345. **Everett, G. L.,** The determination of the precious metals by flameless AAS, *Analyst,* 101, 348, 1976.
1346. **Alcock, N. W.,** A method for between-sample cleaning of the graphite furnace, *At. Absorpt. Newsl.*, 18, 37, 1979.
1347. **Banister, S. J., Sternson, L. A., Repta, A. J., and James, G. W.,** Measurement of free-circulating *cis*-dichlorodiammineplatinum(II) in plasma, *Clin. Chem.*, 23, 2258, 1977.
1348. **Bannister, S. J., Chang, Y., Sternson, L. A., and Repta, A. J.,** AAS of free circulating platinum species in plasma derived from *cis*-dichlorodiammineplatinum(II), *Clin. Chem.*, 24, 877, 1978.
1349. **Dedrick, R. L., Litterst, C. L., Gram, T. E., Guarino, A. M., and Becker, D. A.,** Analysis of Pt in biological material by flameless AAS, *Biochem. Med.*, 18, 184, 1977.
1350. **Le Roy, A. F., Wehling, M. L., Sponseller, H. L., Frianf, W. S., Solomon, R. E., Dedrick, R. L., Litterst, C. L., Gram, T. E., Guriano, A. M., and Becker, D. A.,** Analysis of Pt in biological materials by flameless AAS, *Biochem. Med.*, 18, 184, 1977.
1351. **Standefer, J. C.,** *Direct Assay of cis-Pt in Serum,* Paper 063, Jt. Meet. AACC and CSCC, Boston, July 20 to 25, 1980.
1352. **Hull, D. A., Muhammad, N., Lanese, J. G., Reich, S. D., Finkelstein, T. T., and Fandrich, S.,** Determination of Pt in serum and ultrafiltrate by flameless AAS, *J. Pharm. Sci.*, 70, 500, 1981.
1353. **LeRoy, A. F., Wehling, M., Gormley, P., Egorin, M., Ostrow, S., Bachur, N., and Wiernik, P.,** Quantitative changes in *cis*-dichlorodiammineplatinum(II) speciation in excreted urine with time after i.v. infusion in man: methods of analysis, preliminary studies, and clinical results, *Cancer Treat. Rep.*, 64, 123, 1980.
1354. **LeRoy, A. F.,** Interactions of platinum metals and their complexes in biological systems, *Environ. Health Perspect.*, 10, 73, 1975.
1355. **Pera, M. F., Jr. and Harder, H. C.,** Analysis for Pt in biological material by flameless AAS, *Clin. Chem.*, 23, 1245, 1977.
1356. **Danniston, M. L., Sternson, L. A., and Repta, A. J.,** Analysis of total Pt derived from cisplatin in tissue, *Anal. Lett.*, 14, 451, 1981.
1357. **Priesner, D., Sternson, L. A., and Repta, A. J.,** Analysis of total Pt in tissue samples by flameless AAS. Elimination of the need for sample digestion, *Anal. Lett.*, 14, 1255, 1981.
1358. **Smeyers-Verbeke, J., Detaevernier, M. R., Denis, L., and Massart, D. L.,** Determination of Pt in biological fluid by means of graphite furnace AAS, *Clin. Chim. Acta,* 113, 329, 1981.
1359. **Bel'skii, N. K., Ochertyanova, L. I., and Shubochkin, L. K.,** AA determination of Pt in albumin, *Zh. Anal. Khim.*, 34, 814, 1979.
1360. **Wood, O. L.,** Comparison of naturally occurring Rb and K in human erythrocytes, plasma and urine, *Health Phys.*, 17, 513, 1969.
1361. **Wood, O. L.,** Rubidium and K Concentrations in Human Erythrocytes, Plasma and Urine, Rep. No. C00-119-238, U.S. Atomic Energy Comm., Washington, D.C., 1968.
1362. **Chechan, C., Marchandise, X., and Lekieffre, J.,** Determination of Rb levels in several biological media by AAS. Dangers to avoid, preliminary results, *C.R. Seances Soc. Biol. Ses. Fil.*, 169, 991, 1975.
1363. **Lang, W.,** Flame spectrophotometric determination of serum rubidium, *Z. Ges. Exp. Med.*, 139, 438, 1965.
1364. **Sutter, E., Platman, S. R., and Fieve, R. R.,** AAS of Rb in biological fluids, *Clin. Chem.*, 16, 602, 1970.
1365. **Wood, O. L.,** Determination of Rb in human erythrocytes, plasma and urine by AAS, *Biochem. Med.*, 3, 458, 1970.
1366. **Joseph, K. T., Parameswaran, M., and Soman, S. D.,** Estimation of ^{40}K and ^{87}Rb in environmental samples using AAS, *At. Absorpt. Newsl.*, 8, 127, 1969.

1367. **Lieberman, K. W. and Meltzer, H. L.,** Determination of Rb in biological materials by AAS, *Anal. Lett.*, 4, 547, 1971.
1368. **Clinton, O. E.,** Determination of Se in blood and plant material by hydride generation and AAS, *Analyst*, 102, 187, 1977.
1369. **Tulley, R. T. and Lehmann, H. P.,** *The Determination of Whole Blood Selenium by Flameless AAS*, Paper 503, Jt. Meet. AACC and CSCC, Boston, July 20 to 25, 1980.
1370. **Dickson, R. C. and Tomlinson, R. H.,** Selenium in blood and human tissues, *Clin. Chim. Acta*, 16, 311, 1967.
1371. **Kurahashi, K., Inoue, S., Yonekura, S., Shimoishi, Y., and Tôei, K.,** Determination of Se in human blood by gas chromatography with electron-capture detection, *Analyst*, 105, 690, 1980.
1372. **Hojo, Y.,** Selenium levels in human erythrocyte, saliva, milk, urine, and nail, *Sci. Rep. Kyoto Pref. Univ., Ser. B.* 13, 1980.
1373. **Schroeder, H. A., Frost, D. V., and Balassa, J. J.,** Essential trace elements in man: Se, *J. Chron. Dis.*, 23, 227, 1970.
1374. **Shearer, T. R. and Hadjimarkos, D. M.,** Geographic distribution of Se in human milk, *Arch. Environ. Health*, 30, 230, 1975.
1375. **Shimoishi, Y.,** The gas-chromatographic determination of Se(VI) and total Se in milk, milk products and albumin with 1,2-diamino-4-nitrobenzene, *Analyst*, 101, 298, 1976.
1376. **Lombeck, I., Kasperek, K., Harbisch, H. D., Feinendegen, L. E., and Bremer, H. J.,** The Se state of healthy children. I. Serum Se concentration at different ages; activity of glutathione peroxidase of erythrocytes at different ages; Se content of food of infants, *Eur. J. Pediat.*, 125, 81, 1977.
1377. **Hadjimarkos, D. M. and Shearer, T. R.,** Selenium content of human nails: a new index for epidemiologic studies of dental caries, *J. Dent. Res.*, 52, 389, 1973.
1378. **Poole, C. F., Evans, N. J., and Wibbertley, D. G.,** Determination of Se in biological samples by gas-liquid chromatography with electron-capture detection, *J. Chromatogr.*, 136, 73, 1977.
1379. **Geahchan, A. and Chambon, P.,** Fluorimetry of Se in urine, *Clin. Chem.*, 26, 1272, 1980.
1380. **Watkinson, J. H.,** Fluorimetric determination of Se in biological material with 2,3-diaminonaphtalene, *Anal. Chem.*, 38, 92, 1966.
1381. **Watkinson, J. H.,** Semi-automated fluorimetric determination of ng quantities of Se in biological material, *Anal. Chim. Acta*, 105, 319, 1979.
1382. **Ihnat, M.,** Fluorimetric determination of Se in foods, *J. Assoc. Off. Anal. Chem.*, 57, 368, 1974.
1383. **Ihnat, M.,** Collaborative study of the fluorimetric method for determining Se in foods, *J. Assoc. Off. Anal. Chem.*, 57, 373, 1974.
1384. **Szydlowski, F. J. and Dunmire, D. L.,** Semi-automatic digestion and automatic analysis for Se in animal feeds, *Anal. Chim. Acta*, 105, 445, 1979.
1385. **Hofsommer, H.-J. and Bielig, H. J.,** Determination of Se in foodstuff, *Z. Lebensm.-Unters. Forsch.*, 172, 32, 1981.
1386. **Hoffman, I., Westerby, R. J., and Hidiroglou, M.,** Precise fluorimetric microdetermination of Se in agricultural materials, *J. Assoc. Off. Anal. Chem.*, 51, 1039, 1968.
1387. Analytical Methods Committee, Determination of small amounts of Se in organic matter, *Analyst*, 104, 778, 1979.
1388. **Bem, E. M.,** Determination of Se in the environmental and in biological material, *Environ. Health Perspect.*, 37, 183, 1981.
1389. **Pavlik, L., Kalouskova, J., Vobecký, M., Dědina, J., Beneš, J., and Pařizek, J.,** Selenium levels in the kidneys of male and female rats, in *Nuclear Activation Techniques in the Life Sciences 1978*, IAEA, Vienna, 1979, 213.
1390. **Raie, R. M. and Smith, H.,** The determination of Se in biological material by thermal NAA and AAS, *J. Radioanal. Chem.*, 48, 185, 1979.
1391. IAEA, *Elemental Analysis of Biological Materials. Current Problems and Techniques with Special Reference to Trace Elements*, IAEA, Vienna, 1980.
1392. **Strausz, K. I., Purdham, J. T., and Strausz, O. P.,** X-ray fluorescence spectrometric determination of Se in biological materials, *Anal. Chem.*, 47, 2032, 1975.
1393. **Tam, G. K. H. and Lacroix, G.,** Recovery studies of Se using dry ashing procedure with ashing aid, *J. Environ. Sci. Health*, Part B, B14, 515, 1979.
1394. **Raptis, S. E., Wegscheider, W., Knapp, G., and Tölg, G.,** X-ray fluorescence determination of trace Se in organic and biological matrices, *Anal. Chem.*, 52, 1292, 1980.
1395. **Berti, M., Buso, G., Colautti, P., Moschini, G., Stievano, B. M., and Tregnaghi, C.,** Determination of Se in blood serum by proton-induced X-ray emission, *Anal. Chem.*, 49, 1313, 1977.
1396. **Cutter, G.,** Species determination of Se in natural waters, *Anal. Chim. Acta*, 98, 59, 1978.
1397. **Cappon, C. J. and Smith, J. C.,** Determination of Se in biological material by gas chromatography, *J. Anal. Toxicol.*, 2, 114, 1978.

1398. **Andrews, R. W. and Johnson, D. C.,** Determination of Se(IV) by ASV in flow system with ion exchange separation, *Anal. Chem.,* 48, 1056, 1976.
1399. **Heng-Bin Han, Kaiser, G., and Tölg, G.,** Decomposition of biological materials, rocks, and soils in pure oxygen under dynamic conditions for the determination of Se at trace levels, *Anal. Chim. Acta,* 128, 9, 1981.
1400. **Reamer, D. C. and Veillon, C.,** Determination of Se in biological materials by stable isotope dilution gas chromatography — mass spectrometry, *Anal. Chem.,* 53, 2166, 1981.
1401. **Nakahara, T., Kobayashi, S., Wakisaka, T., and Musha, S.,** The determination of trace amounts of Se by hydride generation — nondispersive flame atomic fluorescence spectrometry, *Appl. Spectrosc.,* 34, 194, 1980.
1402. **Verlinden, M., Deelstra, H., and Adreaenssens, E.,** The determination of Se by AAS: a review, *Talanta,* 28, 637, 1981.
1403. **Tada, Y., Yonemoto, T., Iwasa, A., and Nakagawa, K.,** Graphite-furnace AAS of Se in blood by application of the enhancement effect of rhodium, *Bunseki Kagaku,* 29, 248, 1980.
1404. **Henn, E. L.,** Determination of Se in water and industrial effluents by flameless AAS, *Anal. Chem.,* 47, 428, 1975.
1405. **Manning, D. C.,** Spectral interferences in graphite furnace AAS. I. The determination of Se in an iron matrix, *At. Absorpt. Newsl.,* 17, 107, 1978.
1406. **Alexander, J., Saeed, K., and Thomassen, Y.,** Thermal stabilization of inorganic and organoselenium compounds for direct electrothermal AAS, *Anal. Chim. Acta,* 120, 377, 1980.
1407. **El-Shaarawy, M. I., Maes, R. A., and Otten, J.,** Selenium determination in urine by AAS, *Indian J. Hosp. Pharm.,* 16, 136, 1979.
1408. **Kamada, T. and Yamamoto, Y.,** Use of transition elements to enhance sensitivity for Se determination by graphite-furnace AAS combined with the APDC-MIBK system, *Talanta,* 27, 473, 1980.
1409. **Ihnat, M.,** AAS determination of Se with carbon furnace atomization, *Anal. Chim. Acta,* 82, 293, 1976.
1410. **Ihnat, M. and Westerby, R. J.,** Application of flameless atomization to the AA determination of Se in biological samples, *Anal. Lett.,* 7, 257, 1974.
1411. **Ishizaki, M.,** Determination of Se in biological materials by flameless AAS using a carbon tube atomizer, *Bunseki Kagaku,* 26, 206,1977.
1412. **Ishizaki, M.,** Simple method for determination of Se in biological materials by flameless AAS using a carbon-tube atomizer, *Talanta,* 25, 167, 1978.
1413. **Dillon, L. J., Hilderbrand, D. C., and Groon, K. S.,** Flameless AA determination of Se in human blood, *At. Spectrosc.,* 3, 5, 1982.
1414. **Saeed, K., Thomassen, Y., and Langmyhr, F. J.,** Direct electrothermal AAS determination of Se in serum, *Anal. Chim. Acta,* 10, 285, 1979.
1415. **Nève, J., Hanocq, M., and Molle, L.,** Critical study of the graphite-furnace AA determination of Se in biological samples after extraction with aromatic o-diamines and addition of Ni(II), *Anal. Chim. Acta,* 115, 133, 1980.
1416. **Rail, C. D., Kidd, D. E., and Hadley, W. M.,** Determination of Se in tissues, serum, and blood of wild rodents by graphite furnace AAS, *Int. J. Environ. Anal. Chem.,* 8, 79, 1980.
1417. **Szydlowski, F. J.,** Comparative study of the determination of Se in high carbohydrate nutrient carriers using flameless AAS and fluorimetry, *At. Absorpt. Newsl.,* 16, 60, 1977.
1418. **Vickrey, T. M. and Buren, M. S.,** Factors affecting Se atomization efficiency in graphite furnace AA, *Anal. Lett.,* 13, 1465, 1980.
1419. **Szydlowski, F. J., Peck, E. E., and Bax, B.,** Optimization of pyrolytic coating procedure for graphite tubes used in AAS, *Appl. Spectrosc.,* 32, 402, 1978.
1420. **Verlinden, M., Baart, J., and Deelstra, H.,** Optimization of the determination of Se by AAS: comparison of two hydride generation systems, *Talanta,* 27, 633, 1980.
1421. **Reamer, D. C., Veillon, C., and Tokousbalides, P. T.,** Radiotracer techniques for evaluation of selenium hydride generation systems, *Anal. Chem.,* 53, 245, 1981.
1422. **Meyer, A., Hofer, Ch., Tölg, G., Raptis, S., and Knapp, G.,** Cross-interferences by elements in the determination of traces of Se by the hydride AAS procedure, *Fresenius Z. Anal. Chem.,* 296, 337, 1979.
1423. **Verlinden, M. and Deelstra, H.,** Study of the effects of elements that form volatile hydrides on the determination of Se by hydride generation AAS, *Fresenius Z. Anal. Chem.,* 296, 253, 1979.
1424. **McDaniel, M., Shendrikar, A. D., Reiszner, K., and West, P. W.,** Concentration and determination of Se from environmental samples, *Anal. Chem.,* 48, 2240, 1976.
1425. Analytical Methods Committee, Determination of small amounts of Se in organic matter, *Analyst,* 104, 778, 1979.
1426. **Kamada, T., Shirashi, T., and Yamamoto, Y.,** Differential determination of Se(IV) and Se(VI) with SDDC, APDC, and dithizone by AAS with a carbon-tube atomizer, *Talanta,* 25, 15, 1978.

1427. **Nève, J. and Hanocq, M.,** The determination of traces of Se after extraction with 4-chloro-1,2-diaminobenzene by graphite-furnace AAS. Application to biological samples, *Anal. Chim. Acta,* 93, 85, 1977.
1428. **Vijan, P. N. and Leung, D.,** Reduction of chemical interferences and speciation studies in the hydride generation AA method for Se, *Anal. Chim. Acta,* 120, 141, 1980.
1429. **Cox, D. H. and Bibb, A. E.,** Hydrogen selenide evolution — electrothermal AA method for determining ng levels of total Se, *J. Assoc. Off. Anal. Chem.,* 64, 265, 1981.
1430. **Raptis, S., Knapp, G., Meyer, A., and Tölg, G.,** Systematic errors in the determination of Se in the ng/g-range in biological matrices by hydride-AAS method, *Fresenius Z. Anal. Chem.,* 300, 18, 1980.
1431. **Nève, J., Hanocq, M., and Molle, L.,** Critical study of some wet digestion methods for decomposition of biological materials for the determination of total Se and Se(VI), *Mikrochim. Acta,* I, 259, 1980.
1432. **Meyer, A., Hofer, Ch., Knapp, G., and Tölg, G.,** Determination of Se in the μg/g and ng/g range in inorganic and organic matrices by AAS after volatization in a dynamic system, *Fresenius Z. Anal. Chem.,* 305, 1, 1981.
1433. **Szydlowski, F. J. and Vianzon, F. R.,** Further studies on the determination of Se using graphite furnace AAS, *At. Spectrosc.,* 1, 39, 1980.
1434. **Yasuda, K., Taguchi, M., Tamura, S., and Toda, S.,** Determination of Se in biological samples by solvent extraction — graphite furnace AAS, *Bunseki Kagaku,* 26, 442, 1977.
1435. **McMullin, J. F., Pritchard, J. G., and Sikondari, A. H.,** Accuracy and precision of the determination of Hg in human scalp hair by cold-vapor AAS, *Analyst,* 107, 803, 1982.
1436. **Jordanov, N., Daskalova, K., Khavesov, I., and Toncheva, V.,** AA determination of Se after preconcentration as an organoselenium compound, in *Developments in Analytical Chemistry,* Nauka, Moscow, 1974, 54.
1437. **Lund, W. and Bye, R.,** Flame AAA for Se after electrochemical preconcentration, *Anal. Chim. Acta,* 110, 279, 1979.
1438. **Holen, B., Bye, R., and Lund, W.,** AAS of Se in an air-H_2 flame after electrochemical preconcentration, *Anal. Chim. Acta,* 130, 257, 1981.
1439. **Saeed, K. and Thomassen, Y.,** Spectral interferences from phosphate matrices in the determination of As, Sb, Se, and Te by electrothermal AAS, *Anal. Chim. Acta,* 130, 281, 1981.
1440. **Lo, D. B. and Christian, G. D.,** Microdetermination of Si in blood, serum, urine and milk using furnace AAS, *Microchem. J.,* 23, 481, 1978.
1441. **Fregert, S.,** Silicon in tissues with special reference to skin, *Acta Derm.-Venerol.,* 42 (Suppl.), 39, 1959.
1442. **Sárdi, A. and Tomcsányi, A.,** Fast-neutron activation analysis of Si in sputum, *Analyst,* 92, 529, 1967.
1443. **Bloomfield, J. J., Sayers, R. R., and Goldman, F. H.,** The urinary excretion of silica by persons exposed to silica dust, *U.S. Publ. Health Rep.,* 50, 421, 1935.
1444. **Mehard, C. W. and Volcani, B. E.,** Silicon in rat liver organelles. Electron probe microanalysis, *Cell Tissue Res.,* 166, 255, 1976.
1445. **Frech, W. and Cedergren, A.,** Investigation of reactions involved in flameless AA procedures. VII. A theoretical and experimental study of factors influencing the determination of Si, *Anal. Chim. Acta,* 113, 227, 1980.
1446. **Rawa, J. A. and Henn, E. L.,** Determination of trace silica in industrial process waters by flameless AAS, *Anal. Chem.,* 51, 452, 1979.
1447. **Ortner, H. M. and Kantuscher, E.,** Impregnation of graphite tubes with metal salts for improvement of AAS determination of Si, *Talanta,* 22, 581, 1975.
1448. **Müller-Vogt, G. and Wende, W.,** Reaction kinetics in the determination of Si by graphite furnace AAS, *Anal. Chem.,* 53, 651, 1981.
1449. **Howlett, C. and Taylor, A.,** Measurement of Ag in blood by AAS using the micro-cup technique, *Analyst,* 103, 916, 1978.
1450. **Nakashima, R., Sasaki, S., and Shibata, S.,** Determination of Ag in biological materials by high-frequency plasma-torch emission spectrometry, *Anal. Chim. Acta,* 77, 65, 1975.
1451. **Alexiev, A. A., Bontchev, P. R., and Todorov, I.,** A chemical catalytic method for determination of Ag in human saliva, *Arch. Oral Biol.,* 18, 1461, 1973.
1452. **Buneaux, F. and Fabiani, P.,** Detection and estimation of Ag in biological liquids, *Ann. Biol. Clin.,* 28, 273, 1970.
1453. **Grabowski, B. F. and Haney, W. G., Jr.,** Characterization of Ag deposits in tissue resulting from dermal application of a silver-containing pharmaceutical, *J. Pharm. Sci.,* 61, 1488, 1972.
1454. **Rooney, R. C.,** Determination of Ag in animal tissues by wet oxidation followed by AAS, *Analyst,* 100, 471, 1975.
1455. **Iyengar, G. V., Kasperek, K., and Feinendegen, L. E.,** Retention of metabolized trace elements in biological tissues following different drying procedures. II. Cs, Ce, Mn, Sc, Ag and Sn in rat tissues, *Analyst,* 105, 794, 1980.

1456. **Spector, P. C. and Curzon, M. E. J.,** Strontium concentrations of plaque and saliva, *J. Dent. Res.,* 56, A110, 1977.
1457. **Curzon, M. E. J. and Losee, F. L.,** Strontium content of enamel and dental caries, *Caries Res.,* 11, 321, 1977.
1458. **Montfort, B. and Crlbbs, S. L.,** Determination of Sr in urine by AAS, *At. Absorpt. Newsl.,* 8, 77, 1969.
1459. **Curnow, D. C., Gutteridge, D. H., and Horgan, E. D.,** Determination of Sr in serum and urine of Sr-treated subjects by AAS, *At. Absorpt. Newsl.,* 7, 45, 1968.
1460. **Descube, J., Roques, N., Rousselet, F., and Girard, M. L.,** Estimation of Sr in biological media by AA, *Ann. Biol. Clin.,* 25, 1011, 1967.
1461. **Rousselet, F., El Solh, N., and Girard, M. L.,** The nature and variations of the interactions encountered in AAS in course of the determination of Sr in biological media, *Analusis,* 3, 44, 1975.
1462. **Trent, D. and Slavin, W.,** Factors in the determination of Sr by AAS with particular reference to ashed biological samples, *At. Absorpt. Newsl.,* 3, 53, 1964.
1463. **Helsby, C. A.,** Determination of Sr in human tooth enamel by AAS, *Anal. Chim. Acta,* 69, 259, 1974.
1464. **Tanaka, G., Tomikawa, A., Kawamura, H., and Ohyagi, Y.,** Determination of Sr by AAS, *Nippon Kagaku Zasshi,* 89, 175, 1968.
1465. **Spector, P. C. and Curzon, M. E. J.,** Relationship of Sr in drinking water and surface enamel, *J. Dent. Res.,* 57, 55, 1978.
1466. **Warren, J. M. and Spencer, H.,** Analysis of stable Sr in biological materials by AAS, *Clin. Chim. Acta,* 38, 435, 1972.
1467. **Helsby, C. A.,** Determination of Sr in human tooth enamel by flameless AAS, *Talanta,* 24, 46, 1977.
1468. **Matsusaki, K., Murakami, S., and Yoshino, T.,** Interference effect of chloride on the determination of Sr by AAS with a graphite furnace, *Nippon Kagaku Kaishi,* 7, 1126, 1980.
1469. **Weibust, G., Langmyhr, F. J., and Thomassen, Y.,** Thermal stabilization of inorganic and organically-bound Te for electrothermal AAS, *Anal. Chim. Acta,* 128, 23, 1981.
1470. **Jung, J. D. and Hilderbrand, D. C.,** Enhancement of flameless AA determination of Te in biological materials by means of solvent extraction, *Proc. N. D. Acad. Sci.,* 33, 12, 1979.
1471. **Cheng, J. T. and Agnew, W. F.,** Determination of Te in biological material by AAS, *At. Absorpt. Newsl.,* 13, 123, 1974.
1472. **Greenland, L. P. and Campbell, E. Y.,** Rapid determination of ng amounts of Te in silicate rocks, *Anal. Chim. Acta,* 87, 323, 1976.
1473. **Kamada, T., Sugita, N., and Yamamoto, Y.,** Differential determination of Te(IV) and Te(VI) with SDDC, APDC, and dithizone by AAS with a carbon-tube atomizer, *Talanta,* 26, 337, 1979.
1474. **Havezov, I. and Jordanov, N.,** Separation of Te(IV) by solvent extraction methods, *Talanta,* 21, 1013, 1974.
1475. **Weinig, E. and Zink, P.,** Quantitative mass spectrometric determination of the normal Tl content in the human body, *Arch. Toxicol.,* 22, 255, 1967.
1476. **Cernic, A.,** private communication, in Wall, D. C., *Clin. Chim. Acta,* 76, 259, 1977.
1477. **Geilmann, W., Beyermann, K., Neeb, K. H., and Neeb, R.,** Thallium as a trace element for animals and plants, *Biochem. Z.,* 333, 62, 1960.
1478. **Franke, J. P. and De Zeeuw, R. A.,** DPASV in the detection of acute and chronic Tl toxicity, *Pharm. Weekbl.,* 111, 725, 1976.
1479. **Curtis, A. R.,** Determination of trace amounts of Tl in urine, using the mercury-film electrode and DPASV, *J. Assoc. Off. Anal. Chem.,* 57, 1366, 1974.
1480. **Wall, C. D.,** Determination of Tl in urine by AAS and emission spectrography, *Clin. Chim. Acta,* 76, 259, 1977.
1481. **Fazakas, J.,** personal communication, 1982.
1482. **Fuller, C. W.,** The effect of acids on the determination of Tl by AAS with a graphite furnace, *Anal. Chim. Acta,* 81, 199, 1976.
1483. **L'vov, B. V., Pelieva, L. A., and Sharnopolskii, A. I.,** Eliminating the depressive effect of chlorides during AAA with a graphite furnace by excess lithium additions to the sample, *Zh. Prikl. Spektrosk.,* 28, 19, 1978.
1484. **Chapman, J. F. and Leadbeatter, B. E.,** The determination of Tl in human hair by graphite furnace AAS, *Anal. Lett.,* 13, 439, 1980.
1485. **Curry, A. S., Read, J. F., and Knott, A. R.,** Determination of Tl in biological material by flame spectrophotometry and AA, *Analyst,* 94, 744, 1969.
1486. **Stoeppler, M., Bagschik, U., and May, K.,** Investigations on the extraction of ^{204}Tl for ascertaining optimum working conditions for Tl determinations in biological material, *Fresenius Z. Anal. Chem.,* 301, 106, 1980.
1487. **Savory, J., Roszel, N. O., Mushak, P., and Sunderman, F. W., Jr.,** Measurement of Tl in biological materials by AAS, *Am. J. Clin. Pathol.,* 50, 505, 1968.

1488. **Singh, N. P. and Joselow, M. M.,** Determination of Tl in whole blood by Delves cup AAS, *At. Absorpt. Newsl.*, 14, 42, 1975.
1489. **Morgan, J. M., McHenry, J. R., and Masten, L. W.,** Simultaneous determination of inorganic and organic Tl by AAA, *Bull. Environ. Contam. Toxicol.*, 24, 333, 1980.
1490. **Kambayashi, K.,** Measurement of Tl in urine by AAS, and the effect of penicillamine upon thallotoxicosis, *Igaku to Seibutsugaku*, 80, 309, 1970.
1491. **Shkolnik, G. M. and Bevill, R. F.,** The determination of Tl in urine and plasma by Delves cup AA, *At. Absorpt. Newsl.*, 12, 112, 1973.
1492. **Hagedorn-Goetz, H. and Stoeppler, M.,** Determination of Tl in human head hair in forensic cases with flameless AAS, *Arch. Toxicol.*, 34, 17, 1975.
1493. **Byrne, A. R. and Kosta, L.,** On V and Sn contents of diet and human blood, *Sci. Total Environ.*, 13, 87, 1979.
1494. **Braman, R. S. and Tompkins, M. A.,** Separation and determination of ng amounts of inorganic Sn and methyltin compounds in the environment, *Anal. Chem.*, 51, 12, 1979.
1495. **Horwitz, W.,** Commonly used methods of analysis for tin in foods, *J. Assoc. Off. Anal. Chem.*, 62, 1251, 1979.
1496. Analytical Methods Committee, The determination of small amounts of tin in organic matter. I. Amounts of Sn up to 30 μg, *Analyst*, 92, 320, 1967.
1497. **Englerg, Å.,** A comparison of a spectrophotometric (quercetin) method and an AA method for the determination of Sn in food, *Analyst*, 98, 137, 1973.
1498. **Gorenz, D., Kosta, L., and Byrne, A. R.,** Extractive separation and spectrophotometric determination of Sn in biological materials, *Vestn. Slov. Kem. Drus.*, 21, 9, 1974.
1499. **Byrne, A. R.,** NAA of Sn in biological materials and their ash using ^{123}Sn and ^{125}Sn, *J. Radioanal. Chem.*, 20, 627, 1974.
1500. **Byrne, A. R.,** Activation analysis of Sn at ng level by liquid scintillation counting of ^{121}Sn, *J. Radioanal. Chem.*, 37, 591, 1977.
1501. **Nembrini, P. G., Dogan, S., and Haerdi, W.,** Simultaneous determination of Sn, Pb, and Cu by AC polarography, *Anal. Lett.*, 13, 947, 1980.
1502. **Dogan, S. and Haerdi, W.,** Determination of total Sn in environmental biological and water samples by AAS with graphite furnace, *Int. J. Environ. Anal. Chem.*, 8, 249, 1980.
1503. **Trachman, H. L., Tyberg, A. J., and Branigan, P. D.,** AAS determination of sub-ppm quantities of tin in extracts and biological materials with a graphite furnace, *Anal. Chem.*, 49, 1090, 1977.
1504. **Fritzsche, H., Wegscheider, W., Knapp, G., and Ortner, H. M.,** A sensitive AAS method for the determination of tin with atomization from impregnated graphite surfaces, *Talanta*, 26, 219, 1979.
1505. **Vickrey, T. M., Harrison, G. V., Ramelow, G. J., and Carver, J. C.,** Use of metal carbide coated graphite tubes for the AAA of organotins, *Anal. Lett.*, 13, 781, 1980.
1506. **Hocquellet, P. and Labeyrie, N.,** Determination of tin in foods by flameless AA, *At. Absorpt. Newsl.*, 16, 124, 1977.
1507. **Itsuki, K. and Ikeda, T.,** AAS of Sn using graphite furnace with the aid of La, *Bunseki Kagaku*, 29, 309, 1980.
1508. **Tominaga, M. and Umezaki, Y.,** Determination of sub-μg amounts of Sn by AAS with electrothermal atomization, *Anal. Chim. Acta*, 110, 55, 1979.
1509. **Rayson, G. D. and Holcombe, J. A.,** Tin atom formation in a graphite furnace atomizer, *Anal. Chim. Acta*, 136, 249, 1982.
1510. **Vickrey, T. M., Harrison, G. V., and Ramelow, G. J.,** Treated graphite surfaces for determination of Sn by graphite furnace AAS, *Anal. Chem.*, 53, 1573, 1981.
1511. **Zatka, V. J.,** Tantalum treated graphite atomizer tubes for AAS, *Anal. Chem.*, 50, 538, 1978.
1512. **Hodge, V. F., Saidel, S. L., and Goldberg, E. D.,** Determination of Sn(IV) and organotin compounds in natural waters, coastal sediments and macro algae by AAS, *Anal. Chem.*, 51, 1256, 1979.
1513. **Evans, W. H., Jackson, F. J., and Dellar, D.,** Evaluation of a method for determination of total Sb, As, and Sn in foodstuffs using measurement by AAS with atomization in a silica tube using hydride generation technique, *Analyst*, 104, 16, 1979.
1514. **Vijan, P. N. and Chan, C. Y.,** Determination of Sn by gas phase atomization and AAS, *Anal. Chem.*, 48, 1788, 1976.
1515. **Subramanian, K. S. and Sastri, V. S.,** A rapid hydride-evolution electrothermal AA method for the determination of Sn in geological materials, *Talanta*, 27, 469, 1980.
1516. **Nakashima, S.,** Determination of Sn by long absorption cell AAS following stannane generation, *Ber. Ohara Inst. Landwirtsch. Biol. Okayama Univ.*, 17, 187, 1979.
1516a. **Maher, W.,** Measurement of total Sn in marine organisms by stannane generation and AAS, *Anal. Chim. Acta*, 138, 365, 1982.
1517. **Rigin, V. I.,** AA determination of tin in water and biological materials using electrolytic separation and atomization in the gas phase, *Zh. Anal. Khim.*, 34, 1569, 1979.

1518. **Dabeka, R. W. and McKenzie, A. D.,** AAS determination of tin in canned foods using HNO_3-HCl acid digestion and nitrous oxide-acetylene flame, *J. Assoc. Off. Anal. Chem.,* 64, 1297, 1981.
1519. **Elkins, E. R. and Sulek, A.,** AA determination of Sn in foods: collaborative study, *J. Assoc. Off. Anal. Chem.,* 62, 1050, 1979.
1520. **Glockling, F.,** Organometallic compounds in relation to pollution, *Anal. Proc.,* 17, 417, 1980.
1521. **Byrne, A. R. and Kosta, L.,** Vanadium in foods and in human body fluids and tissues, *Sci. Total Environ.,* 10, 17, 1978.
1522. **Allen, R. O. and Steinnes, E.,** Determination of V in biological materials by RNAA, *Anal. Chem.,* 50, 1553, 1978.
1523. **Gylseth, B., Liera, H. L., Steinnes, E., and Thomassen, Y.,** Vanadium in the blood and urine of workers in a ferroalloy plant, *Scand. J. Work, Environ. Health,* 5, 188, 1979.
1524. **Valentin, H.,** External/Internal Dose-Response Relationship for Hg and V in Occupationally Exposed and Normal Persons, Rep. EUR 6388, Environ. Res. Programme, Comm. Eur. Communities, 1980, 89.
1525. **Sabbioni, E., Marafante, E., Pietra, R., Goetz, L., Girardi, F., and Orvini, E.,** The association of V with the iron transport system in human blood as determined by gel filtration and NAA, in *Nuclear Activation Techniques in the Life Sciences 1978,* IAEA, Vienna, 1979, 179.
1526. **Ishizaki, M., Ueno, S., Fujiki, M., and Yamaguchi, S.,** Determination of V in blood by flameless AAS using a carbon-tube atomizer, *Sangyo Igaku,* 20, 20, 1978.
1527. **Ueno, S., Ishizaki, M., Kataoka, F., Oyamada, N., Murakami, R., Kubota, K., and Katsumura, K.,** Determination of V in biological materials. II. Determination of V in biological materials by N-benzoyl-o-tolylhydroxylamine extraction — flameless AAS, *Ibaraki-ken Eisei Kenkyusho Nempo,* 15, 53, 1977.
1528. **Cornelis, R., Mees, L., Hoste, J., Ryckebusch, J., Versieck, J., and Barbier, F.,** NAA of V in human liver and serum, in *Nuclear Activation Techniques in the Life Sciences 1978,* IAEA, Vienna, 1979, 165.
1529. **Cornelis, R., Versieck J., Mees, L., Hoste, J., and Barbier, F.,** Determination of V in human serum by NAA, *J. Radioanal. Chem.,* 55, 35, 1980.
1530. **Byrne, A. R. and Vrbič, V.,** Vanadium content of human dental enamel and its relationship to caries, *J. Radioanal. Chem.,* 54, 77, 1979.
1531. **Buchet, J. P., Knepper, E., and Lauwerys, R.,** Determination of V in urine by electrothermal AAS, *Anal. Chim. Acta,* 136, 243, 1982.
1532. **Shei, F., Wan, U., and Tze, Y.,** Determination of traces of V in urine and water by catalytic-spectrophotometric method, *Hua Hsueh Tung Pao,* 9, 19, 1980.
1533. **Christian, G. D.,** Catalytic determination of V in blood and urine, *Anal. Lett.,* 4, 187, 1971.
1534. **Ishizaki, M. and Ueno, S.,** Determination of sub-μg amounts of V in biological materials by extraction with N-cinnamoyl-N-2,3-xylyl hydroxylamine and flameless AAS with an atomizer coated with pyrolytic graphite, *Talanta,* 26, 523, 1979.
1535. **Hulanicki, A., Karwowska, R., and Stanczak, J.,** Experimental parameters of V determination by AAS with graphite-furnace atomization, *Talanta,* 27, 214, 1980.
1536. **Myron, D. R., Givand, S. H., and Nielsen, F. H.,** Vanadium content of selected foods as determined by flameless AAS, *J. Agric. Food Chem.,* 25, 297, 1977.
1537. **Barbooti, M. M. and Jasim, F.,** Electrothermal AA determination of V, *Talanta,* 29, 107, 1982.
1538. **Ueno, S. and Ishizaki, M.,** Determination of V in plants and biological samples by extraction with N-cinnamoyl-N-(2,3-xylyl)-hydroxylamine and carbon-furnace AAS, *Nippon Kagaku Kaishi,* 217, 1979.
1539. **Ueno, S., Ishizaki, M., Kataoka, F., Oyamada, N., Murakami, R., Kubota, K., and Katsumura, K.,** Determination of V in biological materials. I. Examination of chelate reagents on the determination of V by solvent extraction — flameless AAS, *Ibaraki-ken Eisei Kenkyusho Nempo,* 15, 47, 1977.
1540. **Dawson, J. B. and Walker, B. E.,** Direct determination of Zn in whole blood, plasma and urine by AAS, *Clin. Chim. Acta,* 26, 465, 1969.
1541. **Chang, T. L., Goover, T. A., and Harrison, W. W.,** Determination of Mg and Zn in human brain tissue by AAS, *Anal. Chim. Acta,* 34, 17, 1966.
1542. **Mazzucotelli, A., Galli, M., Benassi, E., Loeb, C., Ottonello, G. A., and Tanganelli, P.,** AA micro determination of Zn in cerebrospinal fluid by ion-exchange chromatography and electrothermal atomization, *Analyst,* 103, 863, 1978.
1543. **Sohler, A., Wolcott, P., and Pfeiffer, C. C.,** Determination of Zn in fingernails by non-flame AAS, *Clin. Chim. Acta,* 70, 391, 1976.
1544. **Reimold, E. W. and Besch, D. J.,** Determination and elimination of contaminatious interfering with the determination of Zn in plasma, *Clin. Chem.,* 24, 675, 1978.
1545. **Prasad, A. S., Schulert, A. R., Sandstead, H. H., Miale, A., Jr., and Farid, Z.,** Zinc, Fe and N content of sweat in normal and deficient subjects, *J. Lab. Clin. Med.,* 62, 84, 1963.
1546. **Toribara, T. Y. and Jackson, D. A.,** XRF measurement of Zn profile of a single hair, *Clin. Chem.,* 28, 650, 1982.
1547. **Ebdon, L., Ellis, A. T., and Ward, R. W.,** Aspects of chloride interference in Zn determination by AAS with electrothermal atomization, *Talanta,* 29, 297, 1982.

1548. **Whitehouse, R. C., Prasad, A. S., Rabbani, P. I., and Cossack, Z. T.,** Zinc in plasma, neutrophils, lymphocytes, and erythrocytes as determined by flameless AAS, *Clin. Chem.*, 28, 475, 1982.
1549. **Foote, J. W. and Delves, H. T.,** Rapid determination of albumin-bound zinc in human serum by simple affinity chromatography and AAS, *Analyst,* 107, 121, 1982.
1550. **Vieira, N. E. and Hansen, J. W.,** Zinc determined in 10-$\mu\ell$ serum or urine samples by flameless AAS, *Clin. Chem.*, 27, 73, 1981.
1551. **Shaw, J. C. L., Bury, A. J., Barber, A., Mann, L., and Taylor, A.,** A micromethod for the analysis of Zn in plasma or serum by AAS using graphite furnace, *Clin. Chim. Acta,* 118, 229, 1982.
1552. **Stevens, M. D., MacKenzie, W. F., and Anand, V. D.,** A simplified method for determination of Zn in whole blood, plasma and erythrocytes by AAS, *Biochem. Med.*, 18, 158, 1977.
1553. **Arroyo, M. and Plaenque, E.,** Analytical determination of Zn in biological fluids by means of AAS. II. Analytical methods, *Rev. Clin. Esp.*, 134, 227, 1974.
1554. **Butrimovitz, G. P. and Purdy, W. C.,** The determination of Zn in blood plasma by AAS, *Anal. Chim. Acta,* 94, 63, 1977.
1555. **Hackley, B. M., Smith, J. C., and Halsted, J. A.,** A simplified method for plasma zinc determination by AAS, *Clin. Chem.*, 14, 1, 1968.
1556. **Kelson, J. R. and Shamberger, R. J.,** Methods compared for determining Zn in serum by flame AAS, *Clin. Chem.*, 24, 240, 1978.
1557. **Kuleva, V. and Tsalev, D.,** On the dynamics of serum zinc in patients with a chronic kidney insufficiency treated by periodic hemodialysis, *Vatr. Bol.*, 17, 71, 1979.
1558. **Momčilović, B., Belonje, B., and Shah, B. G.,** Effect of the matrix of the standard on results of AAS of Zn in serum, *Clin. Chem.*, 21, 588, 1975.
1559. **Reinhold, J. G., Pascoe, E., and Kfoury, G. A.,** Capillary diameter and rate of aspiration as factors affecting the accuracy of Zn analysis in serum by AAS, *Anal. Biochem.*, 25, 557, 1968.
1560. **Smith, J. C., Jr., Butrimovitz, G. P., and Purdy, W. C.,** Direct measurement of Zn in plasma by AAS, *Clin. Chem.*, 25, 1487, 1979.
1561. **Krčma, V. and Komárek, J.,** Determination of Zn in some body fluids by AAS, *Chem. Listy,* 74, 770, 1980.
1562. **Oiwa, K., Kimura, T., Mikino, H., and Okuda, M.,** Determination of Zn in blood serum and red cells by AAS, *Bunseki Kagaku,* 17, 810, 1968.
1563. **Prasad, A. S., Oberleas, D., and Halsted, J. A.,** Determination of Zn in biological fluids by AAS in normal and cirrhotic subjects, *J. Lab. Clin. Med.*, 66, 508, 1965.
1564. **Nassi, L., Poggini, G., Vecchi, C., and Galvan, P.,** AAS as applied to the determination of total, free and bound Zn in colostrum and human milk, *Boll. Soc. Ital. Biol. Sper.*, 48, 86, 1972.
1565. **Boyde, T. R. C. and Wu, S. W. N.,** Volume exclusion errors and the determination of serum Zn, *Clin. Chem. Acta,* 88, 49, 1978.
1566. **James, B. E. and Macmahon, R. A.,** An effect of TCA on the determination of Zn by AAS, *Clin. Chim. Acta,* 32, 307, 1971.
1567. **Clevay, L. M.,** Hair as a biopsy material. I. Assessment of Zn nutriture, *Am. J. Clin. Nutr.*, 23, 284, 1970.
1568. **Luyten, S., Smeyers-Verbeke, J., and Massart, D. L.,** A comparison of fast destructive methods for the determination of trace metals in biological materials, *At. Absorpt. Newsl.*, 12, 13, 1973.
1569. **Ladefoged, K.,** Determination of Zn in diet and feces by acid extraction and AAS, *Clin. Chim. Acta,* 100, 149, 1980.
1570. Analytical Methods Committee, Determination of small amounts of Zn in organic matter by AAS, *Analyst,* 98, 458, 1973.
1571. **Greger, J. L. and Sickles, V. S.,** Saliva Zn levels: potential indicator of Zn status, *Am. J. Clin. Nutr.*, 32, 1859, 1979.
1572. **Cousins, R. J. and Smith, K. T.,** Zinc-binding properties of bovine and human milk in vitro: influence of changes in Zn content, *Am. J. Clin. Nutr.*, 33, 1083, 1980.
1573. **Eckert, C. D., Sloan, M. V., Duncan, J. R., and Hurley, L. S.,** Zinc binding: a difference between human and bovine milk, *Science,* 195, 789, 1977.
1574. **Giroux, E. L.,** Determination of Zn distribution between albumin and α_2-macroglobulin in human serum, *Biochem. Med.*, 12, 258, 1975.
1575. **Henkin, R. I., Lippoldt, R. E., Bilstad, J., and Edelhoch, H.,** Zinc protein isolated from human parotid saliva, *Proc. Natl. Acad. Sci. U.S.A.*, 72, 488, 1975.
1576. **Henkin, R. I., Mueller, C. W., and Wolf, R. O.,** Estimation of Zn concentration of parotid saliva by flameless AAS in normal subjects and in patients with idiopathic hypogeusia, *J. Lab. Clin. Med.*, 86, 175, 1975.
1577. **Tuompo, H., Raeste, A.-M., and Nuuja, T.,** Ionic Zn in normal human mixed saliva, *J. Dent. Res.*, 56, A 89, 1977.

1578. **Urquhart, N.,** Zinc contamination in a trace-element control serum, *Clin. Chem.,* 24, 1652, 1978.

1579. **Heggie, J.,** Serum Zn analysis: investigation of sources of specimen contamination, *Can. J. Med. Technol.,* 40, 181, 1978.

1580. **Saleh, A., Udall, J. N., and Solomons, N. W.,** Minimizing contamination of specimens for Zn determination, *Clin. Chem.,* 27, 338, 1981.

1581. **Bogden, J. D., Oleske, J. M., Weiner, B., Smith, L. G., Jr., Smith, L. G., and Najem, G. R.,** Elevated plasma Zn concentrations in renal dialysis patients, *Am. J. Clin. Nutr.,* 33, 1088, 1980.

1582. **Gorski, J. E. and Dietz, A. A.,** Contamination-free serum samples for trace analyses, *Clin. Chem.,* 24, 169, 1978.

1583. **Kiilerich, S., Christensen, M. S., Naestoft, J., and Christiansen, C.,** Serum and plasma Zn concentrations with special reference to standardized sampling procedure and protein status, *Clin. Chim. Acta,* 114, 117, 1981.

1584. **Kiilerich, S., Christiansen, M. S., Naestoft, J., and Christiansen, C.,** Determination of Zn in serum and urine by AAS; relationship between serum levels of Zn and proteins in 104 normal subjects, *Clin. Chim. Acta,* 105, 231, 1980.

1585. **Hinks, L. J., Colmsee, M., and Delves, H. T.,** Determination of Zn and Cu in isolated leucocytes, *Analyst,* 107, 815, 1982.

INDEX

A

B

C

D

E

F

G

H

I

J

K

L

M

N

Q

R

S

T

U

V

W

X

Z